Advanced Sciences and Technologies for Security Applications

Indexed by SCOPUS

The series Advanced Sciences and Technologies for Security Applications comprises interdisciplinary research covering the theory, foundations and domain-specific topics pertaining to security. Publications within the series are peer-reviewed monographs and edited works in the areas of:

- biological and chemical threat recognition and detection (e.g., biosensors, aerosols, forensics)
- crisis and disaster management
- terrorism
- cyber security and secure information systems (e.g., encryption, optical and photonic systems)
- traditional and non-traditional security
- energy, food and resource security
- economic security and securitization (including associated infrastructures)
- transnational crime
- human security and health security
- social, political and psychological aspects of security
- recognition and identification (e.g., optical imaging, biometrics, authentication and verification)
- smart surveillance systems
- applications of theoretical frameworks and methodologies (e.g., grounded theory, complexity, network sciences, modelling and simulation)

Together, the high-quality contributions to this series provide a cross-disciplinary overview of forefront research endeavours aiming to make the world a safer place.

The editors encourage prospective authors to correspond with them in advance of submitting a manuscript. Submission of manuscripts should be made to the Editor-in-Chief or one of the Editors.

More information about this series at http://www.springer.com/series/5540

Reza Montasari · Hamid Jahankhani · Richard Hill ·
Simon Parkinson
Editors

Digital Forensic Investigation of Internet of Things (IoT) Devices

Editors
Reza Montasari
Hillary Rodham Clinton School of Law
Swansea University
Swansea, UK

Hamid Jahankhani
London Campus
Northumbria University
London, UK

Richard Hill
Department of Computer Science
University of Huddersfield
Huddersfield, UK

Simon Parkinson
Department of Computer Science
University of Huddersfield
Huddersfield, UK

In 2015, Antonio Mauro, PhD (info@antoniomauro.it) filed a patent named "Forensics Investigation in the Internet of Things (IoT) Devices".

ISSN 1613-5113 ISSN 2363-9466 (electronic)
Advanced Sciences and Technologies for Security Applications
ISBN 978-3-030-60427-1 ISBN 978-3-030-60425-7 (eBook)
https://doi.org/10.1007/978-3-030-60425-7

This Springer imprint is published by the registered company Springer Nature Switzerland AG
The registered company address is: Gewerbestrasse 11, 6330 Cham, Switzerland

Foreword

Watching the progressive rollout of the IOT, it would be easy to form the opinion that we really understand what we are doing and how the network is going to perform, but nothing could be farther from the truth! Reality is, the IOT is a new and evolutionary network form that presents levels of complexity and behavior that we never anticipated and have never seen before. Further, we do not have the tools or abilities to model, characterize, measure, and fully understand the outcomes of our designs and deployments. And along with almost all new systems, security is often omitted completely, or it appears as a weak engineering afterthought. In reality, the IoT is magnifying the attack surface of the planet to the benefit of cybercriminals and rogue states who now see the IOT as a new opportunity and entry window for wider incursions into the networks and facilities of organizations.

> *It is not possible to understate the rapidly growing cyber risks posed by The IoT or indeed the urgency of the address required*

It is, therefore, refreshing to find a book addressing this most important topic with detailed consideration of many of the initial IoT challenges. Primarily, it asks what happens when an IoT attack occurs or failure happens, and how do we locate the point of failure/entry to assess the potential consequences and affect repairs as quickly as possible? In short, the term "forensics" is a perfect fit for what is needed and what is detailed in this first book on the topic. To my mind, it represents a first and vital step in the documentation and development of a new branch of network science and engineering that is urgently required. As an academic, practitioner, and consultant in the field of cyber security, I found the treatment in each chapter refreshing and reassuring with the authors detailing their latest thoughts and research results. Best of all, they opened my mind to new concepts and avenues in the field and left me wanting for more. I, therefore, consider this to be "Volume 1" in the opening salvo of our battle for IoT security supremacy, and the survival of one of our most important components of Industry 4.0 and the realization of sustainable societies.

And so, it is in this context, and with this background that I commend this book to you as a provocative and foundation text in the field. Hopefully, you will find it enlightening and useful, and it might also spur even more innovation.

June 2020

Prof. Peter Cochrane
OBE
University of Suffolk
Ipswich, UK

Contents

Emulation Versus Instrumentation for Android Malware Detection

Anukriti Sinha, Fabio Di Troia, Philip Heller, and Mark Stamp

Abstract In resource constrained devices, malware detection is typically based on offline analysis using emulation. An alternative to such emulation is malware analysis based on code that is executed on an actual device. In this research, we collect features from a corpus of Android malware using both emulation and on-phone instrumentation. We train machine learning models using the emulator-based features and we train models on features collected via instrumentation, and we compare the results obtained in these two cases. We obtain strong detection and classification results, and our results improve slightly on previous work. Consistent with previous work, we find that emulation fails for a significant percentage of malware applications. However, we also find that emulation fails to extract useful features from an even larger percentage of benign applications. We show that for applications that are amenable to emulation, malware detection and classification rates based on emulation are consistently within 1% of those obtained using more intrusive and costly on-phone analysis. We also show that emulation failures are easily explainable and appear to have little to do with malware writers employing anti-emulation techniques, contrary to claims made in previous research. Among other contributions, this work points to a lack of sophistication in Android malware.

A. Sinha (✉) · F. Di Troia · P. Heller · M. Stamp
San Jose State University, San Jose, CA, USA
e-mail: anukriti.sinha@sjsu.edu

F. Di Troia
e-mail: fabioditroia@msn.com

P. Heller
e-mail: philip.heller@sjsu.edu

M. Stamp
e-mail: mark.stamp@sjsu.edu

R. Montasari et al. (eds.), *Digital Forensic Investigation of Internet of Things (IoT) Devices*, Advanced Sciences and Technologies for Security Applications,
https://doi.org/10.1007/978-3-030-60425-7_1

1 Introduction

In 2007, Google launched a mobile operating system (OS) known as Android, which is based on the Linux kernel and other open source software. Android is used primarily on touchscreen devices such as tablets and smartphones. Google distributes Android as an open source platform, which has encouraged the use of smartphones as a computing platform [29]. Android currently dominates the mobile OS market, with more than a billion Android devices having been sold, and more than 65 billion applications (apps) having been downloaded from the Google Play Store. Android devices account for more than 80% of the overall mobile OS market [13].

The prominence of Android has not escaped the attention of malware writers. According to McAfee, more than 3,000,000 Android malware apps were detected in 2017, representing a 70% increase from 2016. Also, in 2017 alone, more than 700,000 malicious apps were removed from the Google PlayStore [20].

In resource constrained devices, such as Android smartphones, malware detection is typically conducted offline, based on emulation. The objective of this research is to explore the effectiveness of malware detection and classification using dynamic features extracted via emulation, as compared to extracting such features via instrumentation (i.e., on-phone analysis). We classify Android apps using a wide variety of machine learning techniques based on these emulator-extracted and "real" features. We find that emulation fails for a significant percentage of apps and that, surprisingly, the failure rate is higher for benign apps than malicious apps. In contrast to claims that appear in the research literature [25], we find scant evidence that such failures are due to anti-emulation techniques being employed by sophisticated Android malware. Instead, the evidence indicates that Android malware writers fail to take advantage of relatively simple techniques that could serve to make the detection problem considerably more challenging [6, 22, 34].

We note that our analysis technique closely follows that in [2]. However, we go beyond the work in [2] in that we consider additional machine learning techniques, we tune the parameters, and in addition to the detection problem, we also consider the classification problem. Furthermore, we show that a simple ensemble technique can provide essentially ideal separation for the malware detection problem. Finally, with respect to the sophistication of Android malware, we draw diametrically opposed conclusions, as compared to previous work such as [25].

The remainder of this paper is organized as follows. Section 2 provides an overview of various feature analysis methods that have previously been used to successfully detect Android malware, along with an overview of selected examples of Android malware research. In Sect. 3, we discuss the methodology used in our experiments. Section 4 gives our experimental results, along with some discussion of the implications of these results. Finally, Sect. 5 concludes the paper and outlines possible directions for future work.

2 Background

In this section, we first discuss the relative advantages and disadvantages of static and dynamic features for malware analysis. Then we briefly consider the potential weaknesses of emulator-based detection for Android malware.

2.1 *Static and Dynamic Features*

Malware detection can be based on static or dynamic features. Features are said to be static if they are collected without executing (or emulating) the code. On the other hand, dynamic features require code execution or emulation. Examples of popular static features include byte n-grams and mnemonic opcodes, while useful dynamic features include opcodes and application programming interface (API) calls that occur when an app executes. In general, static features can be collected more efficiently than dynamic features. The relative advantage of dynamic features is that detection techniques based on such features are often more robust with respect to common obfuscation techniques [7]. In the Android malware literature, both static and dynamic features have been extensively studied [16].

An Android app consists of a package bundled as an Android Package file, which has the file extension `apk`. Among other things, an `apk` file contains a manifest (`AndroidManifest.xml`), class files (`classes.dex`), and external libraries. Figure 1 lists the components of an `apk` bundle, while Fig. 2 gives an example of a typical manifest file.

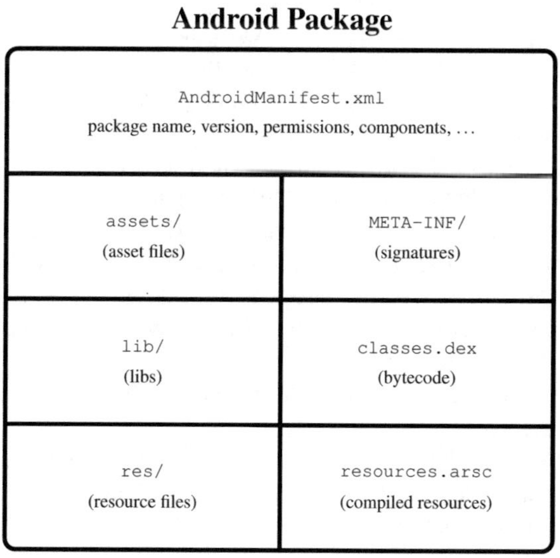

Fig. 1 The parts of an `apk` bundle

```
1  <?xml version="1.0" encoding="utf-8"?>
2  <manifest xmlns:android="http://schemas.android.com/apk/res/android"
3      package="com.novaapps.findevents" android:versionCode="1"
4      android:versionName="1.0"
5      android:installLocation="preferExternal">
6      <uses-sdk android:minSdkVersion="4" />
7      <supports-screens
8          android:largeScreens="true"
9          android:normalScreens="true"
10         android:smallScreens="true"
11         android:resizeable="true"
12         android:anyDensity="true" />
13
14 <uses-permission android:name="android.permission.ACCESS_COARSE_LOCATION" />
15 <uses-permission android:name="android.permission.ACCESS_FINE_LOCATION" />
16 <uses-permission android:name="android.permission.ACCESS_LOCATION_EXTRA_COMMANDS" />
17 <uses-permission android:name="android.permission.READ_PHONE_STATE" />
18 <uses-permission android:name="android.permission.INTERNET" />
19 <uses-permission android:name="android.permission.ACCESS_NETWORK_STATE" />
20
21     <application android:icon="@drawable/icon" android:label="@string/app_name">
22         <activity android:name=".FindEventsGADroidActivity"
23             android:label="@string/app_name" android:configChanges="orientation|keyboardHidden">
24             <intent-filter>
25                 <action android:name="android.intent.action.MAIN" />
26                 <category android:name="android.intent.category.LAUNCHER" />
27             </intent-filter>
28         </activity>
29
30         <activity android:name="com.phonegap.DroidGap" android:label="@string/app_name"
31             android:configChanges="orientation|keyboardHidden">
32             <intent-filter>
33             </intent-filter>
34         </activity>
35     </application>
   </manifest>
```

Fig. 2 Sample `AndroidManifest.xml` file

Many useful static features can be extracted directly from the manifest file. A considerable amount of malware research has focused on static Android features such as permissions (functionality requested by the app). For example, it has been found that number of permissions requested is a surprisingly strong diagnostic [17], with malicious apps requesting more permissions, on average, than benign apps. However, notwithstanding the relative ease and computational efficiency of static analysis, this approach has a significant drawback, as it is relatively easy for malware writers to evade static detection by obfuscating their code. Obfuscation tools are readily available; for example, ProGuard can change data pathnames, variable names, and function names [23].

Dynamic analysis consists of extracting features while code is executing, either on the device for which the code is intended or on an emulator [18]. Some popular dynamic Android features include kernel processes, API calls, and information related to dynamic loading. Dynamic techniques often deal with analyzing internal system calls made by an application at runtime [18]. Previous work has demonstrated the advantage of dynamic features over static features for malware detection [7]. However, the increased efficiency of static feature extraction makes static analysis preferable in cases where it can achieve results that are comparable to dynamic analysis.

To analyze features—static, dynamic, or some combination thereof—researchers can employ a wide variety of machine learning techniques. Examples of such machine learning techniques include *k*-nearest neighbors, hidden Markov models, principal component analysis, support vector machines, clustering, and deep neural networks, among others [30].

From the malware writer's perspective, it is desirable to make a malicious app appear benign under any anticipated analysis. A variety of obfuscation techniques (e.g.. dead code insertion and code substitution) are available to disguise malware and are generally most effective during static analysis. A variety of anti-emulation techniques are available for evading detection by dynamic feature extraction under emulation. These are best understood in the context of the following discussion of emulation.

2.2 *Emulation*

Android malware can access sensitive information such as call history, text and contacts, and can tamper with phone settings. To do this, malicious apps often try to read the background environment via API calls. Examining the result of selected API calls can enable a malicious app to identify the environment on which the code is executing and thereby determine how best to attack the device [2]. Emulators are not entirely faithful to real phone APIs, and malicious apps can use these discrepancies to detect when they are being executed in an emulated environment and therefore should restrict suspicious behavior. An example of an API that can be used to detect emulators is the Telephony Manager API, that is,

```
TelephonyManager.getDeviceId()
```

A call to this API typically returns `000000000000000` when an emulator is executing the code. A real physical device, on the other hand, would not return 0 as the device identifier. This is one of the emulator detection methods that is used by the Pincer family of Android malware [34]. Emulator detection is a significant challenge to security analysis, because most emulators use open source hypervisors such as QEMU, which have detectable identifying functionality [15]. It has been claimed that the Morpheus malware app employs more than 10,000 heuristics to classify its runtime environment [3].

To deal with issues such as these, researchers have attempted to develop improved emulators. Several dynamic analysis tools such as TaintDroid [9], DroidBox [9], CopperDroid [32], Andrubis [19, 35], and AppsPlayground [26] have been developed. In addition, online tools are available for Android application analysis, including SandDroid [27], TraceDroid [33], and NVISO ApkScan [21]. However, these dynamic approaches still rely on emulators or virtualized environments which malware can detect by careful analysis [21].

Since it is possible for Android malware to detect an emulated environment, we might assume that malware would check for emulation and behave benignly when an emulator is detected. Indeed, it has been claimed that such is the case for most Android malware [2]. However, our results indicate that the Android malware datasets used in our experiments are not, in general, using advanced emulation avoidance techniques to any greater degree than benign apps. This observation is based, in part, on the fact that we find that benign apps fail to run in our emulation environment at a higher rate than malware. In addition, we find that these failures are easily explained by the limitations of the emulation environment, rather than advanced anti-emulation strategies.

2.3 Selected Android Malware Research

The authors of the paper [10] study packed Android malware. These authors show that in the time period from 2010 to 2015, about 13% of the Android malware that they consider was packed, and that sophisticated Android malware samples often use (and abuse) custom packers. Similar to code encryption, packing is a well-known technique for defeating signature-based and some other static detection techniques [4]. However, in this paper, we only consider dynamic analysis, which should be unaffected by code packing.

The work [5] considers the problem of detecting privacy leak caused by Android malware. The authors employ a differential analysis technique, here they vary certain key parameters and look for changes in network activity that are evidence of private data leaking. The authors show that their technique is practical and effective. Additional research on the privacy leak problem can be found in [31], where the authors develop and analyze an information flow analysis tool, TAINTART, which can be viewed as an improved version of TaintDroid [37]. Such privacy leakage and information flow work is relevant to the problem consider in this paper, and it serves to illustrate ways that, for example, features could be collected in an Android environment.

The research presented in [12] considers the interesting and challenging problem of detecting Android malware that contains a "logic bomb," which the authors define to be malicious code that only executes under some narrow circumstance. Such code might be used, for example, in an attack that is carefully targeted at a specific user or other entity, and seems to be relatively common in malware developed by nation states. This paper is focused on a narrow and apparently rare class of Android malware, whereas our research considers the general Android malware detection problem.

The main insights in the paper [11] is that Android intents are a stronger feature than permissions. An Android intent is a messaging object that can be used to request an action from another app component [14]. While permissions have been extensively studied in the literature, intents have received much less attention. The authors of [11] also consider a combination of the two feature types—intents and permissions—and show that this yields improved results, as compared to using intents only.

The authors of [25] have developed a tool to extract runtime features, from obfuscated Android malware. For example, encrypted SMS numbers cannot be detected via static analysis and malware that can detect an emulation environment could also hide such data at runtime.

Finally, we note that many research papers claim that it is common for Android malware apps to employ emulation-detection techniques to hide features, while many other research papers implicitly assume that such is the case. Indeed, this assumption is the impetus for considerable research in the Android malware domain. For example, in [25], it is stated that "many malicious applications" use emulation-detection techniques, but no evidence is provided as to the percentage of such applications that actually occur in their malware dataset. Furthermore, the papers cited in [25] as evidence of the supposed widespread use of such detection-avoidance techniques, namely [6, 22, 34], do not provide such numbers either, and instead simply show that it is possible (and, in fact, relatively easy) to implement such feature-hiding capabilities. We return to this issue in Sect. 5.

3 Methodology

This section describes the process we followed to dynamically extract features from Android apps. We extract such features from both benign and malicious apps, using both emulation and on-phone instrumentation. But first, we briefly discuss the datasets used in our experiments before providing details on the feature extraction process.

3.1 *Datasets*

AMGP dataset This dataset is part of the Android Malware Genome Project [38], and it has been used in numerous research papers, including [2]. Of the 2444 apps in the dataset, half are malicious apps from 49 different families, with the remainder being benign apps from McAfee Labs [2]. We use this dataset for binary classification experiments, where we classify samples as either malware or benign.

Drebin dataset We also experiment with 3206 samples from the seven malware families in the Drebin dataset [8]. The list of families and the number of samples from each are given in Table 1. This dataset was used in experiments where we attempted to classify samples into their respective families, as opposed to binary classification (i.e., malware and benign) experiments

Table 1 Drebin data

Family	Apps
FakeInstaller	925
DroidKungfu	667
Plankton	625
Opfake	613
Iconosys	152
Fakedoc	132
Geinimi	92
Total	3206

3.2 Feature Extraction

Feature extraction is a critical aspect of this research, as our approach is based on comparing results from various machine learning techniques, using features collected via emulator versus features collected directly from a phone-based environment. Therefore, feature extraction was performed for both environments, as described below.

Phone environment The Android smartphone used for data collection was configured as follows: Android 5.0 Lollipop, 1.3 GHz CPU, 16 GB internal memory, and 32 GB of external SD card storage. The phone contained a SIM card with activated service to enable 3G data use and outgoing calls. This configuration is consistent with that used in [2]. As discussed in [2], USB 2.0 or 3.0 was used along with the Linux VM so as to avoid the timeout that would result from a USB 1.0 connection with files larger than 1MB.

Emulation environment A Santoku Linux VirtualBox was used to emulate an Android device. The environment was configured as follows: 8 GB of external SD card memory, 2 MB of memory, 4.1.2 Jelly Bean (API level 16, Android version). To more accurately simulate the workings of a real phone, the emulator was enhanced with contact numbers, images, pdf files, and text files. The default IMEI, IMSI, SIM serial number, and phone numbers were altered. After each application was executed, the emulator was re-initialized to ensure the removal of third party apps. This emulation process is consistent with that used in [2].

DynaLog is a dynamic framework that accepts a large number of apps as input, launches them serially in the emulator environment, creates logs of dynamic features, and extract these features for future processing [1]. At the core of DynaLog is the MonkeyRunner API that is able to stimulate apps with random events that are typical of user interactions (e.g.. pressing, swiping, and touching the screen). These simulated actions are designed to stimulate a significant fraction of code functionality.

To extract dynamic features from the phone, we call DynaLog using a Python based tool, as described in [1]. Each app was executed for 15 min during which time we logged and collect dynamic features from the phone, as well as from an emulator

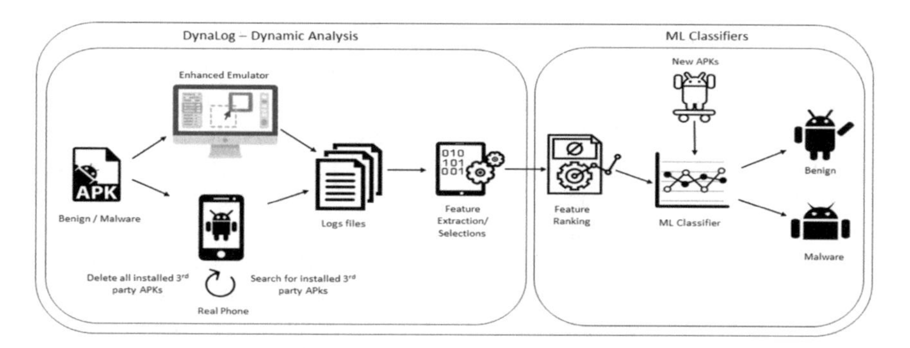

Fig. 3 DynaLog [2]

running the same apps with the same input events. Figure 3 illustrates the use of DynaLog [2].

The data collected from the phone and emulator was saved in files in the `arff` format suitable for feature vector input to machine learning platforms. The 178 features form these vectors were loaded into Weka [36]. The features were then ranked based on information gain (`InfoGain` in Weka) and the top 100 features from each analysis environment (phone and emulation) were then used to test and train the machine learning algorithms considered in this paper.

3.3 Machine Learning Models

In this section, we briefly discuss each of the nine machine learning techniques used in this research. These nine algorithms cover a broad range of techniques, ranging from relatively simple statistical scores to advanced neural networks.

Support Vector Machine A support vector machine (SVM) represents data as points in a high-dimensional space, and computes a hyperplane or manifold that separates points of different classes. The multiclass version of an SVM is known as a support vector classifier (SVC).

Naïve Bayes This approach uses Bayes' theorem to compute probabilities of data points belonging to classes. To simplify computation, features are "naïvely" assumed to be independent of each other even when they are actually dependent.

Simple Logistic Simple logistic is an ensemble learning algorithm that uses multiple simple regression functions to model the training data, computing weights that maximize the log-likelihood of the logistic regression function.

Multilayer Perceptron A multilayer perceptron (MLP) is a feedforward neural network that includes an input layer, an output layer, and one or more hidden layers. MLPs are trained by backpropagation with gradient descent to minimize errors.

IBk This is the Weka implementation of the k-nearest neighbors algorithm, using a Euclidean distance metric to define "nearest." Given an integer k, the algorithm classifies a point in feature space by considering its k nearest classified neighbors.

Partial Decision Trees A partial decision tree (PART) is a simple decision tree that contains branches to undefined sub-trees. In order to develop a partial decision tree, construction and pruning operations are used, with the goal of finding a sub-tree that cannot be further simplified.

J48 Decision Tree An implementation of the C4.5 decision tree algorithm, J48 repeatedly splits on the remaining feature with highest information gain.

Random Forest The random forest (RF) technique relies on a "forest" of decision trees. That is, multiple decision trees are trained, and a majority vote of the trees is used for classification. The RF algorithm uses bagging, whereby subsets of features and samples are selected to construct the component trees. Bagging enables a random forest to greatly reduce the overfitting problem that is inherent in elementary decision trees.

AdaBoost Boosting is a general machine learning technique that can build a strong classifier from a number of weak classifiers. AdaBoost uses a simple adaptive strategy to build such a classifier. The implementation of AdaBoost that we employ is based on decision tree classifiers.

3.4 Evaluation Metrics

From the point of view of this analysis, a positive classification is an identification as malware. We tabulated true/false positive/negative rates for all analyses. Sensitivity and recall are terms that are equivalent to true positive rate. Precision is the ratio of true positives to the number of samples that are classified as positives. Thus, in our binary classification experiments, precision tells us the fraction of samples classified as malware that are actually malware. The primary metric we use in this paper is the F-measure, which is defined as

$$\text{F-measure} = \frac{2 \times \text{precision} \times \text{recall}}{\text{recall} + \text{precision}}$$

By combining both precision and recall into a single statistic, the F-measure provides a useful single value for comparing machine learning approaches.

4 Experiments and Results

This section presents the results of two broad classes of experiments. Our first category of experiments deals with evaluating the effectiveness of Android malware detection based on features extracted via emulation, as compared to features extracted

Table 2 AMGP dataset feature extraction success

Type	Emulator		Phone	
	Number	Percentage	Number	Percentage
Malware	956	78.23	1211	99.09
Benign	807	66.03	1119	91.57
Total	1763	72.13	2330	95.33

Table 3 Drebin dataset feature extraction success

	Emulator	Phone
Number	2598	3201
Percentage	81.03	99.84

directly from a phone. In these experiments, we use the same dataset and feature extraction tools as in [2]. Moreover, we consider additional machine learning techniques, we tune the parameters of the machine learning algorithms,[1] we consider a multi-sensor solution, and we ultimately draw somewhat different conclusions based on our results. We refer to this first set of experiments as malware detection experiments.

Our second set of experiments involves classifying malware samples into families. Again, we consider a variety of machine learning algorithms and we compare the results obtained when using emulator and phone-based features. We refer to this second set of experiments as malware classification experiments.

Before presenting these experimental results, we first discuss the data collection phase in some detail. This is an important issue, since we were not able to extract features from all apps using the automated approach considered here.

4.1 Emulation Versus Instrumentation

Table 2 gives the percentage of apps from the AMGP dataset that we were able to analyze using emulation, as well as the percentage of apps that we could evaluate using on-phone analysis. Table 3 gives analogous results for the Drebin dataset. Recall that the AMGP dataset is evenly split between malware and benign apps, with 1222 in each category; the Drebin dataset contains 3206 malware apps, with the breakdown by family given in Table 1.

Tables 2 and 3 show that nearly 20% fewer malicious Android apps allow for feature extraction using emulation, as compared to the on-phone environment, and this is consistent across both datasets. This has led some researchers to conclude

[1]Based on our experiments, it appears that the authors of [2] consistently used the Weka default settings for their machine learning experiments.

Table 4 Features extracted only from phone environment (AMGP dataset)

Feature	Phone	Emulator
`System;loadLibrary`	212	0
`URLConnection;connect`	15	0
`Context;unbindService`	4	0
`Service;onCreate`	3	0
`BATTERY_LOW`	1	0
`SmsManager;sendTextMessage`	3	0

that anti-emulation techniques must be widely used in Android malware. If such is the case, it is not clear why benign apps would use anti-emulation techniques at an even higher rate than malicious apps—compare the benign and malware results in Table 2. This raises questions as to whether the results for malicious apps are really due to anti-emulation techniques, or whether there might be another explanation.

A more plausible reason why we are able to automatically extract features from more apps using on-phone instrumentation is simply because more APIs can be executed on a phone environment. This is especially an issue for apps that make API calls related to network activity or read incoming and outgoing call activity. Whether such apps are benign or malicious, the phone is able to provide such capabilities and thereby log the relevant API activity, while emulators are not sufficiently sophisticated to simulate all necessary APIs. Manual analysis of a number of apps that fail under emulation reveals that network and call-related issues are indeed responsible for emulation failures for both malicious and benign apps.

Table 4 lists the features that were extracted exclusively from the phone but not by the emulator. For example, the `System.loadLibrary` feature is the API call associated with native code; it is probably not logged under emulation because the emulator does not support native code [2]. The phone based analysis shows a much higher effectiveness in extracting features for analysis; this is clearly an essential benefit for machine learning classification.

4.2 Binary Classification Experiments

In this section, we give the results for binary classification experiments using the AMGP dataset. We consider each of the nine machine learning techniques discussed in Sect. 3.3, and compare the results for features extracted via emulation against results for features extracted via on-phone instrumentation. All experiments were performed using Weka with 10-fold cross validation. The models were fine-tuned over various input parameters, with the following list giving some of the important settings.

Table 5 Results for emulator based features (AMGP dataset)

Model	TPR	FPR	TNR	FNR	F-measure
Simple logistic	0.902	0.097	0.903	0.098	0.901
Naïve Bayes	0.599	0.098	0.902	0.401	0.734
SVM	0.914	0.094	0.906	0.086	0.908
PART	0.902	0.099	0.901	0.098	0.899
J48	0.892	0.116	0.884	0.108	0.886
RF	0.916	0.063	0.937	0.084	0.928
MLP	0.941	0.087	0.913	0.059	0.926
IBk	0.899	0.096	0.904	0.101	0.903
AdaBoost	0.901	0.101	0.899	0.099	0.900

Simple Logistic The ridge estimator for regularization is used to reduce the size of coefficients. The model is trained until it converges.

Naïve Bayes Default values are used for the kernel and for discretization.

Support Vector Machine The complexity parameter C is set to 1.0 and a polynomial kernel is used.

Decision Trees We experimented with various depths for the trees (the `maxDepth` parameter in Weka) and the best accuracy was obtained with a depth of 50. The `noPruning` option was set to `False`.

Random Forest The model yielded the best accuracy with 100 trees and this is what we use in all experiments reported here.

Multilayer Perceptron The number of hidden layers is chosen to be 3.

IBk We use the Euclidean distance with 10 neighbors.

AdaBoost The classifier we use is the decision stump algorithm.

Using features collected from the emulator, we obtain the results in Table 5. From these results, we see that the best accuracy is achieved by a random forest with 100 trees, while an MLP yields a similar F-measure.

For our next set of experiments, we repeat the above analyses, but using features extracted via on-phone instrumentation, with all algorithms parameterized exactly as in the emulation case. Results for these experiments are summarized in Table 6. As with the emulation-based results, the random forest and MLP again perform the best.

We performed another set of experiments on the AMGP dataset using only those apps that successfully executed in both the emulator and on-phone environments. For these apps, the results of testing and training the various machine learning models based on features extracted from the emulator are given in Table 7.

The results in Table 7 again show that random forest yielded the best results. Furthermore, the random forest experiments in Table 7 yielded nearly identical results to those in Table 5. However, for the other techniques, the results are generally

Table 6 Results for phone based features (AMGP dataset)

Model	TPR	FPR	TNR	FNR	F-measure
Simple Logistic	0.923	0.081	0.919	0.077	0.921
Naïve Bayes	0.634	0.119	0.881	0.366	0.748
SVM	0.918	0.090	0.910	0.082	0.914
PART	0.907	0.098	0.902	0.093	0.905
J48	0.929	0.101	0.899	0.071	0.916
RF	0.942	0.074	0.926	0.058	0.934
MLP	0.924	0.082	0.918	0.076	0.925
IBk	0.906	0.086	0.914	0.094	0.910
AdaBoost	0.908	0.087	0.913	0.092	0.906

Table 7 Apps executed in both environments (AMGP data and emulator features)

Model	TPR	FPR	TNR	FNR	F-measure
Simple Logistic	0.887	0.104	0.896	0.113	0.891
Naive Bayes	0.542	0.169	0.831	0.458	0.663
SVM	0.896	0.116	0.884	0.104	0.889
PART	0.896	0.116	0.884	0.104	0.892
J48	0.874	0.088	0.912	0.126	0.894
RF	0.919	0.066	0.934	0.081	0.927
MLP	0.898	0.096	0.904	0.102	0.902
IBk	0.904	0.090	0.910	0.096	0.907
AdaBoost	0.901	0.093	0.907	0.099	0.902

slightly lower than either the exclusively emulator-based or instrumentation-based experiments considered above.

In order to assess the value of analysis with multiple machine learning models, error rates were considered as a function of the number of models. In Table 8 (a), we give results for false negatives (FN), for both the emulator and on-phone features. The row labeled with n in the table gives the number of malware apps that were misclassified as benign by n or more of the nine machine learning techniques considered, based on emulator features (middle column) or on-phone features (last column). Table 8 (b) gives the analogous results for false positives. These results are summarized in the form of line graphs in Fig. 4.

Suppose that we base our classification on a majority vote of the nine machine learning models considered above. Then the numbers in Table 8 (a) and (b) imply that when using the emulator features, we would have only 7 false negatives and 3 false positives, while the corresponding numbers for the on-phone features is 3 false negatives and 0 false positives. The corresponding accuracies and F-measures are

Table 8 Classification errors and machine learning models (AMGP dataset)

Number of models	Features	
	Emulator	Phone
(a) *False negatives*		
1	104	91
2	72	64
3	36	28
4	19	11
5	7	3
6	2	0
7	0	0
8	0	0
9	0	0
(b) *False positives*		
1	73	62
2	46	30
3	21	16
4	9	4
5	3	0
6	0	0
7	0	0
8	0	0
9	0	0

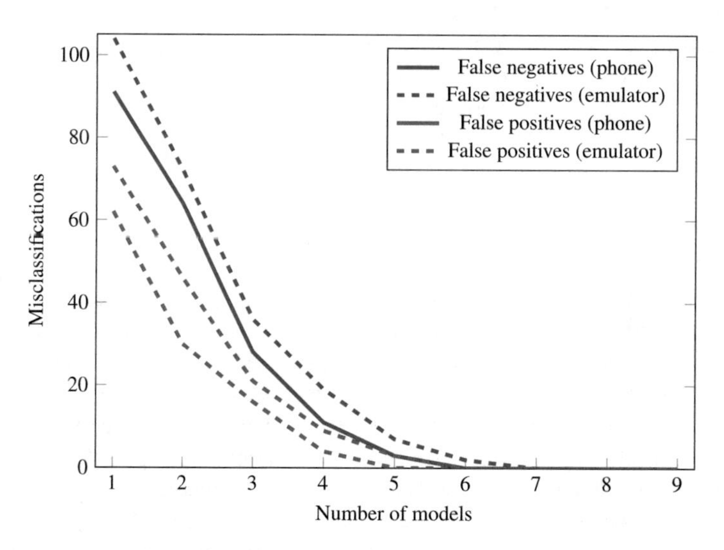

Fig. 4 Misclassifications as a function of the number of models

Table 9 Majority vote of machine learning models (AMGP dataset)

Features	Accuracy	F-measure
Emulator	0.9960	0.9959
Phone	0.9988	0.9988

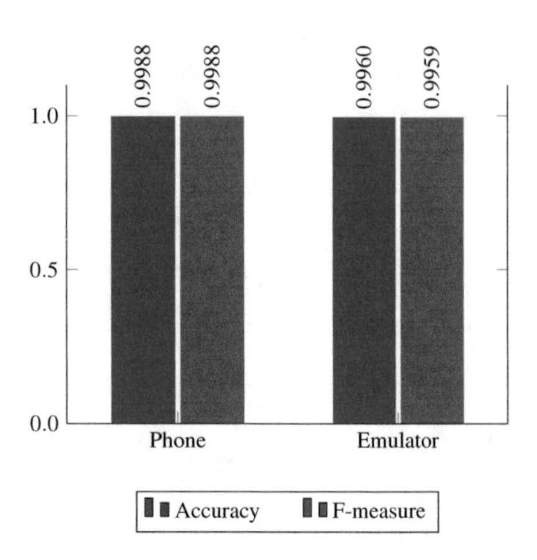

Fig. 5 Majority vote of models

given in Table 9 and in the form of bar graphs in Fig. 5. These results—which are virtually ideal—are far stronger than any of the individual models, and indicate the potential strength of a multi-sensor approach. More sophisticated techniques of combining the output of multiple machine learning models could potentially yield equally strong results with fewer models. For example, in the malware domain, SVMs have been used to combine multiple scores into a single machine learning model [28] and boosting techniques can produce a strong combined classifier [24].

4.3 Multiclass Experiments

In this section we give multiclass results based on a support vector classifier (SVC), which is the mulitclass version of an support vector machine (SVM). For these experiments, we employ the Drebin dataset and we use a linear kernel in all cases. As in the binary classification experiments above, the goal is to compare the performance of models trained on features that have been extracted using on-phone instrumentation with models trained on features extracted via emulation. We expect the multi-family classification problem to be inherently more challenging than the binary classification (malware versus benign) problem due to the larger number of classes.

Table 10 shows the results for our multiclass experiments, with Fig. 6 giving these same results in the form of line graphs. This table and figure include results for both feature extraction environments (emulation and on-phone instrumentation). Note that

Table 10 Family classification results (Drebin dataset)

Families	Combinations	Emulator	Phone
2	21	0.9278	0.9364
3	35	0.9182	0.9276
4	35	0.9113	0.9202
5	21	0.9079	0.9184
6	7	0.8982	0.9064
7	1	0.8890	0.8997

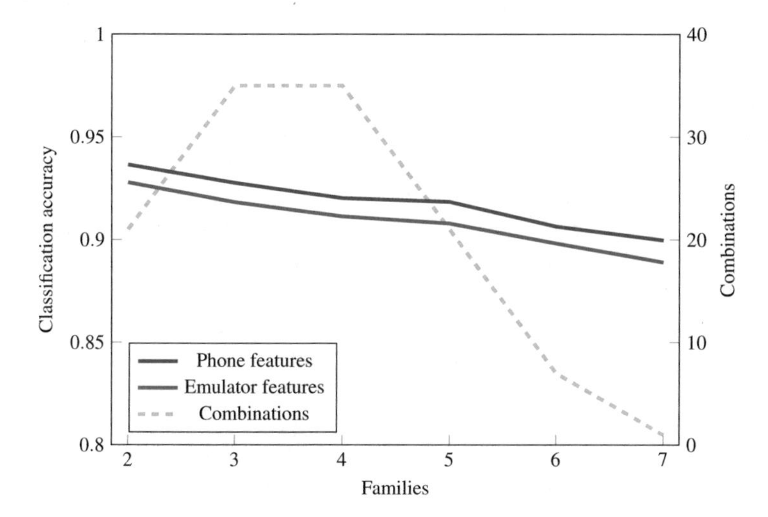

Fig. 6 Family classifications

there are seven families in the Drebin dataset (see Table 1), and we have conducted experiments with each of the 127 nontrivial combinations of these families. The accuracy reported in Table 10 and Fig. 6 for k families is the average of all $\binom{7}{k}$ possible combinations of k families. From these results we see that the on-phone features yield consistently better results than the emulation features, but—as with the binary classification experiments discussed above—the differences are slight. It is interesting that the classification accuracies are so high, which seems to indicate that the families in this dataset may differ substantially from one another.

5 Conclusion and Future Work

In this research, we have considered Android malware detection and classification. Our primary focus was to compare the effectiveness of features extracted on-phone with features extracted using emulation and to consider the implications of these results. In our binary classification experiments we considered nine machine learning techniques (support vector machines, random forest, naïve Bayes, multilayer perceptron, simple logistic, J48 decision tree, PART, IBk, and AdaBoost). We used support vector machines in our classification experiments.

In all cases, we obtained strong results as measured by the F-measure statistic. Although the on-phone features performed marginally better than the emulation features, we conclude that the additional overhead of on-phone analysis is unlikely to be worthwhile in most situations. That is, the incremental reduction in error rates is unlikely to be cost-effective.

A simple majority vote of our nine classifiers yielded essentially perfect detection and F-score results, as given in Table 9. These results exceed those found in previous work, such as [2].

Our results also call into question the oft-stated claim that Android malware frequently uses anti-emulation techniques. Instead, we believe that these experiments offer evidence that Android malware is actually much less sophisticated than is sometimes claimed. In fact, this is easily confirmed by a manual analysis of apps—malware and benign—that fail in the emulation environment. We find that such apps fail simply due to the inability of the emulator to handle call, networking, and similar APIs.

Future work could include a similar analysis on larger and more recent Android malware datasets. While it is not the case that anti-emulation was effectively used by the malware in our datasets, it would not be difficult for a moderately skilled malware writer to generate apps that would be much more challenging to detect. Work involving a more recent dataset would be a way to determine whether Android malware writers have started taking advantage of such techniques.

References

1. Alzaylaee MK, Yerima SY, Sezer S (2016) DynaLog: an automated dynamic analysis framework for characterizing Android applications. In: 2016 international conference on cyber security and protection of digital services, Cyber Security 2016, pp 1–8. arXiv:1607.08166
2. Alzaylaee MK, Yerima SY, Sezer S (2017) EMULATOR vs REAL PHONE: Android malware detection using machine learning. In: Proceedings of the 3rd ACM on international workshop on security and privacy analytics, IWSPA '17, pp 65–72
3. Amos B, Turner HA, White J (2013) Applying machine learning classifiers to dynamic Android malware detection at scale. In: 9th international wireless communications and mobile computing conference, IWCMC 2013, pp 1666–1671
4. Aycock J (2006) Computer viruses and malware. Advances in information security. Springer US
5. Continella A, Fratantonio Y, Lindorfer M, Puccetti A, Zand A, Krügel C, Vigna G (2017) Obfuscation-resilient privacy leak detection for mobile apps through differential analysis. In: 24th annual network and distributed system security symposium, NDSS, 2017. http://www.s3.eurecom.fr/~yanick/publications/2017_ndss_agrigento.pdf
6. Coogan K, Debray S, Kaochar T, Townsend G (2009) Automatic static unpacking of malware binaries. In: 16th working conference on reverse engineering, WCRE 2009, pp 167–176

7. Damodaran A, Di Troia F, Visaggio CA, Austin TH, Stamp M (2017) A comparison of static, dynamic, and hybrid analysis for malware detection. J Comput Virol Hacking Tech 13(1):1–12
8. The Drebin dataset. https://www.sec.cs.tu-bs.de/~danarp/drebin/
9. DroidBox Google archive. https://code.google.com/archive/p/droidbox/
10. Duan Y, Zhang M, Bhaskar AV, Yin H, Pan X, Li T, Wang X, Wang X (2018) Things you may not know about Android (un) packers: a systematic study based on whole-system emulation. In: 25th annual network and distributed system security symposium, NDSS, pp 18–21. https://www.informatics.indiana.edu/xw7/papers/ndss18-paper296.pdf
11. Feizollah A, Anuar NB, Salleh R, Suarez-Tangil G, Furnell S (2017) AndroDialysis: analysis of Android intent effectiveness in malware detection. Comput Secur 65:121–134. http://www0.cs.ucl.ac.uk/staff/G.SuarezdeTangil/papers/2017cosec-androdialysis.pdf
12. Fratantonio Y, Bianchi A, Robertson W, Kirda E, Kruegel C, Vigna G (2016) Triggerscope: towards detecting logic bombs in Android applications. In: 2016 IEEE symposium on security and privacy, SP 2016, pp 377–396. https://sites.cs.ucsb.edu/~vigna/publications/2016_SP_Triggerscope.pdf
13. Global smartphone shipments by OS. https://www.statista.com/statistics/263437/global-smartphone-sales-to-end-users-since-2007/
14. Intents and intent filters: Android developers guide. https://developer.android.com/guide/components/intents-filters
15. Jing Y, Zhao Z, Ahn G-J, Hu H (2014) Morpheus: automatically generating heuristics to detect Android emulators. In: Proceedings of the 30th annual computer security applications conference, ACSAC '14, pp 216–225,
16. Kang H, Jang J, Mohaisen A (2015) Kim HK (2015) Detecting and classifying Android malware using static analysis along with creator information. Int J Distrib Sens Netw 7(1–7):9
17. Kapratwar A, Di Troia F, Stamp M (2017) Static and dynamic analysis of Android malware. In: Mori P, Furnell S, Camp O (eds) Proceedings of the 3rd international conference on information systems security and privacy, ICISSP 2017, Porto, Portugal. SciTePress, pp 653–662, 19–21 Feb 2017
18. Lindorfer M, Neugschwandtner M, Platzer C (2015) MARVIN: efficient and comprehensive mobile app classification through static and dynamic analysis. In: IEEE 39th annual computer software and applications conference, COMPSAC 2015, pp 422–433
19. Lindorfer M, Neugschwandtner M, Weichselbaum L, Fratantonio Y, van der Veen V, Platzer C (2014) Andrubis–1,000,000 apps later: a view on current Android malware behaviors. In: Proceedings of the international workshop on building analysis datasets and gathering experience returns for security, BADGERS 2014, Wroclaw, Poland, Sept 2014
20. McAfee threats report 2017. https://www.mcafee.com/us/resources/reports/rp-quarterly-threats-dec-2017.pdf
21. NVISO Apkscan. https://apkscan.nviso.be/
22. Petsas T, Voyatzis G, Athanasopoulos E, Polychronakis M, Ioannidis S (2014) Rage against the virtual machine: hindering dynamic analysis of Android malware. In: Proceedings of the seventh European workshop on system security, EuroSec '14, pp 5:1–5:6
23. Pincer Android attacks. https://www.fsecure.com/weblog/archives/00002538.html
24. Raghavan A, Di Troia F, Stamp M (2019) Hidden Markov models with random restarts versus boosting for malware detection. J Comput Virol Hacking Tech 15(2):97–107
25. Rasthofer S, Arzt S, Miltenberger M, Bodden E (2016) Harvesting runtime values in Android applications that feature anti-analysis techniques. In: 23rd annual network and distributed system security symposium, NDSS, 2016. https://www.bodden.de/pubs/ssme16harvesting.pdf
26. Rastogi V, Chen Y, Enck W (2013) AppsPlayground: automatic security analysis of smartphone applications. In: Proceedings of the third ACM conference on data and application security and privacy, CODASPY '13, pp 209–220
27. SandDroid—an automatic Android application analysis system. http://sanddroid.xjtu.edu.cn/
28. Singh T, Di Troia F, Visaggio CA, Austin TH, Stamp M (2016) Support vector machines and malware detection. J Comput Virol Hacking Tech 12(4):203–212

29. Smartphone OS market share worldwide 2009–2017. https://www.statista.com/statistics/263453/global-market-share-held-by-smartphone-operating-systems
30. Stamp M (2017) Introduction to machine learning with applications in information security. Chapman and Hall/CRC, Boca Raton
31. Sun M, Wei T, Lui JC (2016) TaintART: a practical multi-level information-flow tracking system for Android runtime. In: Proceedings of the 2016 ACM SIGSAC conference on computer and communications security, CCS '16, pp 331–342. https://www.cse.cuhk.edu.hk/~cslui/PUBLICATION/CCS16.pdf
32. Tam K, Khan SJ, Fattori A, Cavallaro L (2015) CopperDroid: automatic reconstruction of Android malware behaviors. In: NDSS symposium, NDSS 2015, pp 8–11
33. Tracedroid. https://github.com/ligi/tracedroid
34. Vidas T, Christin N (2014) Evading Android runtime analysis via sandbox detection. In: Proceedings of the 9th ACM symposium on information, computer and communications security, ASIA CCS '14, pp 447–458
35. Weichselbaum L, Neugschwandtner M, Lindorfer M, Fratantonio Y, van der Veen V, Platzer C (2014) Andrubis: Android malware under the magnifying glass. Technical Report TR-ISECLAB-0414-001, Vienna Univeristy of Technology, 5
36. Weka 3: machine learning software in Java. https://www.cs.waikato.ac.nz/ml/weka/index.html
37. Yan L-K, Yin H (2012) DroidScope: seamlessly reconstructing the OS and Dalvik semantic views for dynamic Android malware analysis. In: USENIX security symposium, USENIX 2012, pp 569–584. http://www.cs.columbia.edu/~lierranli/coms6998-11Fall2012/papers/droidscope_usenixsec2012.pdf
38. Zhou Y, Jiang X (2012) Android malware genome project. http://www.malgenomeproject.org

Towards a Generic Approach of Quantifying Evidence Volatility in Resource Constrained Devices

Jens-Petter Sandvik, Katrin Franke, and André Årnes

Abstract Forensic investigations of the Internet of Things (IoT) is often assumed to be a combination of existing cloud, network, and device forensics. Resource constraints in many of the peripheral things, however, are affecting the volatility of the potential forensic evidence, and evidence dynamics. This represents a major challenge for forensic investigations. In this chapter, we study the dynamics of volatile and non-volatile memory in IoT devices, with the *Contiki* operating system as an example. We present a way forward to quantifying volatility during the evidence identification phase of a forensic investigation. Volatility is expressed as the expected time before potential evidence disappears. This chapter aims to raise awareness and give a deeper understanding of the impact of IoT resource constraints on volatility and the dynamics of forensic evidence. We exemplify in which way volatility can be quantified for a popular operating system and provide a path forward to generalize this approach. The quantification of the volatility of potential evidence helps investigators to prioritize acquisition and examination tasks to maximize the likelihood of collecting relevant evidence from resource-constrained devices. Our work contributes to establishing a scientific base for evidence volatility and evidence dynamics in IoT devices. It strengthens methods for on-scene triage, event reconstruction, and for assessing the reliability of evidence findings.

J.-P. Sandvik (✉) · K. Franke · A. Årnes
Norwegian University of Science and Technology (NTNU), Trondheim, Norway
e-mail: jens-petter.sandvik@politiet.no

J.-P. Sandvik
National Criminal Investigation Service (KRIPOS), Oslo, Norway

A. Årnes
Telenor Group, Fornebu, Norway

R. Montasari et al. (eds.), *Digital Forensic Investigation of Internet of Things (IoT) Devices*, Advanced Sciences and Technologies for Security Applications,
https://doi.org/10.1007/978-3-030-60425-7_2

1 Introduction

As the Internet of Things (IoT) gains traction, the number of criminal cases involving IoT systems is increasing. The increase in IoT ubiquity in all aspects of daily life will extend both the dependence on these systems and increase the number of devices used for crimes. As more IoT systems will sense their environments, they will also act as new sources of evidence for activities in their environment. From these IoT systems, data and information are in a fast flux, and a crime investigator has to prioritize his or her efforts to collect the relevant data as evidence as long as it exists for the criminal case. A formal approach to volatility quantification requires a well-defined terminology of evidence dynamics and volatility. In this chapter, we are defining key concepts and motivating the formal approach, which will be detailed in the remainder of the chapter.

Challenges introduced by IoT systems for digital forensics are abundant [1]. A subset of IoT forensic challenges that affect volatility are summarized as follows: (i) The ubiquity of their presence, (ii) the resource-constraints, (iii) the lack of interfaces for forensic data collection, and (iv) the data process flow. The data process flow makes data generated by a device hard to locate and to collect, and it can change the data during its lifetime in the system. The set of data that is collected from the system and is used in the investigation is regarded as *evidence*. The changes to data that will be used as evidence are part of the *evidence dynamics*. Evidence dynamics is a term used for all changes a piece of evidence experiences from the creation of the data to the case has been presented in court [2].

Volatility is a term that describes the time interval before evidence disappears, and the term will be defined in this chapter. The disappearance of evidence is a change that happens to it, and it can thus be seen as a subset of evidence dynamics.

It is not only the IoT devices that are resource-constrained but also forensic investigations are limited by resource constraints. This is a double burden. The resources that are available for an investigation are finite, and this includes both manpower, time, and equipment. An investigator with access to several possible sources of evidence needs to prioritize between these to optimize the probability of finding the most valuable evidence. This prioritization task is often referred to as *triage*. Roussev et al. defined triage as "[...] a partial forensic examination conducted under (significant) time and resource constraints" [3].

During triage, the investigator needs reliable and objective sources of information to prioritize the data and evidence collection. Given the available resources for the forensic investigation as well as the evidence dynamics, this prioritization is done to maximize the probability of finding relevant data and evidence. Objective and reliable sources of information about the IoT system can help reduce human errors due to cognitive biases a human investigator is susceptible to.

To overcome the misconception that no evidence can be found in resource-constrained peripheral devices, we are aiming to provide an objective measurement to determine the time window where relevant data is most likely to exist, despite its evidence dynamics. This will increase the confidence of the investigator that evidence exists and where it can be found. The knowledge about the likelihood of some evidence is still present in the system after a given time can be of help when prioritizing between collecting data from two different devices that both might contain evidence, but where one has much lower volatility than the other.[1]

This chapter focuses on the volatility of the evidence, and how we can measure it, which is the objective measure as motivated above. Our contributions are:

- A model of the volatility, to better understand the influencing elements of volatility, with the operating system (OS) Contiki as an example.
- The use of statistical tools to measure the volatility, which borrows from the field of dependability and reliability analysis.

The model of the volatility is a construct to split the analysis of the volatility into smaller, well-defined elements that individually contribute to the volatility for the whole IoT system. The statistical method is a quantification of the contributions of both the individual and the combined elements in the volatility model of the IoT system.

Section 2 describes related work in volatility and how this term has been used. Section 3 introduces the concepts of data volatility and information volatility, together with a model for both data volatility and information volatility. Section 4 introduces the use of statistical methods for measuring the volatility. Section 5 uses the Contiki OS as an example of how the model can be used. Section 6 summarizes and concludes this chapter together with discussions on further work.

2 Related Work

The research in volatility has been focused on the acquisition process, and how to collect evidence in a forensically sound matter, such that the collection process does not change or otherwise overwrite relevant evidence, maintaining evidence integrity. In the case of such changes happening, the acquisition should minimize the number of changes and document what has changed as a part of the chain of custody. This leads to the concept of *order of volatility* (OOV).

The IETF Request For Comments (RFC) 3227, "Guidelines for Evidence Collection and Archiving", is a best practice guide for collecting and preserving evidence from computer systems [4]. In this guide, the *order of volatility* is listed as an important aspect of evidence collection, as the evidence should preferably be collected from the most volatile evidence, and proceed with the less volatile evidence. The order of volatility thus forms the order for prioritizing evidence collection. Examples of evidence in this RFC, ordered in decreasing volatility order, are registers and cache;

[1]This assessment can go both ways. Either prioritize the high volatility device to collect data before it disappears, or prioritize the low volatility one because there is only time to collect data from one of the devices.

routing table, ARP cache, process table, kernel statistics and memory; temporary file systems; disk; remote logging and monitoring; physical configuration and topology; and archival media.

The order of volatility is an assessment tool, and during an investigation, the investigator can decide to rather decrease the risk of overwriting less volatile data on the cost of not collecting more volatile data. As an example, the investigator might want to turn an alarm central off, so that the non-volatile memory won't fill up with warnings, overwriting relevant data from the investigated incident [5].

Ruan and Carthy discuss the order of volatility for cloud providers, and they defined the order of volatility to be, in decreasing order: Service layer artifacts, abstraction layer artifacts, and physical layer artifacts [6]. This is a more generic model than the one defined in RFC 3227 but covers a variety of system architectures.

Dykstra and Sherman researched available tools and the trust challenges that arise with cloud computing and the forensic collection of data from Infrastructure-as-a-Service solutions [7]. The authors described a model with layers of trust in the system where evidence collected from higher layer abstractions such as applications running in a virtual machine need to trust more of the system than, e.g., packet capture at the hardware level. This idea behind the layers in the system as *layers of trust* is similar to the discussion in this paper on the *storage stack layers*, but in this chapter, we focus on the layers from a volatility perspective.

The trust issue and the "changeability" of data have also been the focus for Casey, where he discusses the need for the forensic examiner to detect, quantify, and to compensate for unforeseen changes to the system caused by errors or loss [2].

Some authors have described the challenges with volatile data in systems. Zulkipli et al. point out that the volatility complexity is higher in IoT systems than in other systems, and they see a need for new techniques for filtering and collecting data in IoT environments [8].

Montasari and Hill also discuss the challenges with volatile data in IoT systems, especially with short-lived data in resource-constrained devices together with cloud aggregation and processing of data in the system [9]. The resource-constriction means less memory and, thereby, more volatile data. The cloud aggregation would lead to challenges in the chain of custody, as it will be harder to track the pathway of the data in the system and describe the changes that have happened to it.

In a paper by Sandvik and Årnes, the volatility of the registers keeping the current clock state under low power conditions was discussed. Testing showed that for some devices, the registers kept the state up to 10 s while the processor was connected to a lower voltage than the processor could operate normally under, and it did not have enough power to run the operating system [10]. The results showed the evidence dynamics, as the low power affected registers holding the clock value, which made the clock of the operating system to show the wrong time when power was restored.

3 Volatility in IoT Devices

Any stored data will disappear after an amount of time, whether it is stored in electrically powered circuits or engraved on stone tablets. How fast the data disappears is obviously different for these two technologies, and this can be denoted as the volatility of the data. An intuitive attribute of the volatility is that the faster information disappears, the higher the volatility of the data, but to show how volatility should be defined, we need to go into the details of both what we mean with "disappear" and what we mean with "data". Another intuitive attribute is that the volatility is in some way quantifiable and that we, in general, don't know exactly when the data will disappear, so there are probabilities involved.

From the field of information theory, there is a distinction between the terms *information* and *data*. The information source transmits messages that contain some information, and these messages are encoded in data [11, 12]. This terminology is adopted here, and in our case, we can view the information transmitted as the messages about the events that affect the system, and the data is stored in bits in various physical locations of the IoT system.

With this backdrop, *data volatility* is introduced here as the disappearance of data in the system, and the *information volatility* is introduced as the disappearance of the information about an event, or set of events in the system. Figure 1 shows the difference between these two terms. Data volatility only concerns the specific data and copies of that data found in the system. One example of this might be a file that is stored in a device and then copied to other devices in the system automatically. How fast the data content of this file disappears is data volatility. As the information is stored in the system as data, information volatility is dependent on the data volatility and can be viewed as a superset of data volatility. Information volatility takes into account all data that can be used for reconstructing an event in the system. Even though some data that are stored will disappear, there might still be enough information in the system to reconstruct an event. If data is disappearing, there is at one point in time not enough information to reconstruct an event, given the defined certainty, and this is the point where the information about an event has disappeared.

3.1 Data Volatility

Data volatility is usually what most people refer to when mentioning volatility. The *order of volatility*, which is commonly used as a reference, is a description of the ordering of data lifetime between the various storage types and locations, while the term *volatile memory* is a term used for a type of memory where data disappears as the electric power is removed.

IoT systems can, in many cases, be considered distributed systems, where information about events is stored in many locations. This should also be considered when assessing the volatility of the data. Figure 2 shows a generic IoT system, where data

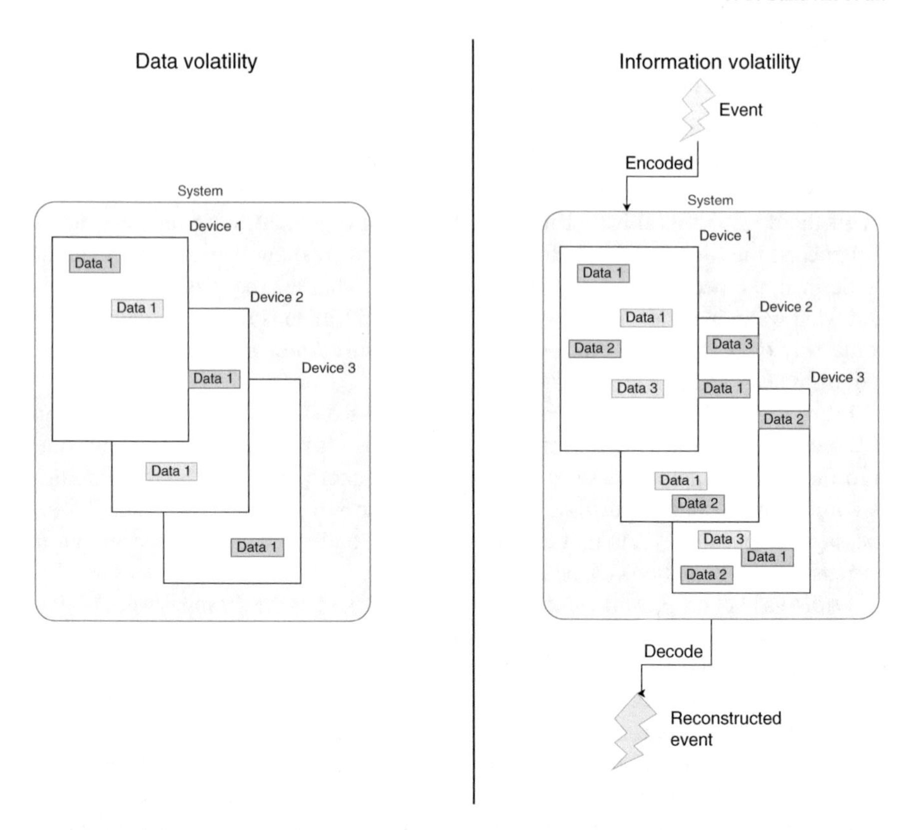

Fig. 1 The difference between data volatility and information volatility. Green signifies existing data, while red signifies inaccessible data. While data volatility focuses on the existence of a specific piece of data, here labeled "Data 1", information volatility focus on all data that can be used for reconstructing an event

can be cloned into several subsystems and several locations in each subsystem. It is important to establish the data locations that are considered in the volatility model, whether we consider data as it is stored in one location or all copies of the same data as it is stored in the system.

The simplest model is where data is contained in one storage location. This can be a timestamp in the metadata of a file or the contents of a file. The time it takes before the particular data is unavailable for the investigator can be considered the volatility. For this, we need to consider the time it takes before a file, or the data is deleted, and the time it takes before the areas in memory (either volatile or non-volatile) containing the data are overwritten. An example of this might be that deletion and erasure of a file. The information about the file and its content is encoded in data found in both RAM and Flash storage, and while the data disappears from the process memory and file system abstraction layer, the contents are still found in a page in the flash memory.

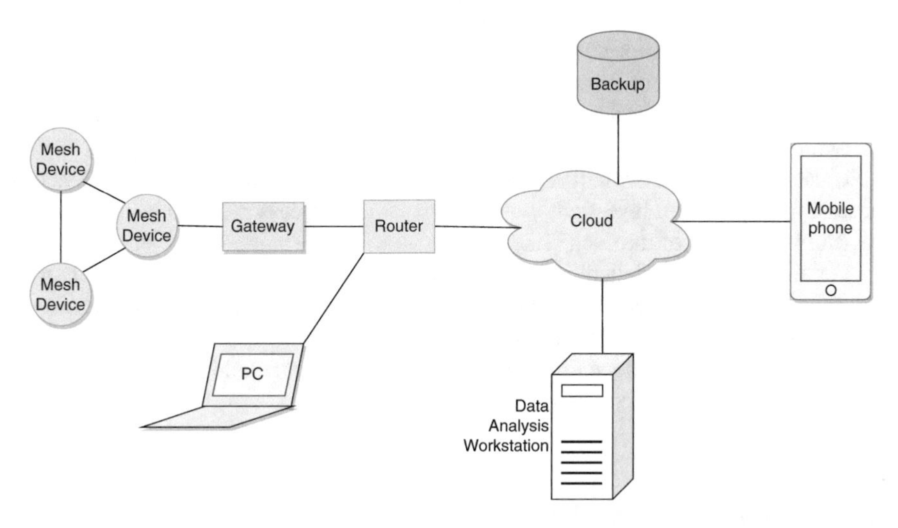

Fig. 2 A generic IoT system with data stored in several locations. Each of the elements in the figure have their own internal system for processing and storing the data in question

A more complex model is where copies of the data are found in several locations in the system. A system in this regard can be a single device consisting of CPU, RAM, and flash memory, or it can be a whole IoT system consisting of several IoT devices, servers, routers, cloud storage, and/or computers. Copies of the data can be found in several places in a system, and it is intuitive to think that the storage location with the lowest volatility is the one contributing the most to the overall volatility. If the contents of a file have been overwritten, there is still a probability for the original pages of the file to be located in the flash memory, as the wear leveling algorithm will write to other pages when the file is modified. Even if the pages are erased with a TRIM command, a command for wiping non-allocated blocks in a flash memory device, the data might still be in a page in a bad block, and therefore not erased. The file can also have been copied to other devices or a cloud service.

If we define the data volatility as the time of disappearance of all copies of the data, the probability for finding at least one of the copies of said data is dominated by the storage location with the lowest volatility. The "disappearance" of data is not a sharp boundary between the existence and non-existence of the data. On the one hand, we can think of disappearance as the point where the data does not exist anymore, or is lost to the mythical place together with single socks from the washing machine. The data is erased, and there are no theoretical methods of reconstructing the data from the storage medium. On the other hand, we can view the disappearance of the data as the point where the data becomes inaccessible for the investigator.

The *inaccessibility* of the data requires another set to describe the volatility, namely the methods and tools available for the investigator for acquisition and examination of the data. There are many ways data can be collected from a system, from documenting status indicators on a user interface to desoldering and reading the flash chip or using

JTAG to dump the memory. Each of these methods has access to a subset of the data that exists in the system. If the investigator does not have the tools for performing a JTAG acquisition, the data is inaccessible for the investigation. The definitions of *order of volatility* as described in Sect. 2, often mentions CPU registers and CPU cache as the most volatile storage locations, but there are no practical ways of accessing these for an investigator, as any use of the processor would overwrite the registers and many of the cache lines.

For an investigator, data disappearance means that data becomes inaccessible for the investigation, and we define the term *data disappearance* as data that becomes inaccessible for a given acquisition method or technique. The data will, therefore, have different volatility depending on where the data is collected from in the system, which translates to the acquisition method.

3.2 A Model for Data Volatility

There are many processes and variables in a system that affect volatility. To split the challenge into more manageable parts, we introduce a model to ease the analysis of data volatility.

The generalized data volatility model, $\mathcal{V}_{\mathcal{D}}$, is introduced here as a 6-tuple given by:

$$\mathcal{V}_{\mathcal{D}} = (L, E, A, M, D, S) \tag{1}$$

where L is the various storage system abstraction layers. E is the set of events that has happened in the system, both internally triggered events and external events. A is the functions of the applications producing, modifying and deleting data, M is the functions of the storage management software and firmware mapping the application data to the physical storage devices, and D is the set of individual memory devices in the system with their physical reliability functions. S is the environment of the data storage devices in the system, including the IoT system with its hardware, software, configuration, the physical environment, and the operational environment. Table 1 shows this model with a short description of the elements.

In short, L is the structure of the data pathway in the system; E is the set of events; A, M, D are functions that operate on the data; and S is the environment in the background of the system. The relationship between these elements of the volatility model is shown in Fig. 3 and is described in more detail in the following subsections. Each of the elements the storage stack consists of contributes to the volatility of the data. The challenge is to find the amount and type of contributions from each element, so the total volatility of the data can be calculated or approximated if we choose to disregard elements with an insignificant contribution.

Table 1 The elements of the volatility model

Model element	Type	Description
Storage abstraction layer, *L*	Physical	Physical, or close to physical storage layers
	Logical	Data structures and access methods for the logical storage layer structure in the system
	Application	The layer that processes the data that are encoded from the events. The application activity functions, *A*, works on this layer
Events, *E*	External	External events affecting the system
	Internal	Internal events in the system, such as delayed response to external events or timed events
Application activity functions, *A*	Applications	The applications that handles the data creation, modification and deletion, together with their rates and probability distribution
Storage management functions, *M*	FTL	Software and firmware mapping the application storage to physical storage
	VMM	
Memory device reliability, *D*	RAM	All devices that contain memory in the system. Each type of memory has its own reliability function and dependencies for failure-free service
	SSDs	
	Tape drives	
	HDDs	
Environment, *S*	Configuration	The current configuration of the system
	Physical environment	The physical environment affecting the system
	Operational environment	The external operational environment, usage patterns, attack intensity, etc.

3.2.1 Storage Abstraction Layers

The storage abstraction layers, L, is the set of storage abstraction layers for the devices in the system. This is in the model for two reasons: It makes it possible to analyze the data loss functions from each layer, it is also used for describing which layer the acquisition method uses together with the volatility.

The OSI model of networking protocols describes the abstraction layers for how data can be transmitted over networks, where each layer has a defined role in the addressing and handling of the data packets. This model is not followed to the letter in contemporary network protocols, but it is still a good tool for analyzing network protocols and for learning the abstraction layers in any network protocol stack. A similar structure for storage abstraction layers can be defined.

The storage abstraction model defined in this work specifies how each level in the storage hierarchy handles the storage of information from the application storage

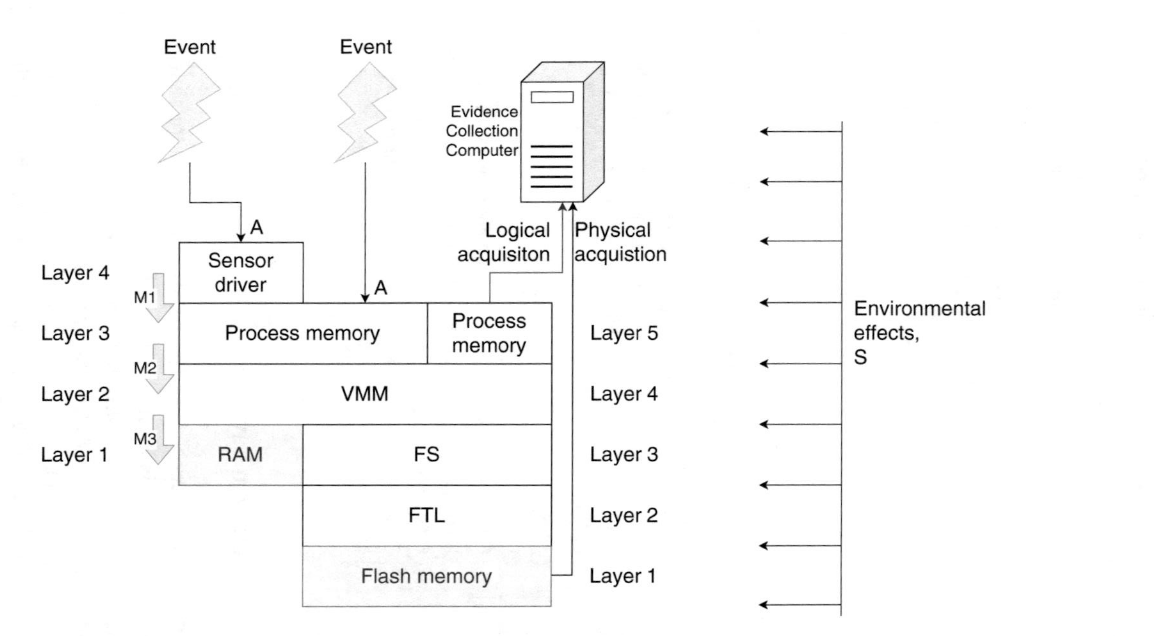

Fig. 3 The relation between the elements of the volatility model. The stack is different depending on both the application layer and the physical storage. Note that there is another stack for the data as it is forensically collected, and this varies with collection method

layer through the file system abstractions and to the storage in the physical medium. The stack will be different for different applications and physical storage locations, as an application may operate on different levels of the storage stack.

The two layers that are always present are the application layer at the top of the stack and the physical layer at the bottom. The *application layer* is the data that is an interpretation and encoding as a response to events. The encoded data is then stored in a data structure, together with other information, the operating system will keep the process data in other structures, and the data will be stored in the *physical layer*, first in physical RAM, before it can be stored in physical flash storage.

These layers define where data can be forensically collected. Physical acquisition is a term used for forensically collecting data from the physical, or a layer close to the physical layer. Sometimes to forensically collect data from a hardware-near layer, such as the flash translation layer (FTL), is called a pseudo-physical acquisition [13]. The forensic collection of data from any of the other storage layers is often referred to as a logical acquisition, regardless of the exact storage layer that is affected.

Various memory technologies have different names and number of abstraction layers, but at least three layers are consistent among the technologies: The physical layer, the addressable-to-physical translation layer, and the application layer. Table 2 shows examples of a DRAM and a Flash memory layer structure. In RAM, the application stores data in the process memory, and the buffers holding the data can be copied otherwise managed during its lifetime. Beneath the in-process memory handling, the Virtual Memory Manager (VMM) handles memory pages and can move the memory pages in and out of a swap file, or other memory modules in case of a NUMA architecture.[2] For the Flash memory layer, the data is first in RAM, and as the file is written to disk, the file system driver will decide where the file is to be written in the address space, and the flash translation layer will move the data from the linear address space to store it in the physical flash pages and keep an index of the corresponding logical address.

It is important to note that the number of layers is dependent on several factors: How many processes or functions are handling these data and the number of abstraction layers within one architecture. The data producer stack can also be different from the collection stack for the same physical data. This can, e.g., be when an IoT device does write directly to flash, but the interface for collection is to connect a computer to the device and logically acquire the data via en Media Transfer Protocol (MTP), which adds a layer of abstraction that the writing did not go through.

3.2.2 Events

The events, E, is the set of events affecting the system, often initiated by some external interface to the system, as an IoT system is typically an open system. Events originating in the system can be events triggered by a timer or triggered by a state change of the system. The events are often what an investigation tries to reconstruct

[2]Non-Uniform Memory Access.

Table 2 Example storage abstraction layers and functions for DRAM and Flash memory

Layer	DRAM	Flash
Application	Sensor reading	Sensor reading
Translation layers	In-process memory management	File system driver
A2P translation	Virtual Memory Manager	Flash Translation layer
Physical	SDRAM cell	Floating gate transistor

from the stored data. External events are events that are triggered from outside the system, while internal events are events originating from the system. The events get encoded into data in the storage abstraction stack and end up in one or more physical storage locations.

3.2.3 Application Activity Function

The application activity function, A, is the set of functions that processes the information and translate events into the storage system. It handles creation, modification, and deletion of data from the top layer of the storage abstraction layer stack: $A : E \rightarrow L_{\text{top}}$. One event can trigger several of these functions. An example of an application activity function is a program running on an IoT device, analyzing sensor inputs, recording sudden changes, and deletes the data after one week.

As the application layer is the data encoding of the external inputs generated by an event, this layer will always be present, and the application activity function will always map events into the application layer. An example if this is log rotation, where the oldest log file is deleted while a new one is created, and the other log files are renamed. Another example is the reading sensor inputs and processing these values. There might also be unexpected events, such as a sudden power failure that affects RAM contents, and non-volatile content that is in the middle of a non-atomic write operation.

3.2.4 Storage Management Functions

The storage management functions, M, are the functions mapping the data between the intermediate layers in the storage abstraction stack: $M : L_x \rightarrow L_y$. One example of this is the flash translation layer, which reorganizes the logical storage address to the flash memory pages. These functions can copy data between locations in a lower layer transparent for the layer above. From the application's view, there is only one occurrence of the data, but it may exist at several physical locations. The deletion of data from the application will also make the data inaccessible for the application layer, but the data might still exist in the physical medium.

The pathway for the data between the application layer and the physical layer can be different depending on the application, the data, and the physical layer, so the storage management functions will not necessarily be the same for all data in a system.

3.2.5 Memory Device Reliability

The memory devices failure probability, M, is the failure function of the physical storage locations of the data. Hardware failures of a memory device will render a part or all of the data inaccessible and thus impact volatility. We can define at least two different aspects of this. The first is the reliability under normal operation; the other is the reliability after an event in the system affecting the memory device has happened. An example of this is the clock registers when losing power. Sandvik and Årnes reported a retention time for the value in the clock register of up to 10 s after an abrupt power loss [10].

Typically, each device will have a failure probability distribution or reliability function. This operates on the lowest level of the storage abstraction layer stack and gives the volatility for the stored data in the absence of other events affecting the data.

3.2.6 Environment

The environment of the system, S is the parts of the system environment that can affect the volatility of the data apart from the direct events. The physical environment can impact the lifetime of the components; the radio environment can affect communications; the digital operating environment shows the attack base rate, or how hostile the digital environment is. The environment can change from one state to another or show cyclical changes over time.

3.3 Information Volatility

From an investigator's point of view, the data loss in a system is not the most critical part of the investigation in itself, but rather that the amount of information available should be enough to reconstruct events with a given confidence. While some data might disappear, there might be other data that can be used for reconstructing the same events. Information about an event is often encoded by several application functions, spread over many pieces of data, and stored in several locations, as shown in Fig. 1.

Information volatility is the probability for enough data to be present to reconstruct events after a period. As data is needed for decoding the information, the data volatility, as discussed, is an important part of, and can be considered a subset of information volatility.

A model for information volatility can be defined as:

$$\mathcal{V}_{\mathcal{I}} = (\mathcal{V}_{\mathcal{D}}, T, C) \tag{2}$$

where $\mathcal{V}_{\mathcal{D}}$ is the set of data that the event is decoded into, T is the threshold for the certainty or confidence, that is needed to reconstruct the event, and C is the decoding function that interprets the data into information about the event.

3.4 Forensic Resources

The volatility model, as described, focuses on the technical part of data and information volatility. For this to be relevant for an investigator, we also have to comment on the socio-technical perspective that is the human and organizational aspects of the investigation. The resource-constraints to the investigation itself is a burden that adds to the resource-constraints of the devices in an IoT system.

To get access to the data or information stored, the investigator is dependent on resources. This can be both personnel, time, acquisition tools, storage space, or knowledge. The available resources affect both the amount and the quality of the collected and examined data. The equipment and knowledge decide the type of acquisition that can be performed, which again decides which memory devices that can be acquired and the layer of the storage abstraction stack that the data can be acquired from. As the volatility of data has to be assessed based on the storage layer from which it is collected, the available resources do affect the perceived accessibility and, thus, the perceived volatility of the evidence. The resources are, however, not a part of the model, but is considered a part of the limitations to the investigation.

4 A Statistical Approach to Data Volatility

The term *volatility* has so far been used to describe the time interval before some data becomes inaccessible to the investigator. To quantify the volatility so it can be used for predicting the probability of finding relevant evidence after a time, a statistical model is needed. We will borrow some methods from reliability and dependability analysis.

We identify that the reliability of a system, or the probability that the system fails within a given period, is similar to the concept of volatility, where the volatility can be viewed as the probability for the evidence to become inaccessible within a given period. The volatility is thus the *reliability of data*, where the reliability analysis' concept of non-repairable failure is the volatility's data inaccessibility. As discussed in Sect. 3, the volatility is dependent on the particular component in the storage stack in which the evidence collection takes place. This means that the probability distribution can vary, depending on the acquisition method, and a volatility function is valid for a particular element in the storage stack.

The reliability function, $R(t)$, is a function that describes the probability that the system has not failed at time t. For a steady state system the reliability function is given by: $R(t) = P(T_F > t)$, where T_F is the time to failure. Instead of *time to failure*, we can use the term *time to inaccessibility* for data and define a volatility function as the probability of the data being inaccessible at time t:

$$V(t) = P(T_I > t) \tag{3}$$

where T_I is the time to inaccessibility. The *mean time to inaccessibility* (MTTI) can then be expressed similar to the *mean time to failure* (MTTF) as:

$$MTTI = \int_0^\infty V(t)\,dt \tag{4}$$

The actual distribution of the volatility function is not generally known, as many variables affect the exact distribution, such as the application's memory and file system operation distribution, encryption, the allocation strategy of the file system, artifacts of the physical storage medium, reboots, or system failures. The model in Sect. 3 does, however, give us some idea about the contributions to the volatility from the various parts of the system. When the data is deleted from the application is dependent on the application function, and the physical layer is dependent on the reliability of the memory chip. The intermediate management layers can also hide or copy data, as we saw in the model description.

To model the volatility, we have to find the corresponding probability distribution function (PDF). Several distributions are used in reliability analysis, and these are candidates for volatility analysis. The distribution can be estimated by empirical testing of the system and matching the PDF, and analytically by assessing all contributing factors to the volatility. Two commonly used PDFs are the Exponential distribution and the Weibull distribution, the former is popular because of its simplicity, the latter because of its flexibility. The distributions and are described in more detail below, together with the motivation of using them.

4.1 Exponential Distribution

The exponential probability distribution is a simple probability function that is popular because of its simplicity but is not always a good approximation of the reliability function [14]. The exponential distribution works well for independent events, has a constant intensity, and a memoryless property, which can be true for external events in the volatility model described in Sect. 3.2.2. The probability distribution function has one parameter, λ, which is the rate of the data deletion. The distribution is given by:

$$f(t, \lambda) = \lambda e^{-\lambda t} \tag{5}$$

The cumulative distribution function is given by:

$$F(t) = 1 - e^{\lambda t} \tag{6}$$

Figure 4 show examples of the PDF and the CDF for the exponential distribution for various values of λ. A λ of 0.1 means a rate of 0.1 events per unit of time, and 0.01 means 0.01 events per unit of time. This can be, e.g., an internal event like a garbage collection routine, happening every 100 s on average, which gives a λ of 0.01 events/s. The mean of this distribution is λ^{-1} and the reliability function for this distribution is given by $R(t) = e^{-\lambda t}$. This can be a good description for the events

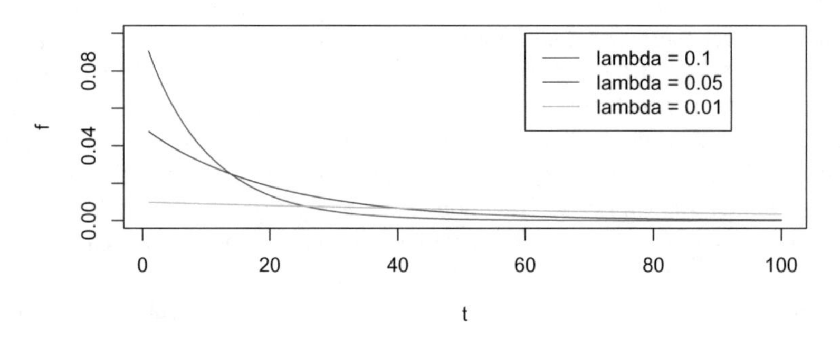

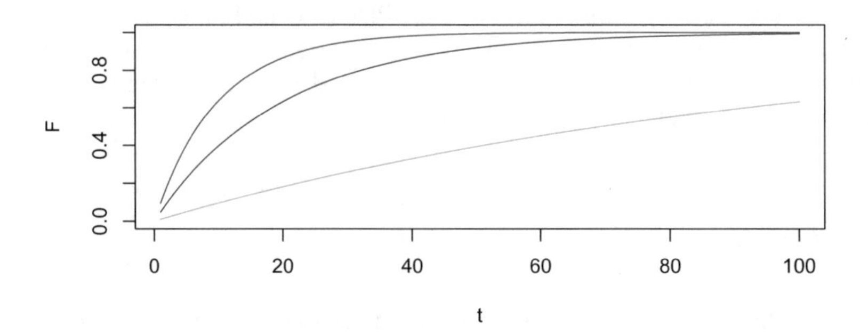

Fig. 4 Examples of the exponential probability and cumulative distribution functions

encoded by the application activity function, as external events might be modeled as happening at a constant intensity, triggering modifications and erasure of data in the file system.

4.2 Weibull Distribution

The Weibull distribution is another common distribution used for modeling reliability [15]. It has two parameters that can adjust the shape of the distribution, α and γ, which are called the scale parameter and the shape parameter, respectively. The flexibility of the function lets it approximate several other distributions. Depending on the shape parameter, it can model falling ($\gamma < 1$), steady ($\gamma = 1$), or rising ($\gamma > 1$) failure rates.

This distribution is often used to model physical components' failure rate, as the failure from wear of the components is not constant over time, but changes with the age of the component. Flash memory in SSD storage devices is an example of a failure distribution closely resembling a Weibull distribution [16]. This can, therefore, fit the physical reliability function in the volatility model, as described in Sect. 3.2.2.

The probability distribution can be parameterized in several ways, two of them are given below:

$$f(t, \gamma, \alpha) = \frac{\gamma}{t}\left(\frac{t}{\alpha}\right)^{\gamma} e^{-\left(\frac{t}{\alpha}\right)^{\gamma}} \tag{7}$$

$$f(t, k, \lambda) = k\lambda(\lambda t)^{k-1} e^{-(\lambda t)^{k}} \tag{8}$$

When $\gamma = 1$, this distribution is identical to the exponential distribution, with $\lambda = \alpha^{-1}$.

The cumulative distribution function is given by:

$$F(t, \gamma, \alpha) = 1 - e^{\left(\frac{t}{\alpha}\right)^{\gamma}} \tag{9}$$

Figure 5 shows examples of the Weibull distribution and the cumulative distribution function for a few values of γ and α. For $\gamma = 1$, the plot is equal to an exponential distribution. As the shape and scale parameters can't easily be decided analytically, the parameters often are estimated by empirical observations and the probability distribution fitted to the data. See also Sect. 4.4. The mean of the distribution is given by $\alpha\Gamma\left(1 + \gamma^{-1}\right)$, where $\Gamma(x)$ is the Gamma function, which for natural numbers is $\Gamma(N) = (N - 1)!$, and a slightly more complex definition for non-natural numbers: $\Gamma(z) = \int_0^{\infty} x^{z-1} e^{-x}\, dx$, where $z \in \mathbb{C}$ and $\Re(z) > 0$. The reliability function for this distribution is given by $R(t) = e^{-\left(\frac{t}{\alpha}\right)^{\gamma}}$.

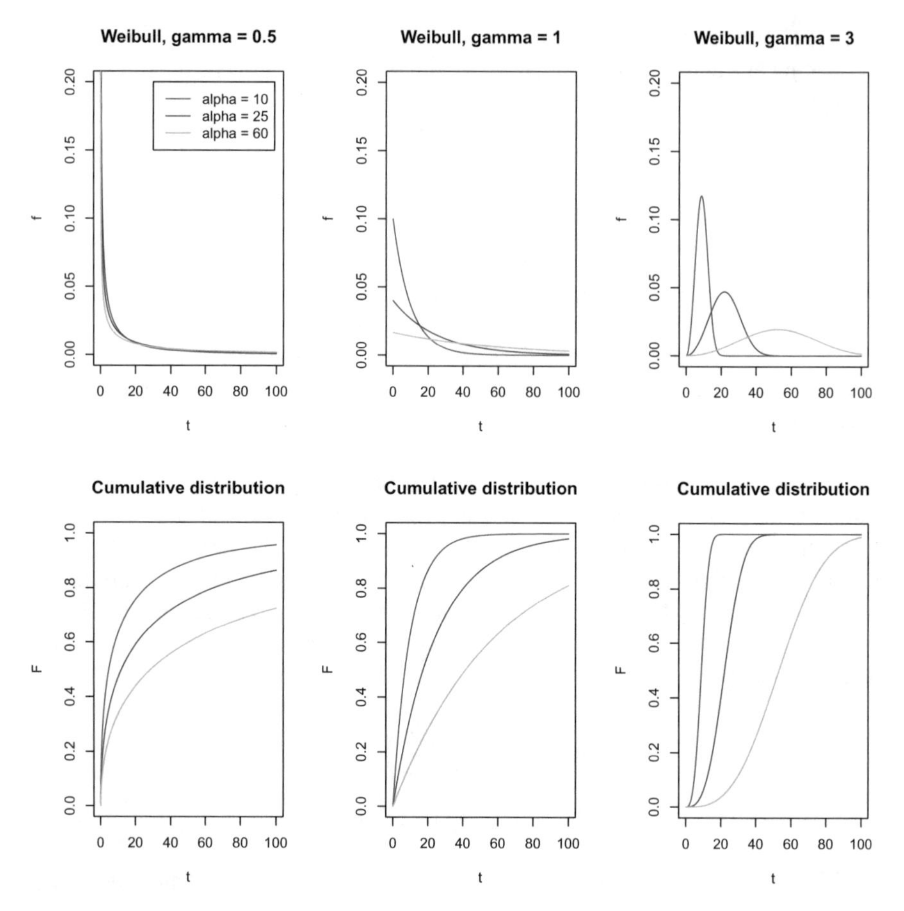

Fig. 5 Examples of the Weibull distribution and cumulative distribution functions for some values of γ and α

4.3 *Series and Parallel Systems*

The storage system can be seen as a system consisting of elements connected both in series and parallel. This is similar to the block model used for calculating reliability in a compound system [15]. The system can be viewed as such a block model to ease the analysis. Figure 6 shows such a block model of components attached in series and parallel. The volatility function for these connections is given by:

$$V_{\text{series}}(t) = \prod_i V_i(t) \tag{10}$$

$$V_{\text{parallel}}(t) = 1 - \prod_i (1 - V_i(t)) \tag{11}$$

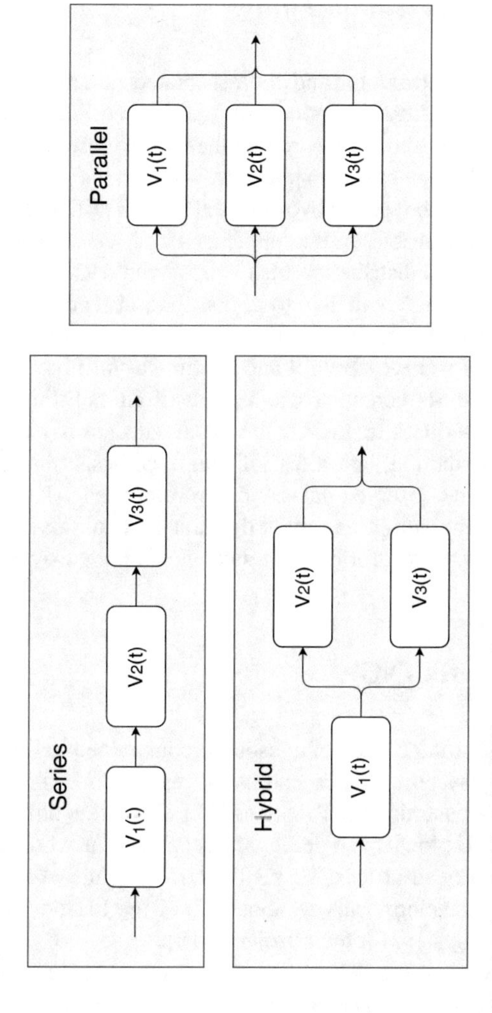

Fig. 6 Components in series, in parallel, and hybrid connection

The storage management functions are, in most cases, connected in series. If data is duplicated, there will be a parallel structure in the data pathway. The system's volatility can thus be calculated by combining Eqs. 10 and 11 following the block structure of the system components.

4.4 Probability Distribution Fitting

Finding a probability distribution and the associated parameters that best fit the observed data, and measuring the goodness of fit, is known as probability distribution fitting. The fitting process should ensure that the model closely fits the observations and that the model selection can be explained.

Distribution fitting can be seen as two different tasks. The first is to find the optimal parameters for a given distribution that matches the observed data, and the other is determining how good the distribution fits the observed data.

Several software packages can help to fit distributions and select optimal parameters. One such software package is the library *fitdistrplus* for the statistical computation and graphics software *R*.[3] Both R and the fitdistrplus library is free[4] software. Sagemath, Matlab, and Mathematica also have distribution fitting functionality.

To measure the goodness of the fit, the observed data has to be compared to the hypothesized distribution, and a test of this hypothesis, that the hypothesized distribution explains the observed data is therefore needed. There are several tests for this, each with their assumptions about the data being observed. Among the tests often encountered, are χ^2-test for discrete data and Kolmogorov-Smirnov test.

5 Example: Contiki-NG

As an example of the storage stack of a resource-constrained IoT device, we can use the Contiki operating system. Contiki and its successor, Contiki-NG, is an operating system for resource-constrained IoT devices [17]. Contiki is built around an event-driven kernel, and utilize loadable modules and services. The whole operating system is about 100 kB, and need at least 10 kB of RAM to run.[5] According to Eclipse Foundation's annual developer survey, about 5% of the IoT developers were using the Contiki operating system for their projects [18].

For embedded devices, the acquisition of running RAM can be a challenge in investigations, but non-volatile memory is easier to acquire, as the contents still are present after the power has been removed. The non-volatile storage is managed by the Coffee File System, a file system that is both minimalist and designed for flash

[3]https://www.r-project.org, visited 2020-07-01.

[4]Free as in beer and speech.

[5]https://github.com/contiki-ng/contiki-ng/wiki, visited 2020-07-08.

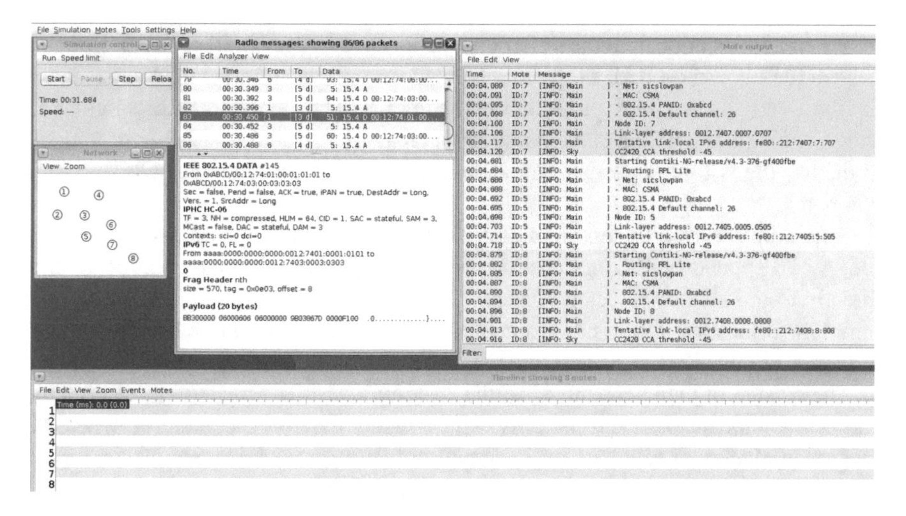

Fig. 7 Cooja running a simulation

memory devices. This section, therefore, focuses on the file-system specific part of Contiki. From a volatility perspective, the theory is similar for RAM data, but the specific memory allocation methods need to be taken into account.

Some advantages from a research perspective of using Contiki as a case study is that the Cooja simulator that comes with the OS can simulate various types of networks and configurations and also dynamic environments. It can both emulate specific micro-controllers and run native code on the host architecture. Contiki/Cooja also implements 6LoWPAN and other protocols that are used in IoT systems. Figure 7 shows a screenshot of a running simulation.

The Coffee file system has been designed to run on resource-constrained nodes and to include wear-leveling techniques. Because of the resource constraints, the file system has simplified many operations that we take for granted in a general-purpose file system. It does not contain much metadata, the actual file size is not among the metadata but has to be calculated, and there are no timestamps among the metadata.

Flash memory works differently than old-fashioned hard disks or RAM. Writing to flash memory can only be done by writing a whole page [5]. The page size is specific to a particular chip and is often 512, 1024, or 2048 bytes. A bit can only be set or flipped from ‘1’ to ‘0’, and to flip the bit back to ‘1’, a whole erase block has to be erased. Erase-blocks consist of several pages. This means that modifying data in flash memory involves writing new versions of the data rather than overwriting existing data. A file system operating on a flash memory need to either take this into account or introduce a flash translation layer that mimics an addressable read-write memory area for the operating system, while it hides the data shuffling happening in the background. The coffee file system is designed for operating on a flash device and does not use a flash translation layer.

When a file is allocated, the default is to allocate 11 pages for the file, and when a page in the file is modified, a log file is created, recording a number of changes in the file until there are no more pages left in the log file. The next modification will then trigger a new file with the same name to be created. The default number of updates in the log file is four pages, and the default page size is 0x100 bytes, which equals 256 bytes.

Appending data to the end of an existing file does not create an entry in the log file. As an append operation doesn't modify existing data, the data can be written directly to the already allocated pages. When an append operation has reached the initial file allocation size, a new copy of the file will be made, and twice the number of pages will be allocated.

The erasure of data happens when the write-pointer in the file system reaches the end of the addressable flash memory. This triggers a garbage collection, where all free erase blocks, or sectors as it is named in Coffee, are erased and are free to be reused. The file system does not move existing data to free up more pages.

The simplicity of both the file system and Contiki as an operating system removes many of the storage abstraction layers and simplifies the model considerably. From the top, the device has an application running, where it receives inputs, and reacts to this. The application function acts on the events; as an example, it can store a value that is read from a device interface in a file. As the file containing these records is sent to a central server by an application, the log file is deleted by the application. This all happens at the top layer in the stack. Beneath this layer, the data is held in data structures in RAM and temporarily stored in the physical RAM. The file write will open a new stack toward the flash memory and write the files there.

The storage management functions are different between RAM and Flash, and the function will trigger both copies of the data to be spread over the physical Flash memory, and keep track of the unused memory such that new writes can be done at the right location. In the physical flash, the data can be held for a long time, until the flash memory chip breaks, or the flash cell stops working. This is the domain of the memory device's failure function.

Figure 8 demonstrates a simplified block model of an app running on a Contiki device. To calculate the volatility of the system, we can use the equations from Sect. 4.3 to calculate the volatility for the acquisition methods:

$$V_{\text{physical}}(t) = V_1(t) \times V_2(t) \times V_4(t) \times (1 - (1 - V_{5,1}(t))(1 - V_{5,2}(t))) \quad (12)$$

$$V_{\text{RAM}}(t) = V_1(t) \times V_2(t) \times V_3(t) \quad (13)$$

$$V_{\text{logical}}(t) = V_1(t) \times V_2(t) \times V_4(t) \times V_5(t) \times V_6(t) \times V_7(t) \quad (14)$$

Each of the individual volatility functions in Fig. 8 has to be assessed, to find $V_i(t)$ used in the above calculation. In our example, the data was duplicated when stored in flash memory, therefore the parallel combination in V_5. For the logical acquisition, $V_{\text{logical}}(t)$, the duplicated data in the flash memory is not visible, and not used by the file system when reading, so the whole pathway is in series.

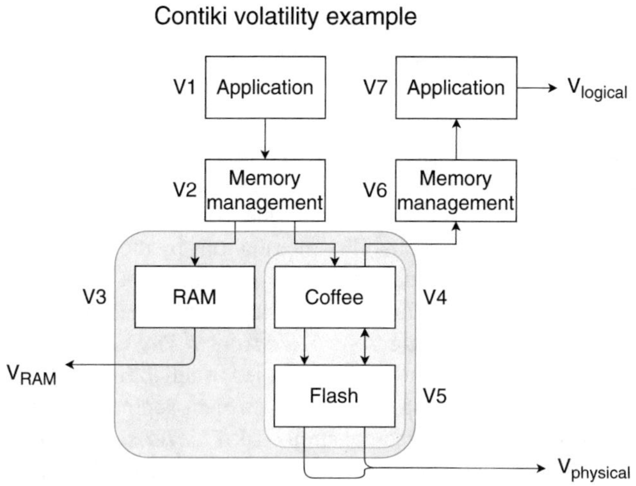

Fig. 8 A simplified model of an application running on a Contiki device and 3 different acquisition methods

Another storage stack includes the transfer of the file to a central server. Here the storage management functions translate the data to network packets that are sent and stored in RAM of the gateway and other network equipment until it reaches the central server. This is outside the bounds of the Coffee file system, though.

To acquire data, the investigator can perform a logical acquisition, some devices allow the connection through USB, and might use a Media Transfer Protocol for transferring a subset of the files in the operating system. This is shown in Fig. 3. The storage stack is different between the application layer that process events and the application layer for the MTP server that transfer the data to the forensic collection computer. If the investigator instead uses a chip-off method, the data will be collected from the physical flash storage layer, at level 1.

6 Summary and Conclusions

In this chapter, a model of data and information volatility is introduced. This volatility model is used to analyze the elements in IoT systems that affect the volatility of data and of information in the IoT system. The emphasis in this chapter is on the data volatility. The data volatility model consists of (i) an abstraction for a storage layer stack; (ii) events and system states affecting data; (iii) and functions for how data is transformed between the application and the physical storage. From this data volatility model, all individual components contribute to the overall volatility. Based on the volatility model, we also introduce a quantitative measure for volatility. It allows for a more objective assessment of volatility. The volatility measure is dependent on the forensic method used, as each forensic method collects data from different components in the storage layer stack. We show that the IoT system can have several volatility measures, one for each element in the storage stack used to collect forensic data.

We derive a statistical measurement method from dependability and reliability analysis. This statistical measuring method is used to calculate the volatility contribution in each storage layer component. Quantifying the volatility can thus be established by combining the probability distributions by the data pathway connections in the storage stack. The data pathways can be a mix of series and parallel connections. For the volatility function, two commonly used probability distributions are described, and the *Mean Time To Inaccessibility* is introduced. We have modeled this after the Mean Time To Failure used in reliability analysis.

The model is exemplified using *Contiki-NG*, an open-source, minimalist, and real-time operating system for resource-constrained IoT devices. By using Contiki-NG as an example, we can focus on the core model without the added complexities of more advanced operating systems and storage management functions. In addition, Contiki contains a powerful simulator that we can use for our study and for volatility analysis.

Our study described in this chapter is a step towards establishing a scientific base for measuring the data and information volatility in an IoT system. As IoT systems, in particular, and other computer networks in general, become more and more complex, an objective and reliable assessment of the systems' volatility is becoming more crucial. It works toward a forensically sound acquisition of data with a high evidential value.

With our study, we aim to establish a theoretical foundation toward a scientific base for volatility analysis. Further empirical studies are needed to reveal each element's specific volatility contributions in the volatility model, and to translate this theoretical model into working procedures for forensic practitioners. The similarities and differences between IoT systems and individual devices regarding the components and their volatilities are open for further studies. Information volatility is introduced in this chapter but deserves to be focused in more detail elsewhere.

Our objective is that this research will leverage better tools for the forensic investigator and the forensic community at large, to objectively and reliably be able to quickly prioritize the data collection during the triage process such that the quality of the investigation can be upheld. In the courtroom, enough evidence with high quality is a necessity for a fair trial. By enabling the investigator to collect and analyze more relevant data, the court has a better understanding of the facts to make a just judgment, decreasing the chance of a miscarriage of justice.

Acknowledgements The research leading to these results has received funding from the Research Council of Norway program IKTPLUSS, under the R&D project "Ars Forensica – Computational Forensics for Large-scale Fraud Detection, Crime Investigation & Prevention", grant agreement 248094/O70.

References

1. Conti M, Dehghantanha A, Franke K, Watson S (2017) Internet of things security and forensics: challenges and opportunities. Future Gener Comput Syst
2. Casey E (2002) Error, uncertainty, and loss in digital evidence. Int J Digit Evid 1(2):45
3. Roussev V, Quates C, Martell R (2013) Real-time digital forensics and triage. Digit Investig 10(2):158–167
4. Brezinski D, Killalea T (2002) RFC 3227: guidelines for evidence collection and archiving. RFC 3227
5. Årnes A, Flaglien A, Sunde IM, Dilijonaite A, Hamm J, Sandvik J-P, Bjelland P, Franke K, Axelsson S (2017) Digital forensics. Wiley, Ltd
6. Ruan K, Carthy J (2013) Cloud forensic maturity Model. In: Digital forensics and cyber crime. ICDF2C 2012. Lecture notes of the institute for computer sciences, social informatics and telecommunications engineering, vol 114. Springer, Berlin, Heidelberg. pp 22–41
7. Dykstra J, Sherman AT (2012) Acquiring forensic evidence from infrastructure-as-a-service cloud computing: Exploring and evaluating tools, trust, and techniques. Digit Investig 9(SUPPL.)
8. Zulkipli NHN, Alenezi A, Wills GB (2017) IoT forensic: bridging the challenges in digital forensic and the internet of things. In: Proceedings of the 2nd international conference on internet of things, big data and security, pp 315–324, Jan 2017
9. Montasari R, Hill R (2019) Next-generation digital forensics: challenges and future paradigms. In: Proceedings of 12th international conference on global security, safety and sustainability, ICGS3 2019
10. Sandvik J-P, Årnes A (2018) The reliability of clocks as digital evidence under low voltage conditions. Digit Invest 24:S10–S17
11. Shannon CE (1948) A mathematical theory of communication. Bell Syst Tech J 27(3):379–423
12. Cover TM, Thomas JA (2006) Elements of information theory, 2nd edn. Wiley, Inc., Hoboken, New Jersey
13. Klaver C (2010) Windows mobile advanced forensics. Digit Investig 6(3–4):147–167
14. Murphy KE, Carter CM, Brown SO (2002) The exponential distribution: the good, the bad and the ugly. A practical guide to its implementation. In: Proceedings of the annual reliability and maintainability symposium, pp 550–555
15. Billinton R, Allan RN (1992) Reliability evaluation of engineering systems. Springer US, Boston, MA
16. Meza J, Wu Q, Kumar S, Mutlu O (2015) A large-scale study of flash memory failures in the field. Perform Eval Rev 43(1):177–190
17. Dunkels A, Grönvall B, Voigt T (2004) Contiki—a lightweight and flexible operating system for tiny networked sensors. In: Proceedings—conference on local computer networks, LCN, pp 455–462
18. Eclipse Foundation (2019) Eclipse IoT developer survey 2019. Technical report, Eclipse Foundation, Apr 2019

Application of Artificial Intelligence and Machine Learning in Producing Actionable Cyber Threat Intelligence

Reza Montasari, Fiona Carroll, Stuart Macdonald, Hamid Jahankhani, Amin Hosseinian-Far, and Alireza Daneshkhah

Abstract Cyber Threat Intelligence (CTI) can be used by organisations to assist their security teams in safeguarding their networks against cyber-attacks. This can be achieved by including threat data feeds into their networks or systems. However, despite being an effective Cyber Security (CS) tool, many organisations do not sufficiently utilise CTI. This is due to a number of reasons such as not fully understanding

R. Montasari (✉) · S. Macdonald
Hillary Rodham Clinton School of Law, Swansea University, Richard Price Building, Sketty Ln, Sketty, Swansea SA28PP, United Kingdom
e-mail: Reza.Montasari@Swansea.ac.uk
URL: https://www.swansea.ac.uk/staff/law/montasari-r/

S. Macdonald
e-mail: s.macdonald@swansea.ac.uk
URL: http://www.swansea.ac.uk

F. Carroll
Cardiff School of Technology, Cardiff Metropolitan University, Llandaff Campus, Western Avenue, Cardiff CF52YB, United Kingdom
e-mail: FCarroll@cardiffmet.ac.uk
URL: http://www.cardiffmet.ac.uk

H. Jahankhani
Information Security and Cyber Criminology, Northumbria University, 110 Middlesex Street, London E17HT, United Kingdom
e-mail: hamid.jahankhani@northumbria.ac.uk
URL: http://www.london.northumbria.ac.uk

A. Hosseinian-Far
Faculty of Business and Law, University of Northampton, Waterside Campus, University Drive, Northampton NN15PH, United Kingdom
e-mail: Amin.Hosseinian-Far@northampton.ac.uk
URL: https://www.northampton.ac.uk/

A. Daneshkhah
School of Computing, Electronics and Mathematics, Coventry University, Priory Street, Coventry CV15FB, United Kingdom
e-mail: ac5916@coventry.ac.uk
URL: https://www.coventry.ac.uk/

R. Montasari et al. (eds.), *Digital Forensic Investigation of Internet of Things (IoT) Devices*, Advanced Sciences and Technologies for Security Applications,
https://doi.org/10.1007/978-3-030-60425-7_3

how to manage a daily flood of data filled with extraneous information across their security systems. This adds an additional layer of complexity to the tasks performed by their security teams who might not have the appropriate tools or sufficient skills to determine what information to prioritise and what information to disregard. Therefore, to help address the stated issue, this paper aims firstly to provide an in-depth understanding of what CTI is and how it can benefit organisations, and secondly to deliver a brief analysis of the application of Artificial Intelligence and Machine Learning in generating actionable CTI. The key contribution of this paper is that it assists organisations in better understanding their approach to CTI, which in turn will enable them to make informed decisions in relation to CTI.

Keywords Cyber security · Threat intelligence · Artificial intelligence · Machine learning · Cyber physical systems · Digital forensics · Big Data

1 Introduction

Cyber threats are constantly growing in frequency and complexity [1–4]. Through the use of intrusion kill chains, campaigns and customised tactics, techniques and procedures, cyber criminals are able to bypass organisations' security controls [5–7]. Cyber Security (CS) breaches and outages have been widely covered in the media, and statistics concerning the number of cyber-attacks are available in a variety of sources [8–11]. However, despite many CS breaches, there is little expert analysis of the areas that organisations should prioritise in order to increase their effectiveness in addressing known threats while also minimising the risk from evolving attacks [12]. One of the ways to help mitigate security breaches is by developing and implementing robust CTI. CTI is focused on analysing trends and technical developments in three areas of CS, Hacktivism and Cyber Espionage. CTI is used by nations states as an efficient solution to devise preventive CS measures in advance and as a result to uphold international security.

CTI is a branch of CS that concerns the contextual information surrounding cyber-attacks, i.e. the understanding of the past, present, and future tactics, techniques and procedures (TTPs) of a wide variety of threat actors. It is actionable and timely and has business values in that it can inform the security teams in organisations of adversarial entities so that they can prevent them. CTI is also a proactive security measure that involves the gathering, collation and analysis of information concerning potential attacks in real time so as to prevent data breaches and subsequent adverse consequences. Its primary objective is to deliver detailed information on the security threats that pose a higher risk to an organisation's infrastructure and simultaneously guide the security teams on preventative actions.

By providing continuously updated threat data feeds, CTI can enable security teams to defend against cyber-attacks before they can enter their networks or detect already malicious activities on enterprise networks. For instance, CTI can assist the teams in gaining a detailed understanding of the adversary and their modus operandi. This, in turn, enables them to improve their protection against specific attack methods known to be used by the adversary, and helps produce actionable information that can enable decision makers to comprehend their operational risks and better prioritise and allocate resources. Therefore, to be effective, CTI must be able to provide context and to be understood by decision makers. While CTI's main focus is on traditional IT systems, industrial control system (ICS) and network operators could also benefit from this capability given that many of the threats to ICS are facilitated by traditional IT networks. A CTI network can be considered as a combination of regular updating and learning feeds that develop the basis of powerful layered network security. Such threat feeds enable individual devices and networks to take advantage of the intelligence of numerous devices to safeguard their endpoints and networks.

Considering the above, many organisations attempt to include threat data feeds into their networks or systems without fully understanding how to deal with a daily flood of data filled with extraneous information across their security systems. This adds an additional layer of complexity to the tasks performed by security analysts who might not have the appropriate tools to determine what information to prioritise and what information to disregard. Therefore, to address the stated issues, this paper aims firstly to provide an in-depth understanding of what CTI is and how it can benefit organisations, and secondly to analyse the application of Artificial Intelligence (AI) and Machine Learning in generating actionable CTI. The key contribution of this paper is that it assists organisations in better understanding their approach to CTI, which in turn will enable them to make informed decisions in relation to CTI.

The remainder of this paper is structured as follow: Sect. 2 provides a brief overview of CTI and its benefits. Section 3 discusses phases of our recommended six-phase CTI Cycle (CTIC) and how each phase can be utilised to provide intelligence, help to guide decisions, shorten the information aggregation and dissemination timelines, and assist organisations in protecting their networks from cyber-attacks. Section 4 analyses the application of Artificial Intelligence (AI) and Machine Learning (ML) in producing actionable CTI. In Sect. 5, a discussion is provided, and finally the paper is concluded in Sect. 6.

2 Cyber Threat Intelligence

2.1 Overview of CTI

CTI is an ambiguous concept with numerous definitions attributed to it that are based on different procedural viewpoints and competitive imperatives. One definition that provides a comprehensive description is provided by McMillan [13], who defines CTI as:

evidence-based knowledge, including context, mechanisms, indicators, implications and actionable advice, about an existing or emerging menace or hazard to assets that can be used to inform decisions regarding the subject's response to that menace or hazard.

Despite its ambiguity, CTI should have three main characteristics including, (1) evidence based: cyber threat evidence may be acquired from malware analysis to ensure that the threat is valid, (2) utility: there must have some utility for organisations to have a positive impact on security incidents, and (3) actionable: the gathered CTI must drive not only data or information but also security control action [14]. It must include the combination of information detailing possible threats with a solid insight into network structure, operations, and activities. In order to produce this evidence-based knowledge, information on the mechanisms and indicators, i.e. threat feeds, will need to be put into context by contrasting it with the core knowledge of network activity. The process of gathering and collation of threat feeds will result in threat intelligence, "which then informs 'security analytics' to improve chances of detection" [15]. Security analytics in a network defence environment often consists of one of the following two forms, both of which are informed by CTI: 'Big data' platform processing large amounts of network data to determine trends, and 'Security information and event management (SIEM) infrastructure' to automate the detection of anomalous activities.

CTI is collected by continuously analysing large quantities of threat data with the aim of organising and adding context to cyber threat activities, trends and attacks. It can be derived from external threat feeds, internal networks, analysis of historical attacks, and research. For instance, it can be generated through the aggregation of fused, heterogeneous and highly reliable sources of data such as security networks, web crawlers, botnet monitoring service, spam traps, research teams, the open web, dark web, deep web, social media, and collected historical data about malicious objects. All the aggregated data is then carefully examined and processed in its entirety (often in real-time) through several pre-processing techniques, including statistical criteria, expert systems (such as sandboxes, heuristics engines, similarity tools, behaviour profiling etc.), security analysts' validation and whitelisting verification.

2.2 *Types of Threat Intelligence*

CTI can be classified into four main types as depicted in Fig. 1 in relation to information assortment, knowledge analysis and intelligence consumption. These consist of Tactical, Technical, Operational and Statistical threat intelligence [16]. The followings describe each type.

Tactical Cyber Threat Intelligence Tactical CTI (TaCTI) focuses on the techniques and procedures of threat actors such as methodologies, tools, and tactics, relies on sufficient resources and includes certain specific measures against malicious actors

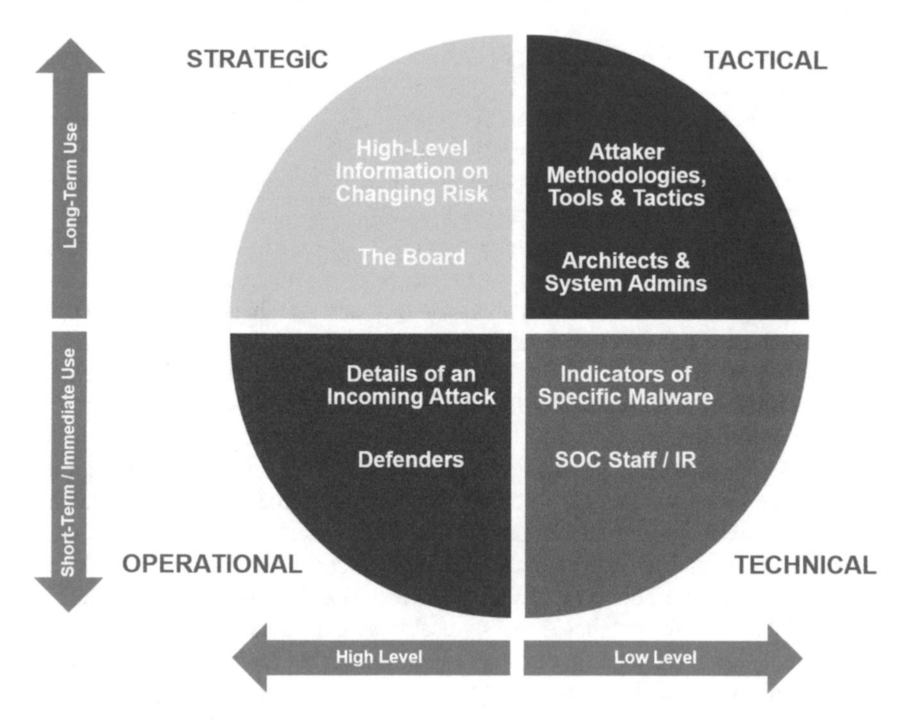

Fig. 1 Types of threat intelligence

attempting to infiltrate a network or system. TaCTI should be used to evaluate real-time events, investigations, and activities, and to provide support for day-to-day operations and events such as the development of signatures and indicators of compromise (IOCs). It must be aimed at the immediate future and identifies simple IOCs (such as malicious IP addresses, URLs, file hashes and known malicious domain names). If implemented properly, it can provide organisations with a deeper understanding of how they have been previously attacked and how they can mitigate such attacks. TaCTI is often automated and machine-readable enabling security products to ingest it through feeds or API integration. It is considered to be the easiest type of intelligence to be produced, and as a result, it can be found through open source and free feeds. It must be noted that TaCTI has a short lifespan given that IOCs can become outdated in a short period of time.

Technical Cyber Threat Intelligence Technical CTI (TeCTI) should focus on the technical clues that are indicative of a CS threat such as the subject lines to phishing emails, fraudulent URLs or specific malware. TeCTI enables security analysts to determine what to look for, rendering it valuable for analysing social engineering attacks. However, in the financial sector such as the banking sector, penetration testing no longer appears to be sufficient to shield sensitive business sectors. Considering this, the UK Financial Authorities have recommended several steps which can be found in [17] to protect financial institutions from cyber threats.

Operational Cyber Threat Intelligence Operational CTI (OCTI) pertains to details of specific events associated with the cyberattack in order to facilitate an understanding of the nature, severity, timing, and intent of specific attacks. OCTI involves cybersecurity professionals learning about threat actors and is focused on addressing the 'attribution' elements of CTI, such as 'who', 'why' and 'how' questions. In this context, 'who' refers to threat actors, 'why' addresses the motivation or intent, and 'how' consists of tactics, techniques and procedures (TTPs) that adversaries use to carry out attacks. The attribution elements offer context, and context, in turn, provides insight into how attackers plan, conduct, and sustain campaigns and operations. Such an insight is considered to be operational intelligence which cannot be produced by machines alone. If implemented properly, OCTI will be able to provide highly specialised and technically focused intelligence to guide and assist with the response operations.

Thus, OCTI should be based on details of the specific incoming attack and evaluation of an organisation's capability in determining future cyber-threats. It must be able to assess specific attacks associated with events, investigations and malicious behaviour, and provide an understanding that can guide and support response to specific incidents. This type of CTI requires Cyber Security Analysts who can convert data into a format that is readily usable by end-users. Despite the fact that OCTI necessitates more resources than that required by TaCTI, it offers a longer valuable lifespan. This is due to the fact that attackers will not be able to alter their TTPs in the same way that they could easily change their tools. OCTI is often most beneficial for those cybersecurity specialists operating in security operations centers (SOCs) who are in charge of conducting routine operations. Professionals operating in CS branches such as Vulnerability Management, Incident Response and Threat Monitoring are the main customers of OCTI as it assists them with becoming more capable and effective at their assigned tasks [16, 18].

Strategic Cyber Threat Intelligence Strategic CTI (SCTI) must be aimed at long-term issues and be based on high-level information on CS modus operandi, threats, details concerning impact of fund on different cyber activities, attack tendencies, and the effect of high-level business assortments. Therefore, SCTI must be employed (1) to evaluate disparate pieces of information to establish unified views, and (2) to develop an overall picture of the intent and capabilities of cyber threats (such as the actors, tools and TTPs) through the identification of trends, patterns, and evolving threats with a view to inform decision makers. An effective SCTI should also be able to enable time alerts of threats against organisations' important assets such as IT infrastructure, employees, customers, and applications. This information should be in the format of reports, whitepapers, policy documents, or publications in the industry and must then be presented to high-level executives, such as Chief Information Security Officers (CISO) for the purposes of decision making.

Furthermore, SCTI can be used as a means to understand how global events, foreign policies, and other long-term national and international movements can influence the CS of an organisation. This understanding can assist decision-makers in understanding cyber threats against their organisations more effectively. In turn, this knowledge can enable them to make CS investments that safeguard their organisations and are aligned with its strategic priorities [18]. SCTI is the most challenging type of intelligence to produce as it entails human collection and analysis that require an in-depth knowledge of both CS and global geopolitical situation. To this end, often, senior leadership is required to perform critical evaluations of cyber threats against their organisations.

2.3 Benefits of Cyber Threat Intelligence

If implemented effectively, CTI provides substantial benefits as threat information can be shared in machine-readable formats that can be promptly obtained and imported for immediate use by security incident and event management (SIEM) tools and CTI platforms (CTIPs). CTI can enable the development of a focused defence against specific threats as well as the insight to apply the appropriate CS tools and solutions to protect organisations. Furthermore, CTI can provide organisations with context such as intelligence about the attackers, their motivation and capabilities and indicators of compromise (IoCs) in their system to investigate. This information will enable organisations to make informed decisions about their security. Based on its classification, described in the previous section, CTI offers four types of tactical, technical, operational, and strategic benefits as shown in Table 1.

In addition to the above, CTI can contextualise threat information that is more meaningful for the end-user. This, in turn, reduces ambiguity, enhances situational awareness, and results in more informed risk management and security investment. Furthermore, CTI can assist vulnerability management teams in prioritising the most vital susceptibilities more accurately with access to the external understandings enabled by CTI. Similarly, comprehending the existing threat landscape (comprising key insights on threat actors and their modi operandi) that CTI provides can augment other high-level security processes such as fraud prevention and risk analysis. As well as assisting organisations to protect their networks, CTI can also enable them to regulate costs of sustaining their network security and provide the security teams with the knowledge they require to concentrate on what really matters.

3 Cyber Threat Intelligence Cycle

To produce intelligence (the final product of the CTI cycle), organisations would firstly require to collect raw data. This raw data represents simple facts that are available in large quantities such as IP addresses or logs. On its own, the raw data has limited usefulness until it is converted to information through data processing for the purposes of producing a valuable output. An example of information is a collated series of logs that display an increase in suspicious activities. Intelligence can then be produced by processing and analysing this information, which must be able to inform

Table 1 Benefits of tactical, technical, operational and strategic CTI

Types of CTI	Benefits
Tactical CTI	• enables organisations to develop a proactive cybersecurity posture and to strengthen overall risk management policies • informs better decision-making during and after the detection of a cyber-attack • assists with a cybersecurity posture that is predictive • facilitates enhanced detection of advanced threats
Technical CTI	• connects details associated with attacks rapidly and accurately • provides rapid response to new indicators • enables security analysts to determine what to look for rendering it valuable for analysing social engineering attacks
Operational CTI	• provides context and relevance to a large amount of data that enable organisations to gain better insight into how threat actors plan, carry out, and sustain offensives and major operations • enables organisations to detect and respond to cyber-attacks more swiftly and assisting them in preventing future incidents•enables organisations to detect and respond to cyber-attacks more swiftly and assisting them in preventing future incidents
Strategic CTI	• provides a more in-depth situational awareness • assists decision-makers in understanding the risks posed by cyber threats to their organisations • enables decision-makers in making cybersecurity investments that effectively defend their organisations and are aligned with its strategic priorities • produces an organisational situational awareness that will help existing and future security strategies

decision making. As an example, the collated data is placed in context along with prior incident reports in relation to similar activities that enable the development of a strategy to reduce cyberattacks [19]. Figure 2 represents a useful model that visualises the processing of raw data into a complete intelligence product.

The Cyber Threat Intelligence Cycle (CTIC), that produces intelligence, must be a methodical, continuous process of analysing potential threats to detect a suspicious set of activities that can threaten organisations' systems, networks, information, employees, or customers. It must visually represent and evaluate a number of specific intrusion sensor inputs and open source information to determine specific threat courses of action [20]. Therefore, the CTIC should be a process whereby raw data and information are identified, collected and then built into a complete intelligence for use by decision makers. The model must also be able to support organisations' risk management strategies and the information security teams' decision-making. To exploit the benefits of CTI, it is essential to define both appropriate objectives as well as relevant use cases. Since the CTIC is intended to produce intelligence, the information security teams must be able to formulate new questions and identify gaps in knowledge during this lifecycle. In turn, this should result in the requirements development. Furthermore, in order for a CTIC to be an effective intelligence

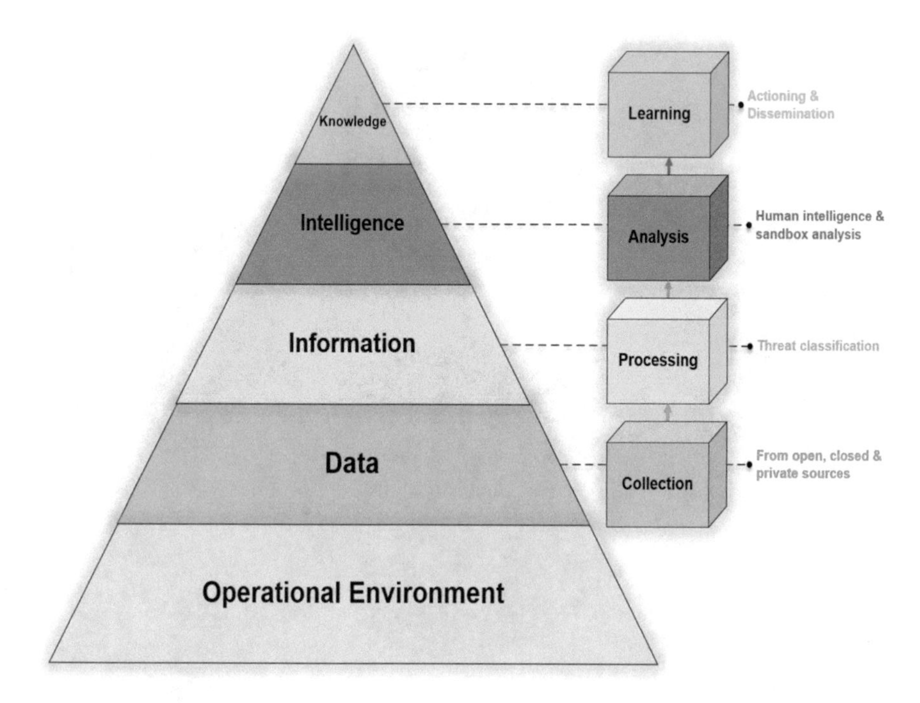

Fig. 2 Processing of raw data into a complete intelligence product

scheme, it must be based on iterative phases that can become more sophisticated over time.

Therefore, considering the discussion above, we recommend a six-phase cycle consisting of the following stages: Planning and Direction, Data Collection, Data Processing and Exploitation, Data Analysis and Production, Dissemination and Integration, and Feedback. All steps in the cycle must also incorporate an Evaluation process and a Review process that must be performed simultaneously throughout the entire six phases so as to ensure that the necessary materials are being processed accurately and that the original questions are being addressed effectively. Figure 3 represents our recommended CTIC along with the description of each phase.

3.1 Planning and Direction

The CTI's production life cycle starts with requirements or questions unique to the end-user that should be answered. After the CTI requirements have been identified and prioritised, a data collection plan comprising identification and evaluation of information sources should be created. Planning and Direction is the first phase of the CTI, that is intended to produce actionable threat intelligence based on a set of

Fig. 3 Six-phase cyber threat intelligence cycle

accurate questions that should enable the development of actionable threat intelligence. These questions must focus on a single fact, event, or activity as opposed to broad, open-ended questions [21]. A key aspect of this phase should be understanding who will consume and benefit from the complete product. Next, individuals involved with planning and direction should be able to establish the precise requirements of the consumer, called intelligence requirements (IRs), and prioritise intelligence requirements (PIRs). These IRs and PIRs must be based on certain factors such as how closely they comply with organisations' core values and must determine what data and information are required and how it should be collected. This output is often systematised in an intelligence collection plan (ICP) [19]. It is important that this phase involves substantial interaction between the consumer and producer.

3.2 Collection

The next step in the CTI is the Collection phase, that involves gathering raw data [22–25]. This data must be meaningful to the organisation and able to address the initial CTI requirements established in the first phase. Raw data can be gathered from a wide variety of sources such as internal ones including network event logs and records of past incident responses and external ones from the open web, the dark web, and technical sources [21]. The data Collection phase must be timely and accurate, as well as being applicable to deal with incidents that can occur or are occurring. Understanding which sources are likely to generate the desired information, be reliable, and provide information that can be used in a timely manner is a complex process that necessitates thoughtful and robust planning and direction to assist in isolating the signals from the noise.

Instances of CTI data sources consist of traditional Security Information and Event Management (SIEM) tools (such as network monitors, firewalls, intrusion detection systems), dedicated CTI data feeds, vulnerability and malware databases, and the system users. It is through these data sources that indicators of compromise (IoCs) can be identified, documented, and further analysed. IoCs which represent threat data concerns measurable events that can be classified as either network-based or host-based events. Examples of network-based IOCs comprise email addresses, subject line and attachments, connections to specific IP addresses or web sites, file hashes, and fully qualified domain names utilised for botnet command and control server connections. Instances of host-based IoCs consist of the presence of filenames on a local drive, programs and processes that are running on a machine, and creation or modification of dynamic link libraries (DLLs) and registry keys [26]. Furthermore, IoCs can also include vulnerability information, such as the personally identifiable information of customers, raw code from paste sites, and text from news sources or social media.

3.3 Processing and Exploitation

Processing and Exploitation is the third phase in the CTCI, that involves converting the raw data into intelligence. The raw data that have been collected from multiple data sources must be integrated and sorted in order to produce more consistent, accurate, and useful information than that provided by any individual data source. To achieve this, one needs to sort and fuse it with other data sources by organising it with metadata tags and filtering out redundant information or false positives and negatives [21]. During this phase, both human and machine capabilities are needed to address the IRs for the engagement while complying with the tenets of intelligence. Given that data is collected from millions of log events and indicators every day, processing such data manually is extremely cumbersome. Thus, collecting data must be automated in order to extract meaningful intelligence from it. One of the best ways to achieve this is to deploy solutions such as SIEM since it facilitates structuring and correlating event data with rules that can be established for various use cases (even though it can only deal with a limited number of data types). See Sect. 4 for details on more powerful data processing solutions.

3.4 Analysis and Production

Analysis and Production is the next phase in the CTIC, where analysts will need to make sense of the processed data. The objective of this phase is to look for possible security threats and inform the relevant audience in a format that achieves the intelligence requirements defined in the Planning and Direction phase [21]. The analysis must be determined based on three elements of actors, intent, and capability, with consideration given to their tactics, techniques, and procedures (TTPs), motivations, and access to the intended targets. By examining these three elements, it is often possible to make informed, forward-leaning strategic, operational, and tactical assessments. Furthermore, during this phase, analysts must be able to produce intelligence products, i.e. the answers to the questions posed earlier during the requirements gathering, and identify connection between the technical indicators, attackers, their motivations and aims, and information related to the target [27]. This should then result in informative and proactive decision-making. To do so, analysts will need to employ a wide range of quantitative and qualitative analytical techniques to evaluate the significance and implications of processed information, merge contrasting items of information to find patterns, and then interpret the meaning of any newly developed knowledge.

Additionally, they will need to apply a variety of approaches to assess the reliability of the sources and the material collected and to ensure accurate and unbiased evaluations that need to be predictive and actionable. It is also vital for any potential ambiguities to be handled properly, for instance, by determining how the questions have been addressed. Analysis phase must be accurately documented and efficiently implemented to assist organisations in utilising the collected data more effectively. This should be followed by a timely dissemination of intelligence to internal and external audiences in a format understandable to them such as threat lists and peer-reviewed reports.

3.5 Dissemination and Integration

Dissemination phase should involve communicating and distributing the complete product in a suitable form to its intended consumers. In order for CTI to be actionable, it must be delivered to the right audience at the right time, i.e. the occurrence of dissemination should correspond to the time period on which the content is based. For instance, operational material requires to be regularly conveyed whilst strategic content will be more sporadic. The Dissemination phase must also be traceable in order that there is continuation between one CTIC and the next and that the learning is not lost. One of the ways in which this can be achieved is by utilising ticketing systems that integrate with the consumers' other security systems to trace each stage of the CTIC. Everytime a new intelligence request is made, tickets can be submitted, written up, reviewed, and fulfilled by different audience in one place. By obtaining feedback and refining existing IRs or creating new ones, the CTI cycle can commence again [21].

3.6 Feedback

The Feedback is the final stage in CTIC, in which a complete intelligence has been developed linking it to the original Planning and Direction phase. During this phase, individual/s who made the original request reviews the complete intelligence product to establish whether their questions have been addressed. This assists in informing the objectives and procedures of the next CTI cycle, once again highlighting the importance of documentation and continuation.

4 Application of Artificial Intelligence and Machine Learning in Producing Actionable CTI

AI and ML are two promising fields of research that can significantly improve CS measures. For instance, CS applications using AI and ML can perform anomaly detection on a network more effectively than those performed by traditional methods. With rapid pace of development and the desire for more effective countermeasures, AI and ML come as a natural solution to the problem of coping with the ever-growing number of cyber-attacks. This interdisciplinary endeavour has created a joint link between computer specialists and network engineers in designing, simulating and developing network penetration patterns and their characteristics. Some of these diverse methods are directed towards: Multi-Agent Systems of Intelligent Agents, Neural Networks, Artificial Immune Systems and Genetic Algorithms, Machine Learning Systems, including: Associative methods, Inductive Logic Programming, Bayes Classification, Pattern Recognition Algorithms, Expert Systems, and Fuzzy Logic.

Examples of AI and ML applications that can be used in CS solutions include: Spam Filter Applications, Network Intrusion Detection and Prevention, Fraud Detection, Credit Scoring and Next-Best Offers, Botnet Detection, Secure User Authentication, Cyber Security Ratings, and Hacking Incident Forecasting, etc. For instance, by determining certain distinctive features, AI and ML systems can be trained to analyse and distinguish between a normal software and malware. These features can comprise: accessed APIs, accessed fields on the disk, accessed environmental products, consumed processor power, consumed bandwidth, and amount of data transmitted over the internet. By utilising these distinct features, the system is developed. Once a test software is fed to the system, it can then determine whether the software is a malware or not by analysing these distinct features [28].

In the specific context of CTI, organisations can utilise AI and ML methods to automate data acquisition and processing, combine with their existing security solutions, absorb unstructured data from disjunctive sources, and then link information from different places by adding context on compromise and modi operandi of malicious actors. This is particularly important in the context of Big Data, due to the scales of which its processing necessitates automation to be comprehensive. This processing should comprise the fusion of data points from a wide range of sources such as open web, deep web, dark web, and technical sources in order to draw up the most robust strategy. This can help to convert these large quantities of data into actionable CTI. Furthermore, by means of AI and ML techniques, data can be structured into categories of entities based on their names, properties, relationships to each other, and events by separating concepts and assembling them together. This will facilitate robust searches on the categories, enabling the automation of data sorting as opposed to sorting data manually [29]. In addition, AI and ML techniques can be applied for the purposes of structuring text in many languages through Natural Language Processing (NLP). For instance, NLP can be exploited to analyse text from almost infinite unstructured documents across a wide range of languages and categorise them by means of language-independent groups and events [21].

Moreover, ML techniques can be developed to categorise text into groups prose, data logs, or code, and remove ambiguities between entities with the same name through the use of contextual clues in the surrounding text. ML and statistical methodology can be implemented to sort entities and events even further based on significance, for instance by evaluating risk scores to malicious entities. Risk scores can be calculated by the ML trained on an already examined dataset. Classifiers such as risk scores deliver both a judgment and context describing the score since different sources verify that this IP address is malicious. Automating risk classification saves substantial time by sorting through false positives and determining what to prioritise. In addition, ML can be used to predict events and entity properties by producing predictive analysis models more accurately than those created by humans based on deep pools of data that have been previously mined and categorised [21, 29]. It is also likely that ML techniques could function as active sensors that feed data into a common threat intelligence network that can be employed by the entire user base. The above said, the process of applying ML and AI methods at the different levels of CTI is at very different stages. For instance, studies in Operational Intelligence type are still in the experiment and research stage and as a result necessitate substantial resources.

5 Discussion

Cyber threats are constantly growing in frequency and complexity, and the threat landscape is continually evolving. Through the use of various customised TTPs, cyber criminals are able to bypass organisations' security controls. As a result, organisations are under constant pressure to manage security vulnerabilities. One of the ways to help address security vulnerabilities is by developing and implementing robust CTI. CTI is based on traditional intelligence gathering and processing activities used to track, analyse and counter CS threats. The information collected through CTI can enable the security teams to identify, prepare, and impede cyber-attacks that can pose risk to the data integrity. CTI feeds can assist organisations in this process by identifying common IoCs and suggesting required steps to stop cyber-attacks. The most common IoCs consist of [30]:

- IP addresses, URLs and Domain names: An example is malware that targets an internal host that is communicating with a known threat actor.
- Email addresses, email subject, links and attachments: An example is a phishing attempt that depends on a user clicking on a link or attachment and starting a malicious command.
- Registry keys, filenames and file hashes and DLLs: An example is an attack from an external host which has already been flagged or that is already infected.

Robust CTI feeds could potentially have millions of computers functioning as security sensors which feed CTI to the entire users subscribing to that feed. At the same time, millions of security updates can automatically and seamlessly take place on the daily basis to end users and networks.

It is important to note that in order for organisations to be able to access CTI when needed, they will need to incorporate it into their broader security model as an essential component that enhances every other function (as opposed to a separate function). Incorporating CTI into security solutions that organisations already employ reinforces their security postures. Such an integration can enable security operations teams to respond to and process the alerts more effectively by helping automatically to prioritise and sieve through security threats. It is also imperative that there will be a clear distinction between threat data and threat intelligence. Without intelligence, data will not be able to provide the predictive knowledge required to detect threats before they can enter organisations' networks.

6 Conclusion

CTI can add significant values to organisations' security functions as well as to every level of government entity such as Chief Information Security Officers (CISOs), police chiefs, policy makers, information technology specialists, law enforcement officers, security officers, accountants, and terrorism and criminal analysts. If implemented properly, CTI can facilitate better understanding into cyber threats, enabling a faster, more targeted response and resource development and allocation [31]. It can enable decision makers to define acceptable business risks, create controls and budgets, make equipment and staffing decisions, provide insights that guide and support incident response and post-incident activities (operational/technical intelligence), and advance the use of indicators by validating, prioritising, specifying the length of time an indicator is valid for (tactical intelligence). Likewise, when timely, relevant, and actionable, CTI can enable organisations to operate more efficiently and effectively by gaining the advantage they require to combat cyber-attacks prior to loss being incurred. Furthermore, by utilising CTI, organisations will be able to update their endpoint and network security proactively in real-time without the need to update their network security environments manually. For instance, in cases where one endpoint device faces a threat, that intelligence will be able to update the larger CTI network automatically. This enables organisations to stay ahead of cyber threats and attackers consistently and ensure that they are safeguarded against the latest cyber-attacks.

As security vendors compete with each other to deal with the consumer demand for assistance with the increasing number of threats, the market is now providing a wide range of CTI tools. However, not all tools are developed equal. For a successful implementation of security at this level to function effectively, the tool must be able to search through the vast and miscellaneous stretch of online content for potential security threats at every second. Therefore, a CTI security solution must be customizable and capable of providing clear and complete investigation with advanced analytics such as AI and ML that can be adapted to specific behavioural activities [30].

It is envisioned that over the next few years the inclusion of CTI into organisations' and governments' operations will become increasingly vital, as all levels and employees are forced to respond to the cyber threats. It is also envisaged that in the near future, cloud-based network security and secure web gateways fed by threat intelligence replace legacy firewalls, appliances, software and much of the resources required to patch and update in traditional environments [32]. As a future research direction, one area of CTI that has remained underexplored concerns the application of Multi-Agent Systems (MASs) in Tactical Cyber Threat Intelligence (TCTI). Therefore, experiments should be performed with the application of MASs to determine whether it can be an appropriate method for the needs of the CTI.

References

1. Montasari R, Hill R (2019) Next-generation digital forensics: challenges and future paradigms. In: 2019 IEEE 12th international conference on global security, safety and sustainability (ICGS3). IEEE, pp 205–212
2. Montasari Reza (2017) A standardised data acquisition process model for digital forensic investigations. Int J Inf Comput Secur 9(3):229–249

3. Montasari R (2018) Testing the comprehensive digital forensic investigation process model (the cdfipm). In: Technology for smart futures. Springer, pp 303–327
4. Montasari R, Hill R, Carpenter V, Montaseri F (2019) Digital forensic investigation of social media, acquisition and analysis of digital evidence. Int J Stratg Eng(IJoSE) 2(1):52–60
5. Montasari R, Hosseinian-Far A, Hill R (2018) Policies, innovative self-adaptive techniques and understanding psychology of cyber security to counter adversarial attacks in network and cyber environments. In: Cyber criminology. Springer, pp 71–93
6. Montasari R, Hill R, Parkinson S, Peltola P, Hosseinian-Far A, Daneshkhah A (2020) Digital forensics: challenges and opportunities for future studies. Int J Organ Collect Intell (IJOCI) 10(2):37–53
7. Montasari R, Hosseinian-Far A, Hill R, Montaseri F, Sharma M, Shabbir S (2018) Are timing-based side-channel attacks feasible in shared, modern computing hardware? Int J Organ Collect Intell (IJOCI) 8(2):32–59
8. Farsi M, Daneshkhah A, Hosseinian-Far A, Chatrabgoun O, Montasari R (2018) Crime data mining, threat analysis and prediction. In: Cyber criminology. Springer, pp 183–202
9. Montasari R (2017) An overview of cloud forensics strategy: capabilities, challenges, and opportunities. In: Strategic engineering for cloud computing and big data analytics. Springer, pp 189–205
10. Montasari R, Hill R, Montaseri F, Jahankhani H, Hosseinian-Far A (2019) Internet of things devices: digital forensic process and data reduction. Int J Electr Secur Digital Forensics
11. Montasari R, Peltola P (2015) Computer forensic analysis of private browsing modes. In: International conference on global security, safety, and sustainability. Springer, pp 96–109
12. Pescatore J (2019) SANS top new attacks and threat report, 2019. SANS Institute Cyber Security Report
13. McMillan R (2013) Definition: threat intelligence. Accessed 29 March 2019
14. Johansen G (2017) Digital forensics and incident response: an intelligent way to respond to attacks. Packt Publishing
15. CERT-UK (2015) An introduction to threat intelligence, 2015. CERT-UK. TLP White
16. NCSC (National Cyber Security Centre) (2016) Vulnerability management: guidance to help organisations assess and prioritise vulnerabilities. https://www.ncsc.gov.uk/guidance/vulnerability-management. NCSC. Accessed 05 March 2020
17. CBEST (2016) CBEST intelligence-led testing: CBEST implementation guide, 2016. CBEST. Version 2.0
18. Crowd Strike (2019) Cyber threat intelligence. https://www.crowdstrike.com/epp-101/threat-intelligence/. Crowd Strike. Accessed 27 Feb 2020
19. CREST (2019) What is cyber threat intelligence and how is it used? CREST. CTIPS (CREST Threat Intelligence Professionals)
20. KimeB (2016) Threat intelligence: planning and direction. SANS Institute. White Paper
21. Recorded Future (2020) What is threat intelligence? https://www.recordedfuture.com/threat-intelligence/. Crowd Strike. Accessed 17 Feb 2020
22. Montasari R, Peltola P, Evans D (2015) Integrated computer forensics investigation process model (icfipm) for computer crime investigations. In: International conference on global security, safety, and sustainability. Springer, pp 83–95
23. Montasari R (2016) Review and assessment of the existing digital forensic investigation process models. Int J Comput Appl 147(7):41–49
24. Montasari R (2016) Formal two stage triage process model (ftstpm) for digital forensic practice. Int J Comput Sci Secur 10:69–87
25. Montasari R (2016) An ad hoc detailed review of digital forensic investigation process models. Int J Electron Secur Digit Forensics 8(3):205–223
26. Stephen D, Mason R, Robert M, Matthew S (2016) Applying cyber threat intelligence to industrial control systems. J Cyber Secur Inf Syst 7(2)
27. Shackleford D (2015) Who's using cyberthreat intelligence and how? SANS Institute. Accessed 24 Jan 2018

28. NormShield (2020) Cyber threat intelligence. https://www.normshield.com/cyber-security-with-artificial-intelligence-in-10-question/. Recorded Future. Accessed 24 Feb 2020
29. Pokorny Z (2018) 4 ways machine learning produces actionable threat intelligence. https://www.recordedfuture.com/machine-learning-threat-intelligence/.NormShield. Accessed 25 Jan 2020
30. Forcepoint (2020) What is threat intelligence?: Threat intelligence defined and explored. https://www.forcepoint.com/cyber-edu/threat-intelligence.Forcepoint. Accessed 29 Feb 2020
31. Intel & Analysis Working Group (2020) What is cyber threat intelligence? https://www.cisecurity.org/blog/what-is-cyber-threat-intelligence/. CIS (Centre for Internet Security). Accessed 26 Jan 2020
32. Avast (2020) What is threat intelligence?. https://smb.avast.com/answers/threat-intelligence. Avast. Accessed 07 March 2020

Drone Forensics: The Impact and Challenges

S. Atkinson, G. Carr, C. Shaw, and S. Zargari

Abstract Unmanned aerial vehicles (UAV) have surged in popularity over the last few years. With this, crime involving drones has also dramatically increased. Therefore, there is a dire need of successful Drone programmes that significantly would lower the amount of crime being committed involving Drone devices. Drone forensics is a concept that is less well known or documented. Research has shown that there have been Drone Forensic programmes to support the forensics investigations, however, many have failed for a few reasons such as the lack of understanding of the technology or other limited resources. It is also known within the Digital Forensics community that Anti-Forensics techniques are constant threats and hinder investigations, resulting in less convictions. This study aims to ascertain exactly what data can be extracted from UAV devices (Drones), the usefulness of this data, and whether consumers are able to obfuscate the data in efforts to evade detection (i.e. Anti-forensics techniques). A number of primary and secondary datasets have been utilised in this research. Primary data includes carrying out a flight using a UAV device and consequently analysing the resulting data and an interview with a qualified Digital Forensic Analyst. Secondary data was gained from VTO Labs, recommended by NIST which was able to be interrogated in order to deliver interesting results. This study found that Drones have the ability to hold a wealth of evidence that could potentially be very useful to assist forensics investigations. This included the flight path of the Drone, date and time of flight, altitude, home-point and alerts to inform whether the Drone was near restricted airspace such as airports (No Fly Zones). Moreover, it was found that it is possible for the manufacturers to

S. Atkinson · G. Carr · C. Shaw · S. Zargari (✉)
Faculty of Science, Technology and Arts, Sheffield Hallam University, Sheffield, UK
e-mail: S.Zargari@shu.ac.uk

S. Atkinson
e-mail: sianatkinson15@gmail.com

G. Carr
e-mail: georgecarr97@gmail.com

C. Shaw
e-mail: callumshaw12@hotmail.co.uk

R. Montasari et al. (eds.), *Digital Forensic Investigation of Internet of Things (IoT) Devices*, Advanced Sciences and Technologies for Security Applications,
https://doi.org/10.1007/978-3-030-60425-7_4

build in Anti-Forensics software into their devices, but it would not be possible for a consumer to utilise such techniques.

Keywords Digital forensics · Drones · UAV · Anti-forensics · Mobile forensics · Drone forensics

1 Introduction

An Unmanned Aerial Vehicle (UAV) or drone, is a pilotless aircraft that is controlled via flight software and a remote pilot. The first UAV was a quadcopter built in 1907 [16]. By 1917 developments were being made to create the Ruston Proctor Aerial Target, which is used by the army to fly bombs into enemy territories [16]. Such developments resulted in military drones that are used today. The use of unmanned aerial vehicles (drones) has soared in recent years across the UK. PWC recently reported that by 2030 there could be up to 76,000 drones operating in the UK's skies, with 628,000 jobs created within the UK economy involving drones [37].

However, this increased availability has resulted in tremendous growth of drone crime over recent years [32]. Even as far back as 2014, there were 283 drone crimes reported in the UK [44]. Some of these criminal acts include drug and weapon delivery into prisons, as well as being used to stake out homes for burglaries [12]. Drone technology is continuously evolving and is 'part of a complex digital ecosystem' [32]. This is due to the use of controllers and connected devices, meaning that it can be challenging to keep up with their many uses. This correlated with the lack of understanding within law enforcement agencies [47] about the technology.

Drones provide quick aerial views via remote pilot. The functionality and accessibility of drones has increased drone usage in various sectors such as construction, filming, photography, and estate agents who are more frequently hiring drone operators to help in commercial activities.

There has also been an increase in the recreational use of drones by hobbyist's [19]. Users fly drones remotely to capture aerial images, create films and record their experiences. Recreational drones can be purchased for approximately £400 in the UK and can be used immediately by the user after unpacking and charging the drone [42].

Due to the accessibility and usability of drones, criminals have taken advantage of their abilities to commit a vast range of crimes. In 2018, police forces across the UK reported that they had received 2,435 reports of incidents involving drones [34]. This was up 2% from previous year and a dramatic 42% higher than incidents reported in 2016 [34]. Perhaps the most high-profile case involving drones occurred in December 2018, whereby Gatwick airport closed in response to drone sightings within the surrounding airspace [19]. Flights were cancelled and delayed during the 36 h of the closure of the airport, EasyJet reported that in total the cost of compensation reached £15 m and a total of 82,000 customers were affected [19].

1.1 Unmanned Aerial Vehicles

Rouse [40] defines a drone as unmanned aerial vehicles (UAVs) or unmanned aircraft systems (UAS) [9]. Rouse [40] also states that essentially, a drone is a flying robot that can be remotely controlled or fly autonomously through software-controlled flight plans in their embedded systems, working in conjunction with onboard sensors and GPS.

Recently, UAVs were most often associated with the military, where they were used initially for anti-aircraft target practice, intelligence gathering and then, more controversially, as weapons platforms [40]. Drones are now also used in a wide range of civilian roles ranging from search and rescue, surveillance, traffic monitoring, weather monitoring and firefighting, to personal drones and business drone-based photography, as well as videography, agriculture and even delivery services [40].

There are various brands and designs of drones which have different components to fit the job they have been created for. Drones generally have the same or similar components to be able to function, some of which are:

- Propellers (Can be made of plastic or carbon fibre depending on the specific drone)
- Motor (The better the motor, the better that battery life of the drone)
- Receiver (Radio signals to the drone through the controller)
- Transmitter (Radio signals from the controller to the drone)
- GPS Module (Responsible for longitude, latitude, and elevation points)
- Battery (Allows the drone to fly)
- Camera (Can be inbuilt or detachable)
- Electronic Speed Controllers (Controls the speed of the drone) [1]

Similarly, any data gathered from the drones can be stored differently depending on the drone. Some drones have internal storage which varies in size from 4 to 8 GB. It is worth noting that the internal drone data will be overwritten when it has reached capacity, so it is wise to regularly check the memory and extract any data you may need. Drones also use SD or Micro SD cards, giving the user more storage capacity. There are also options such as extracting drone data to an external hard drive or even the cloud, depending on whether the data is something the pilot wants to keep.

1.2 Current Threats and Impact of UAVs

The misuse of drones poses threats to public safety, organisations, and national security due to such incidents where drones have been flown in no-fly zones. Such threats are reasons why drone forensic programs are essential in aiding the understanding of drone technology and in reducing the crime rate, especially when successful in gathering evidence for a case. However, even with the benefits a drone forensics program will bring to law enforcement agencies, this is still an area that remains relatively unexplored [5]. Drone crime can vary in threat levels, and there needs to be a more precise understanding by the public as well as law enforcement about these threats posed by drones, as drone crime is 'only limited by the imagination of the criminal' [21].

UAVs are getting more popular and accessible with consumers which makes it easier for criminals to take advantage of the technology for nefarious reasons. Drones have been used for a range of crimes from smuggling, spying/stalking, criminal damage, and even theft of card details from ATM's [34]. With all technology there is always the potential for misuse, leading to issues and negative impacts on individuals but also the community as people start to fear what the technology is capable of. With drones being adaptable and so varied, it makes the job of law enforcement even harder as they must tackle newly emerging crimes and disruptions, possibly without the necessary legislation in place for guidance.

1.3 Legal Implications

Depending on the context the drone is to be used for, there are legislations in place to outline the correct use of the drone and rules to be adhered to such as ensuring the correct licences are held for a drone. As drones were never designed for criminal use, it is impossible to create legislation to cover all possible drone crimes, as there are endless possibilities. What can be done, is to create legislation to cover the general misuse of drones; anything that causes harm or distress to the person, property, or community. This way there is a clearer line of what is unacceptable by law, therefore likely to lead to consequences. The NPCC's lead for drone crime said, 'those who choose to use drones for a criminal purpose should be in no doubt that they face serious consequences and police will use all available powers to investigate and prosecute them' [34].

Current regulations of drone fall under the Civil Aviation Act 1982 and the Air Navigation Order 2016 (amended in 2018), covering appropriate drone usage with flight restriction zones [25]. In 2019, it was proposed that registration of all drones be mandatory with possible competency tests. The idea being that law-abiding citizens would register their drones making it easier for law enforcement to track down criminal drone use [25]. Whilst this concept proves successful with newly registered drone users, it is not possible to ensure all drones purchased prior to the new regulations will be registered. Unfortunately, with all technology and the regulations put in place, there will be a loophole found by criminals to continue with their criminal activities.

1.4 Motivation

With the influx in use by the public, there have been reports of Drone devices being involved in criminal activity. Therefore, this study will carry out research into various crimes that have been recorded that involve such devices and how evidence gained from them aided the investigation. Additionally, this project will examine Drone devices in order to ascertain exactly what data/evidence can be extracted from these devices and how useful they could be to an investigation.

Additionally, Anti-Forensics techniques are in existence and are at times used on devices such as PCs. This study will also look into the various techniques that are available and ascertain whether or not these techniques can be used on Drone devices.

1.5 Research Aims and Objectives

There are various aims and objectives to this study due to the collaboration of multiple projects. The research objectives are as follow:

1. Gain access to Drone data and extract using popular forensics extraction software to ascertain the usefulness of evidence.
2. Gain in-depth drone knowledge to carry out successful drone data analysis.
3. Gain information regarding the current processes used by Police Forces and other authorities to analyse the data and prevent drone crimes and how they are investigated.
4. Gain Knowledge of the various Anti-Forensics techniques that are currently available and in use by users who want to obfuscate data and determine whether these methods can be used on Drone devices.

2 Existing Research

This section focuses on multiple elements relating to drones to provide a broad coverage of how technology, tools, crime, and devices affect drone technology. There are more and more research topics relating to drones, therefore, this section provides a small insight into what route the research can take into the technology. There will also be an emphasis on challenges facing the technology, whether by legislation in place, restrictions or nefarious means.

2.1 The Internet of Things (IOT)

The internet of things is described as a world where many otherwise ordinary devices are uniquely identifiable, addressable, and contactable via the internet [20]. Hegarty et al. [20] split IOT challenges into four stages: Identification, Preservation, Analysis and Presentation. Whilst Hegarty et al. [20] identify the issues IOT forensics faces, their paper does not provide insights how to deal with challenges. A further limitation of Hegarty et al. [20] paper is that it only covers the basic challenges of IOT forensics in 2014 and thus the findings are likely to be outdated. However, their study does suggest that more frameworks need to be introduced for forensic examinations of the internet of things [20].

The findings presented more recently by Conti et al. [10] are congruent with preceding work conducted by [20]. Conti et al. [10] claim the main challenges which IOT forensics face are: Evidence Identification, Collection and Preservation, Analysis and Correlation and Attribution. However, they fail to propose a framework to meet the requirements of IOT which states the challenges that forensic examiners are facing in 2018.

A review of the literature has indicated there are gaps in the knowledge surrounding IOT frameworks and procedures and thus, more work needs to be done to present a framework which considers a range of IOT devices.

2.2 UAV Devices

Whilst the definition of a drone remains constant, there are several different types of drones, which vary in size, weight, capabilities, appearance, brand, and features.

Flynt [17] breaks drones down into size; ranging from very small drones, small drones, medium drones and large drones. Additionally, Flynt [17] states drones vary with the range at which they can be flown with from some drones only being able to be flown from 5 km away whereas others can be flown from as far as 650 km. Drones vary in size and shape depending on their function as a drone, for whom they are targeted to be used for and the tasks they can complete.

- Quadcopters—The most popular model on the market uses four rotors positioned in each corner of the square body. This type of drone will be considered in this study as it resembles a popular choice for smuggling contraband into prisons and is used for lots of other criminal activities.
- GPS Drones—Drones which are linked to satellites via GPS, the flight direction of the drone will depend on the satellite.
- Photography Drones—These are drones which have a camera attached to the main body, they are used to take HD pictures and videos and are popular among hobbyists.
- Racing Drones—Small, fast and agile which are streamlined for speed and free of excess weight that can reach speeds of up to 60 mph.

2.3 UAV Offences

The UK Civil Aviation Authority (CAA) set the rules for drone use in the UK [46].

1. Always keep your drone in sight.
2. Stay below 400 ft (120 m) to comply with the *drone code*.
3. Every time you fly your drone you must follow the manufacturer's instructions.
4. Keep the right distance from people and property—150 ft (50 m) from people and properties and 500 ft (150 m) from crowds and built up areas.
5. You are responsible for each flight.
6. Stay well away from aircraft, airports and airfields when flying any drone—It is illegal to fly inside the airport's flight restriction zone without permission.

Drone crimes generally are carried out by repurposed larger drones [49] that have longer flight times, transmission distance as well as have self-adaptive flight systems that adjust flight parameters based on different payloads, such as the DJI Matrice 600 [14].

Smuggling contraband into prisons is not a new concept, however, as drones provide an effective way to smuggle more dangerous and larger items into prison grounds. Many prisons have taken an approach of non-technical solutions such as barbed wire and perimeter nets, and there are the technical solutions of installing jammers inside and on the perimeter of the prisons [41]. Agencies that have a drone forensics program will be able to use their skills to see if they can determine if any suspect drones found near prison property were carrying any contraband.

UAVs have provided criminals with new ways to carry out their crimes, such as spying and scoping potential houses to burgle [6]. Not only are drones able to store the camera footage onto a connected SD card, but there is also the option for live streaming footage, which could lead to further distress to the victim(s). An investigator needs to be able to know how to read the data they are presented with in order to provide enough evidence to support claims of the crime taking place, especially for sensitive cases.

2.4 Data Storage in Drones

Due to the various brands and designs of drones, there are different storage methods. The two main areas are the drone's internal memory, which varies in capacity depending on the drone, and SD or Micro SD slots on the drone. Depending on the drone, there is also the potential for data to be stored on the remote controller and the connected mobile devices. The descriptions below are based on the DJI Spark drone.

Drones Internal Memory—To extract the data from the drone itself, the drone needs to be connected to the laptop/virtual machine via the USB connector or the UFED Device Adapter depending on the software used for extraction. The software can either be DJI Assistant 2 or UFED 4PC. To carry out the analysis on the drone's internal memory, UFED Physical Analyser, Csv View and DatCon were used.

External SD Card—The SD that was inserted into the drone during the flight was copied to the laptop via NUIX Evidence Mover and analysed through FTK imager, where a physical image of the SD card was taken.

Remote Controller—Looking at the controller data that is held in the .DAT files, it is clear to see that there are only slight changes to the data depending on whether the drone was flown when connected to the RC or a connected mobile device. The data provided when the drone was connected to the RC provides information regarding the 'Rudder' and 'Throttle' as well as a 'Connected' identifier. The RC does not hold data that would be useful in an investigation. The examiners would need to have access to either the drone itself or the connected mobile device to collect adequate evidence to aid their investigation.

Connected Mobile Devices—For the extracted data held on the connected mobile device, the mobile device was connected to the laptop via a USB connector. This allows for viewing the file structure within the DJI GO 4 App.

2.5 Process of Drone Forensics

Drone forensics consists of various elements that allow law enforcement/private agencies to build a larger picture of how drones have been used, what evidence they hold and even make new discoveries about drone data. There are three evidence categories which need to be considered during the drone forensic process. The first category being the physical evidence, such as the aircraft, mobile devices, battery, radio controller and laptop/computer. This can relate to sensors, data links to ground stations and the flight controller. The second category is the digital evidence that relates to a drone, such as the SD/Micro SD cards, the drone itself, laptop/computer and mobile device OS (e.g. Linux, Windows, Android). This category includes file systems, media storage and firmware. The third category can be classed as miscellaneous to cover all other evidence artifacts which relate to the forensic process. This category includes social media, purchase records and even fingerprints [31].

2.5.1 Forensics Tools

Below are descriptions of various tools/software that can be used to extract and analyse drone data. Some tools/software are recognised forensic tools whereas others are free offline apps.

DJI Assistant 2—Upon opening the DJI Assistant 2 software, there is an information box, which informs the examiner that the data will be uploaded to DJI's server before starting your data extraction. With DJI Assistant 2 the flight log data can be uploaded to a local drive, which is uploaded as .DAT files. These files are unintelligible and need to be converted to a .CSV file via DatCon to read the data.

DJI Assistant 2 has a section that provides the examiner with data held in the drone black box. When extracted, this data is encoded and can only be decoded by the DJI Company. This is due to the security of the data. The type of data held in the Black Box is used by DJI regarding user enquiries leading to investigating any issues they had during their drone flights. The data being encoded means that it will not be tampered with to affect an investigation taking place by DJI themselves.

DatCon—DatCon has the option to convert any .DAT files into a .CSV file or a .LOG file. Once the .DAT flight logs from the drone internal memory are converted into a .CSV file, it will open as an excel workbook with readable data. This data output is the most detailed flight log format found during the extraction of all the components. The data provided as part of the flight log include the 'GPS', 'Motor' and 'AirCraftCondition'.

DJI GO 4 App—The DJI GO 4 App is required to be able to control the drone from a smartphone. The app itself does provide some details about the flight taken, however not as much as that found in the drone's internal memory. The app provides the pilot with a series of interchangeable screens. It was found that the DJI GO 4 App cache held the most data out of all the components analysed. Within the mobile phone, the DJI cached files were contained in 'My Files > Internal Storage > DJI > dji.go.v4′. While the app cache holds useful data, there is a significant amount of data that is encoded. Some files can be decoded by simply converting the file format, while others are unable to be decoded. Further examination and research would be needed to determine why this is the case.

CsvView—CsvView allows .DAT, .txt, .csv or .tsv files to be uploaded into the software to identify the data held within these files. This is particularly helpful when the original files are encoded. For example, when a 'FlightRecord.txt' file is uploaded to the software the user is able to see the initial upload page, which provides data regarding the 'droneType', 'aircraftName', 'appType' and more specifically the 'aircrafSn' (serial number). This information is useful for verifying what type of drone the investigators are looking for, matching the drones given name to those held in DJI's databases and determining that an android phone is a connected device related to the drone.

Cellebrite (UFED 4PC)—The UFED 4PC software comes with a kit containing the cables and connector tips that may be needed to extract data from a range of sources. To be able to carry out the extraction, the examiners will need a Cellebrite Device Adapter as well as the Cable A with the back-tip T-100 for a DJI Spark drone.

Cellebrite (UFED Physical Analyser)—Once the extracted files are uploaded into the software, it took approximately 2 min for all the data to be processed and decoded. The timings will vary depending on the amount of data extracted from the drone. UFED Physical Analyser displays the data in a way that lets the examiner know where the data was found. The most crucial evidence found during the analysis of a DJI Spark drone was the battery and the aircraft serial numbers. From this small amount of evidence, the investigators would be able to get in contact with DJI to request access to any data linked with the found serial numbers, providing they have identified enough evidence to support the claim of a crime. Physical Analyser also creates a timeline of the flight. This is the first time during the analysis that a timeline has been identified; however, the timeline only displays the latest flight that was conducted, meaning that potentially important data is not detected.

2.6 Challenges in Drone Forensics

This section discusses some of the current challenges in drone forensics.

2.6.1 Anti-Forensics

As the digital age is in constant development and is an essential part of modern life, it is easy to assume that it is difficult to carry out a crime without the involvement of a digital device somewhere along the timeline. Thus, the community of digital forensics has become an integral part of criminal investigations in recent years. However, individuals have come across and created a number of applications and methods of erasing/hiding data over the years. Such methods are referred to as 'Anti-Forensics Techniques'. Kesler states in their paper [30] that the term 'Anti- Forensics' is defined as '*Viewed generically, anti-forensics (AF) is that set of tactics and measures taken by someone who wants to thwart the digital investigation process*'. As of the time of writing this chapter there are a number of different methods of carrying out Anti-Forensics; listed below are some examples:

- **Artifact Wiping**
  - Using tools such as 'Eraser' and 'BC Wipe' to clear the slack/unallocated space
- **Data Hiding**
  - Relocation of data- transferring data to portable device
  - Steganography
  - Altering file extensions

    Signature analysis catches this out
- **Trial Obfuscation**
  - Modification of Metadata

    Altering timestamps
- **Attack on Computer Forensics Tools (CFT) and processes**
  - Forensic tools are well known and well documented

An attacker could gain a copy of the tool and learn the ins and outs of the software, also learn its flaws

– DoS attack

For a number of years there have been techniques used by criminals to try and obfuscate or delete evidence from devices in order to avoid detection. Jaon and Chhabra constructed a paper on the analysis of anti-forensic techniques [26] in which they document a number of well-known and widely used techniques with the intent of delaying or the destruction of investigations. The paper describes in detail each technique and gives examples of how it could be used by criminals and what types of data/evidence can be tampered with. Techniques such as Artefact Wiping, Data Hiding, Trial Obfuscation and Attacks on Computer Forensics Tools are all thoroughly described.

Although the study provides examples and clear descriptions of various anti-forensics techniques, unfortunately there is no mention of UAV devices within the paper; instead simply stating 'Computers' and 'Digital Devices'. This is a weakness of the paper as it does not go into depth of anti- forensics in different devices, instead simply 'computers'. The term 'computers' in technical terms refers to anything electronic that carries out a calculation, so the authors could be talking about any device. However, due to the references they make (e.g. 'different files in computer are identified by their file extensions'), it is assumed that they are referring to PCs. It is unfortunate that the study does not include a section dedicated to UAV anti- forensics techniques as it would have been very useful to be included within this study.

While this paper does have flaws, it does include a lot of interesting and very useful information regarding the various anti- forensics techniques that can be utilised by the general public.

A Digital Forensic Analyst was interviewed on their knowledge regarding Digital Forensics and UAV forensics. In relation to Anti-Forensics in UAV devices, the analyst felt that it was unlikely that users would be able to implement Anti-Forensics techniques on such a device, unless the manufacturers installed one as default. A feature such as encrypting all data on a UAV device, which would mean the data could not be extracted and analysed by the investigations team. Nonetheless, a lot of the data captured and analysed when investigating a UAV device is extracted from the mobile app used to control the device. Flight paths for example are stored there. It is possible to download a scheduler on Android devices that wipes the data from apps at designated times. Thus, it could be possible for a user to install such an app and set it to delete everything from the drone app if it has not been opened in 'x' amount of days. This would count as an Anti-Forensic technique and is something that could be carried out by an end user. Nevertheless, it is unlikely that a standard or computer illiterate user will know about advanced features such as this; it would take an advanced 'tech savvy' user to carry out the technique. However, this can be said for almost any Anti-Forensic technique. The user would have to be aware of the presence of data in order to hide/delete it.

2.7 UAV Legislation

Due to the popularity of UAV devices, it is important that their use is governed, and guidelines are set to ensure that they are used safely and securely without endangering others. The CAA [43] have issued such guidelines within The Air Navigation Order 2016 (amended March 2019) [45] in which all UAV device users must comply. The CAA collaborated with NATS [7] to develop a website called 'Drone Safe' [8]. The site includes an array of features including a copy of the 'Drone Code' (the guidelines that must be adhered when flying such devices) [43], information regarding training opportunities for beginner fliers and general resources regarding UAV devices such as the names of approved retailers and safety checklists.

In addition to the CAAs efforts, UAV device manufacturers such as DJI offer guidance on global and regional legislation [13]. The 'Fly Safe' page of their site allows users to select a region and country e.g. Europe, United Kingdom and information regarding current legislation will be displayed.

2.7.1 No Fly Zones

As a result of the continued disruption being caused by UAV devices around airfields, authorities have made the decision to extend the no fly zones to a total of three miles, as opposed to the previous 0.6 mile radius. Therefore, from 13th March 2019 it will be a criminal offence to fly a UAV device within three miles of an airport [4]. Failure to comply with the new law may result in the end user being charged and sent to prison for up to five years [7]. The 'Drone Code' on the Drone Safe UK site has been updated to include the updated law regarding the extension of the no fly zone surrounding airports.

In addition to the 'Drone Code' document on the site, there are other available features that can be utilised to help end users with locating restricted/no fly zones.

It has been reported that in the U.S, drone manufacturers such as DJI have hard coded 'No Fly Zones' into their devices [48]. In the referenced example from 2017, a TFR (Temporary Flight Restrictions) had been established around an area where President Trump was to be residing. In addition, it is stated in the article that various airport airspaces have also been hard coded into the devices meaning that they would be unable to fly in that airspace; the device would either stop in its tracks and hover or would descend automatically. Nevertheless, there are reports of the existence of software constructed by Russian developers that modify a device's GPS software in order to gain access to 'no fly zones' [29]. CopterSafe is the name of the company that is able to provide members of the public with such modifications, currently a modification board for a DJI Phantom 4 stands at $200 [11]. In addition to CopterSafe, another company has been found to provide similar services when an internet search was carried out [36]. NLD appear to offer a software client that will permit users access to a number of modifications that are compatible for an array of models. However, this company charges significantly less money than the CopterSafe hardware counterpart, only requesting $34.99.

2.7.2 Airfield Restrictions Maps

Present on the Drone Safe site is a page which displays a Google map of the UK with pinpoints of the unauthorized flying zones (Fig. 1) [8]. In addition, there are smaller maps present which display the 'no fly zones' surrounding the major airports of the

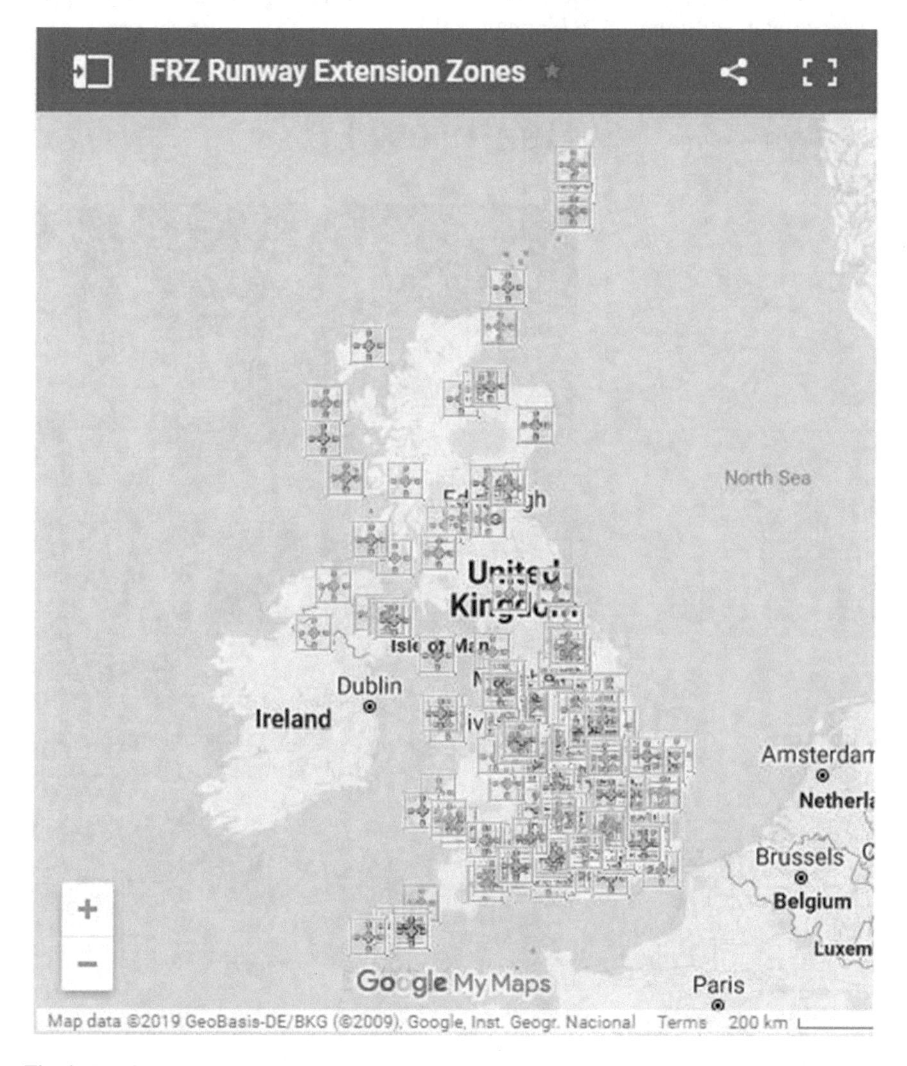

Fig. 1 UK flight restrictions map

UK, such as Heathrow and Manchester (Figs. 2 and 3). This is a very useful resource for any UAV device flier, especially a beginner as they may not be aware of all the no fly zones and may accidently get into trouble. Trouble which would be much greater after March 2019 with the introduction of the new heavier sentences.

Nevertheless, it is not only the Drone Safe UK site that offers these maps, there are a variety of other sites online that offer similar services. UAV manufacturer DJI for example offer such a service [13] along with other sites such as 'No Fly Drones' [28], however these are not governed by a professional body such as the CCA like the data from Drone safe UK; however according to the 'Contact' page of the 'No Fly Drones' site [27], the sources of information regarding the rules and airspace must be obtained from the CAA directly [19].

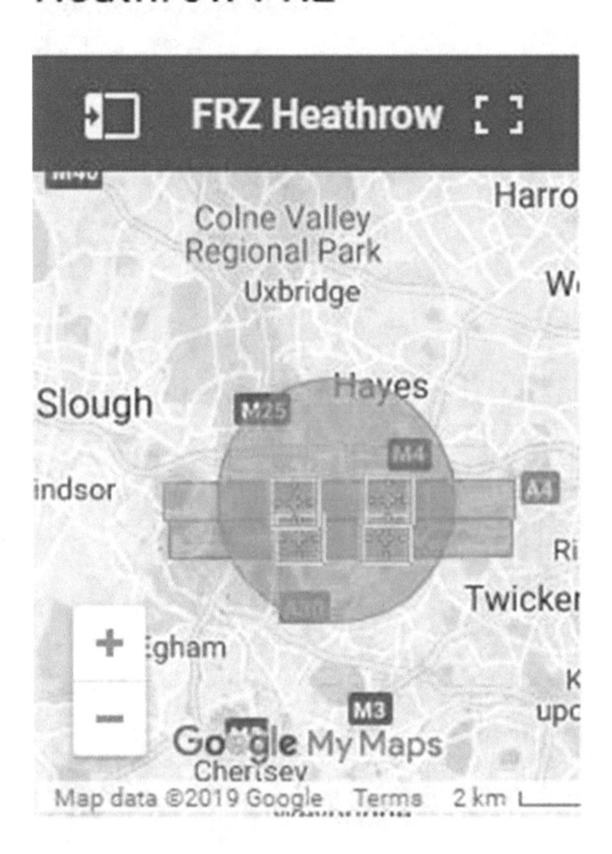

Fig. 2 DroneSafeUK no fly zone—Heathrow Airport

Fig. 3 DroneSafeUK no fly zone—Manchester Airport

2.7.3 Useful Mobile Safety Apps

A further feature present on the DroneSafeUK site is the description of a mobile app developed by NATS named 'Drone Assist' [46]. This app acts in a similar manner to the maps feature present on the website but it has enhanced features such as 'Area Report', to inform the user of an overview of the risks associated with flying in a specific area. Also present in the app is a feature that provides the user with weather information. This is very useful to ensure that the flight remains safe and lowers the risk of an incident.

Furthermore, the application allows users to collaborate with each other in the form of the 'Fly Now' feature. By utilising this function users are able to share the current location of their UAV device with other users of the app. Naturally this reduces incidents as users are aware of other UAV traffic in the vicinity [46].

As stated, the app allows its users to gain detailed information regarding no fly zones. Figure 4 is an example of a 'High Risk' zone as it is surrounding Manchester Airport which naturally contains a lot of air traffic. Usefully available is a description

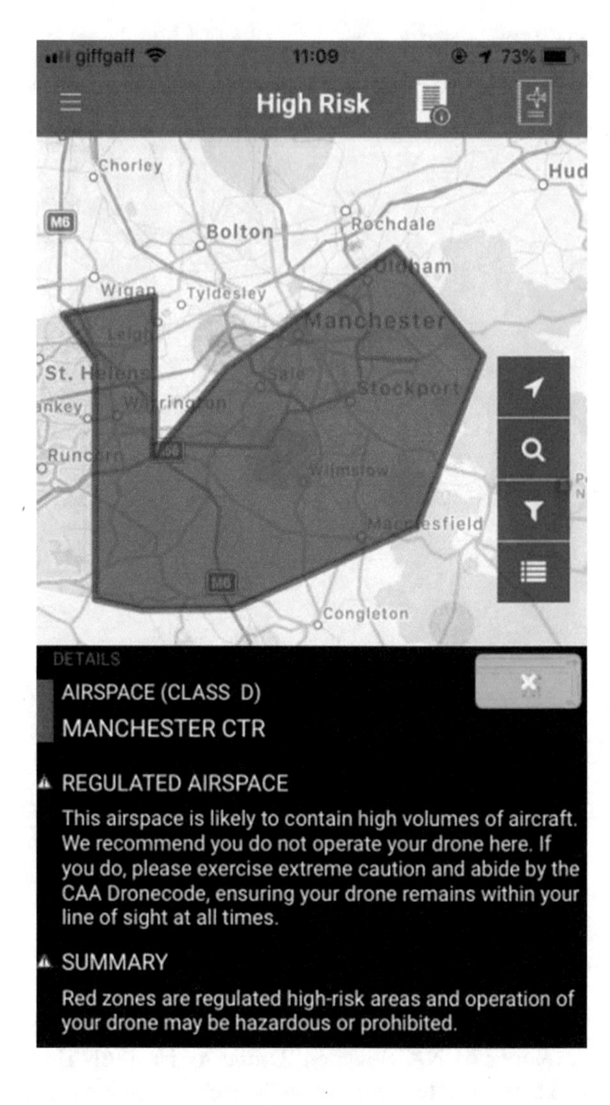

Fig. 4 Drone Assist app 'High Risk' area over Manchester Airport

of why the area is classified as 'High Risk'. In this figure the reason being the air space is in the vicinity of Manchester Airport. There are a number of different classifications of 'zones' (Fig. 5) is a screenshot taken from the app which clearly describes each zone.

As useful as this application is, it does have some issues that should be addressed in the near future. It has been discussed earlier in this chapter that UAV legislation has been scrutinised and new laws will come into place from March 2019. However, this app does not make any mention to this. Upon loading the app for the first time, a message box appears stating a change in law, in July 2018 (Fig. 6). The new laws being that it is an illegal act to fly a drone device above 400 feet without prior permission from the CAA, and that it is now illegal to fly a drone device closer than

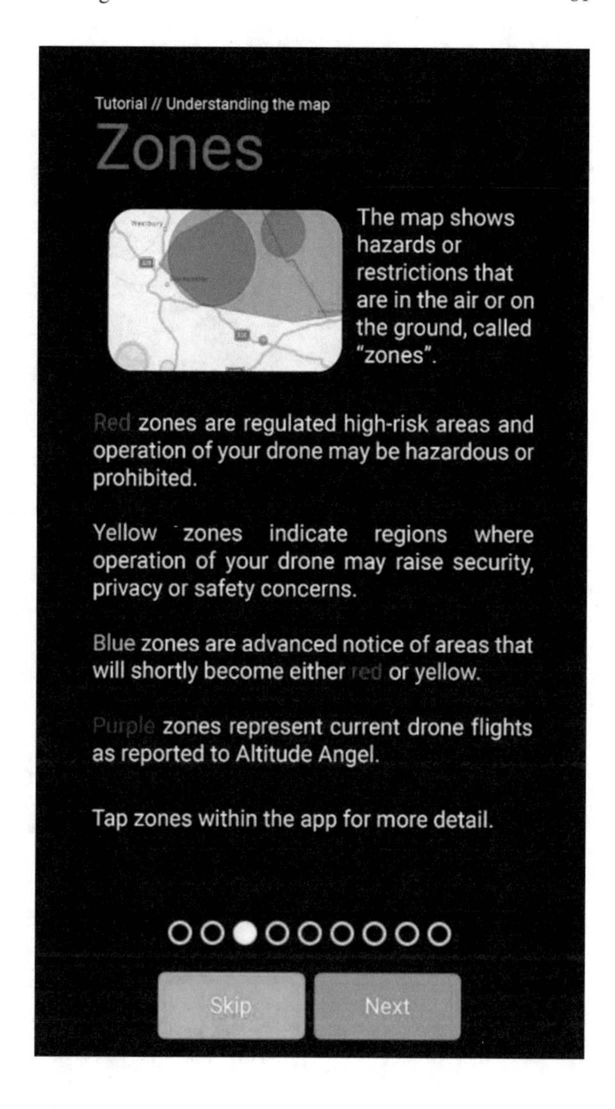

Fig. 5 The different classification of 'Zones' on the Drone Assist app

1 km from the boundary of aerodromes. As the new laws surrounding the extended 'no fly zones' does not come into place until March 2019, the content of this message box is factually correct. However, stating that from March 2019 there will be a further change in the law regarding 'no fly zones' would be a good addition; also, that higher penalties will be enforced.

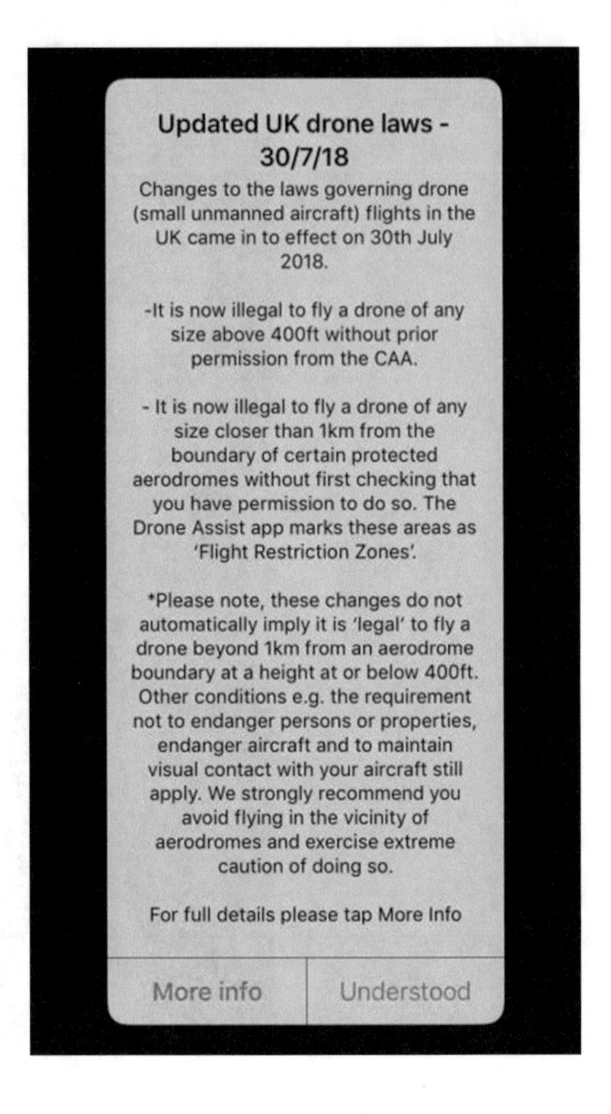

Fig. 6 Information on updated Drone laws

2.8 Summary

In summary, there exists vast amounts of legislation and laws that must be adhered to when flying UAV devices. These laws ensure the safety of members of the public by restricting the areas where UAV device flying is permitted. It is also demonstrated in this chapter how easy the information relating to these laws and legislation are made available to UAV pilots. By distributing the free to download 'Drone Assist' app and the Drone SafeUK website, fliers simply have to quickly check online before flying to ensure they are not about to break any legislation/no fly zones quickly and easily. In essence, with the relatively easy access to this information there is not really any excuse for fliers to accidentally break laws/legislation.

In addition, it has also been stated within this chapter that UAV devices have been commercially sold in the U.S with hard coded 'No Fly Zones' included as default. Areas such as airports etc. The article in question was published in 2017. If this technology had been applied to all devices in the U.K at the same time, it is a possibility that the disruption to airports in 2018 would not have occurred. It is unknown by this study why the devices have not been distributed with such technology in the U.K.

Although it is clear that authorities and manufacturers are attempting to restrict users in how and where they fly their devices, it is demonstrated in this chapter that roguish beings will try their hardest to combat these efforts and break the law.

Due to the expanding nature of drones and the constant updating and release of new models, this study will focus on the examination of the DJI Phantom 4 [15] of which no current research papers could be located. A review of the literature has indicated there are limited studies which focus on the examination of drone forensics. Thus, this study will focus on the examination of the DJI Phantom 4 to contribute to further understanding regarding drone forensics. As a result of this study, a process of examination on a DJI Phantom 4 will be proposed, in an attempt to streamline investigations. There is currently limited research relating to how drone data to proving crimes despite the rise in crimes committed with drones.

Due to the increase of drone technology, it is vital for government agencies to be able to deal with the increasing demands of the devices and have the knowledge of how to use and gather evidence from UAVs, which would benefit in investigations as well as become a useful tool to be used by the agencies. There have been several Drone Forensic Programs in the USA and UK, and while some have been successful, there have been a vast majority that have been terminated, even though drone programs would increase public safety [18]. There are two types of the programs; the first being the use of UAVs as a tool by law enforcement to aid their work, the second being where law enforcement agencies have departments dedicated to extracting evidence from UAVs used in criminal activity.

The use of UAVs as a tool by law enforcement is outside the scope of this work as it excludes essential evidence extraction and analysis skills that would be relevant to a police investigation. Therefore, this study will focus on drone data used as evidence for a police investigation. One of the reasons for these programs failures is due to a lack of understanding of UAV technology. This study will highlight a first-hand drone extraction and analysis on a DJI Spark to determine the complexity behind these processes as well as data interpretation. The study will also highlight some of the ways that drones are used for criminal activities to show how the data can support or refute criminal claims. As there is a lack of understanding in this field, there needs to be a real-life application to the analysis, to allow law enforcement agencies to know how to correctly interpret the data gathered to be successful in their investigation.

3 Research Methodology

This section focuses on the research methods used as part of the projects, to show the different ways drone data can be used to draw varying conclusions. There will be focus on the questionnaire and experimental work.

3.1 Research Method

As there are a vast amount of UAV device models currently on the market, it is not feasible to gain primary flight data from all of them. However, NIST (National Institute of Science and Technology) allow access to a total of thirty-two UAV images created by VTO Labs [35]. This allows for the analysis of a much wider range of devices in order to gain a clear understanding of any differences that may be present in different models and manufacturers.

Nevertheless, in order to gain a clear and concise understanding of what forensic investigators are challenged with; it is important that evidence from at least one UAV device is manually examined and analysed. Therefore, a UAV device (DJI Spark) will be taken on a number of test flights in an array of locations in order to generate good quality data to analyse. Cellebrite UFED 4PC will then be used to extract data from the device and Cellebrite Physical Analyser used to examine the data. As the UFED 4PC software houses features that exclusively extract data from UAV devices, it seems the perfect software to use for the extraction and analysis.

In addition to the use of Cellebrite software, IEF will also be used to analyse data by means of a comparison between the two tools. The comparison being an experiment of the amount of data returned by each toolkit and whether one of the tools appears to have missed some data during the extraction or decoding processes.

The extraction experiments are scheduled to take place after the interview with the forensic investigator; the reasoning being that the outcome of the interview will provide an understanding of how forensic investigation units deal with UAV devices and which methods are utilised to extract evidence from them. Thus, the extraction and handling of the test flight data as well as the data set evidence would mirror that of a real life scenario. Due to the experiments using both the VTO Labs datasets and the manual test flight data, a mixture of both primary and secondary data will be used during analysis.

During the examination of the images, it will be deduced exactly what types of evidence can be obtained from UAVs, how useful this evidence could be in given scenarios as well as ascertain if one tool is superior to another regarding analysis.

3.1.1 Freedom of Information Request

The literature has indicated that the crimes committed by drone usage has increased due to the rise in drone accessibility and functionality. A freedom of information request was made to West Yorkshire Police which asked for the number of incidents made by unmanned drones. Quantitative data was attained which was categorised by incident type since the data began being recorded and the calendar year the crime was committed. As a result, the data provided an understanding as to whether there had been an increase in the number of incidents year on year, what type of categories these incidents were classified as and whether there was a need for further research in this area (Appendix 1).

3.1.2 Ethical Consideration

Due to the nature of this project, the subject of ethics must be considered. Two main forms of methodology are used to gain suitable research and test data: Interviews and UAV Extractions.

In relation to the interview methodology, it is absolutely vital that the name(s) of all participants are not revealed along with the name of the police force they are associated with. As several of the questions that are to be asked are 'open ended' and open to interpretation, it is important that this is adhered to avoid any repercussions.

Concerning to the examination of the UAV data, it is important that privacy and confidentiality is maintained. As a feature of most UAV models is that of a high resolution camera, it is more than likely that individuals may have been inadvertently captured by the device. As it is more than possible for permission not to have been granted by these individuals, it is vital that any content analysed is not to be made public and be used in a purely academic manner. Nevertheless, the possibility of capturing a criminal event using the on board camera has to also be considered. Although it is highly unlikely for a criminal act to be inadvertently captured on video, it is a serious issue. Although by providing officials with the footage would be a breach of privacy, it could be a major advantage to a current investigation. If a crime such an indecent assault on a minor was captured on the device, although it would have to be reported to the Police, it could be seen as distributing child pornography. However, if the footage were not reported to the Police, the owner of the footage could be tried for the creation and possession of indecent images of minors. Nevertheless, this theory is quashed by the content of the various related legislations. Although the Protection of Children Act 1978 states that it is an offence 'to distribute or show indecent images' and 'have in his possession indecent images' [23], the Sexual Offences Act 2003 furthers the content of the 1978 act and creates a defence to this 'offence' by stating in Section 46.1A that 'the defendant is not guilty of the offence if he proves that it was necessary for him to make the photograph or pseudo- photograph for the purposes of prevention, detection or investigation of crime, or for the purposes of criminal proceedings' [24].

Therefore if a member of the public were to take content from their UAV device to a Police official stating the content and their wish for it to be investigated, it is unlikely that they would be prosecuted as they would be seen as 'distributing' the images they 'made' in the public interest and their intent to get the crime investigated. In addition to this, it would be the CPS (The Crown Prosecution Service) who would prosecute such offences and only do so if prosecution against an individual is in the public interest. It is unlikely that prosecution of an individual such as the one in this example would not be in the public interest.

3.2 *Questionnaire*

In order to gain a good and relevant understanding of how investigators examine UAV devices and the current protocols associated with such examinations, conducting an interview with a qualified forensic investigator from a police force is a worthy way to gain an insight into the workings of a forensic investigation unit. A mixture of qualitative and quantitative approaches will be used during the interview. By answering the questions asked, the interviewee will be able to provide a clear and concise impression of the depth of their knowledge on the subject and how useful their input will be to this study.

Before the interview with the forensic investigator commences, the interviewee must complete, sign and date the consent form to ensure that they are fully aware of the nature of the project, the risks and benefits as well as the fact that their answers and opinions would be completely anonymous.

3.2.1 Wording of Questions

In order to gain quality information from the interview, it is important that open-ended questions are asked as well as closed questions. By utilising this approach, it will allow the interviewee to provide clear, factual information such as statistics and information regarding current protocols used within forensic investigation units as well as being able to expand and express their own opinions regarding the current situation regarding the use of UAV devices and the way that the digital forensic community is adjusting to include such devices into the various protocols and investigation methods used.

3.3 *Experimental Work*

The initial flights that took place were to be used for the primary analysis. These flights are used to identify the extraction and analysis techniques that need to be used by the examiner, as well as to determine what different data can be identified with varying methods of control. A flight was carried out via the connected mobile device, and the following flight was conducted via the DJI controller. Forensic tools and the free offline apps were installed on a LENOVO Laptop to carry out the analysis. The extraction processes were conducted through a virtual machine. By using first-hand data and not using secondary data ensures more control over the reliability of the results, as well as accounting for any variables that may have affected the data.

The main experiments were focused around two main drone crimes: smuggling and spying. The first experiment regarding smuggling was to determine how the motor data and battery data may be affected by various payloads. And how the drone movements and gimbal directions can be used in support of a 'spying' claim.

3.3.1 Experiment One

There have been many cases where drones have been used to smuggle phones, weapons, and drugs into prisons. Therefore, the payload tests are to determine whether it is possible to identify whether a drone has had a payload during a flight. For this scenario, various payloads were added to the drone to see what effect the added weight had on the drone's motor and battery data. The payloads of 146, 18 and 10 g will be compared to that of a flight without a payload. The flights conducted consisted of take-off and hovering at the default take-off height of 2 m for a desired time of 3 min.

3.3.2 Experiment Two

In an investigation where a claim has been made by a victim that they have been spied on by a drone, investigators want to be able to look at the data extracted from the drone and find the evidence to support or refute the claims made. For this scenario, the focus will be on the rotor movements and the camera. For the DJI Spark, the camera is located at the front of the aircraft with the gimbal having an 85° tilt range. When looking at the rotor movements, it can be determined which direction the camera was facing. The only issue is that currently, it is not possible to determine whether the camera was facing forward or tilted by the gimbal. This likely would only be determined by actual camera footage if it was taken and kept by the suspect or looking at the gimbal settings in DJI GO 4 app, which only provides limited details, provided that the settings haven't been changed since the flight in question.

Similarly, when there is evidence to show that the suspect's drone was in fact in the area of the crime but not necessarily the suspect themselves, there can be claims that there was no footage taken, or that the camera wasn't facing towards the victim or property. The SD containing the .jpg and .mp4 files from the camera can be examined to determine if this was the case. In the instance that the camera files have been deleted, the investigators need to be able to determine whether the camera was facing the victim or property.

3.4 Data Analysis

This section looks at how the free offline apps were used to carry out the analysis of the drone data. There will be a focus on two data components: the drone internal memory and the connected mobile device.

Internal Drone Memory—Using DatCon, the .DAT flight logs from the drone's internal memory are converted into .CSV files, which opens as an excel workbook. This data output is the most detailed flight log format found during the extraction of all the components. Below is an example of the data from the workbook (Fig. 7).

Connected Mobile Device—The DJI GO 4 App is required to be able to control the drone from a smartphone. The app itself does provide some details about the flight taken, however not as much as that found in the drone's internal memory. The app provides the pilot with a series of interchangeable screens. In the case of a drone forensics program, the app would not be as helpful to an investigation as to have access to the data the mobile device needs to be connected to the drones WIFI. To protect the integrity of the data, the drone needs to remain disconnected to any external devices, until safe in the lab for analysis. If law enforcement were to attain the mobile device and not the drone, they would be missing the data from the app. However, the examiners would also be missing the data that is provided in the DJI GO 4 App's cache on the connected mobile device.

The DJI GO 4 App cache held a large amount of useful data. The 'DJI_RECORD' folder holds any video recording taken during the flight (.mp4). The folder contains '.info' files for each recording within the cache folder, which provides relevant information that would be helpful in an investigation, such as the UUID for the drone.

The UUID is the Unique User Identification code related to each drone pilot's DJI account, as well as the pilot's identification and flight information. The '.info' file also holds the GPS longitude and latitude locations of where the recording was taken, again providing the evidence to prove the suspect's location at the recorded time and date at the top of the .info file (Fig. 8).

The 'FlightRecord' folder also holds the sub-folder of 'SyncResults' which holds .txt files. As expected, this file contains the data for each synchronisation that took place. However, this file also holds the email address linked with the drone's pilot account, providing investigators with evidence of proof of ownership, as they will be able to see if the email matches that in the DJI database (Fig. 9).

The 'FlightRecord' folder also holds individual .txt files which are unintelligible. These .txt files were uploaded to CsvView to convert to a .CSV file to read the data. The initial upload page provides data regarding the 'droneType', 'aircraftName', 'appType' and more specifically the 'aircrafSn' (serial number). This information is useful for verifying what type of drone the investigators are looking for, matching the drones given name to those held in DJI's databases as well as determining that an android phone is a connected device related to the drone (Fig. 10).

The option 'GeoPlayer' opens a new window showing a map of the flight path (Fig. 11).

	A	B	C	D	E	F	G	H
1	Tick#	offsetTime	IMU_ATTI(0):Longitude	IMU_ATTI(0):Latitude	IMU_ATTI(0):numSats	IMU_ATTI(0):barometer:Raw	IMU_ATTI(0):barometer:Smooth	IMU_ATTI(0):accel:X
2	63058508	14.013	-0.755856901	54.12667755	10	172.95943	172.80284	-0.10461119
3	63212969	14.047	-0.75583693	54.12667757	10	173.02264	172.80731	-0.10159523
4	63365302	14.081	-0.755855956	54.12667759	10	172.74942	172.80855	-0.10321622
5	63515734	14.115	-0.755856989	54.12667761	10	172.87584	172.80861	-0.11125354
6	63669049	14.149	-0.755856953	54.12667763	10	172.81874	172.8056	-0.108064
7	63820479	14.182	-0.755856931	54.12667764	10	172.70149	172.80357	-0.10609724
8	63977828	14.217	-0.755856912	54.12667766	10	172.72699	172.80203	-0.101901665
9	64132567	14.252	-0.755856897	54.12667767	10	172.73515	172.79897	-0.10333833
10	64292799	14.287	-0.75535688	54.12667769	10	172.72292	172.7974	-0.10080014

Fig. 7 Drone internal memory, flight log example

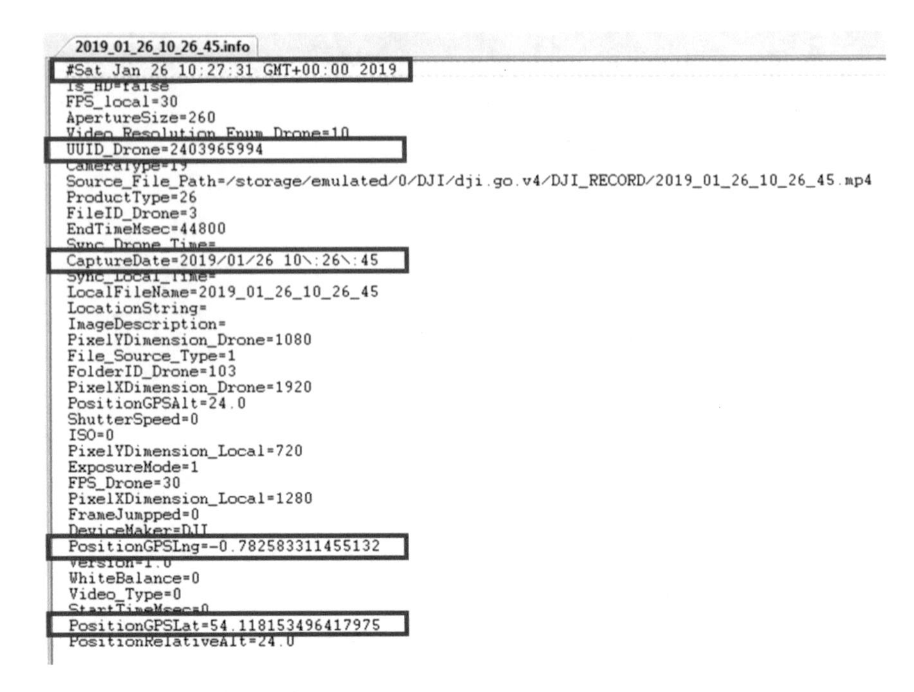

```
2019_01_26_10_26_45.info
#Sat Jan 26 10:27:31 GMT+00:00 2019
Is_HD=false
FPS_local=30
ApertureSize=260
Video_Resolution_Enum_Drone=10
UUID_Drone=2403965994
CameraType=19
Source_File_Path=/storage/emulated/0/DJI/dji.go.v4/DJI_RECORD/2019_01_26_10_26_45.mp4
ProductType=26
FileID_Drone=3
EndTimeMsec=44800
Sync_Drone_Time=
CaptureDate=2019/01/26 10\:26\:45
Sync_Local_Time=
LocalFileName=2019_01_26_10_26_45
LocationString=
ImageDescription=
PixelYDimension_Drone=1080
File_Source_Type=1
FolderID_Drone=103
PixelXDimension_Drone=1920
PositionGPSAlt=24.0
ShutterSpeed=0
ISO=0
PixelYDimension_Local=720
ExposureMode=1
FPS_Drone=30
PixelXDimension_Local=1280
FrameJumpped=0
DeviceMaker=DJI
PositionGPSLng=-0.782583311455132
Version=1.0
WhiteBalance=0
Video_Type=0
StartTimeMsec=0
PositionGPSLat=54.118153496417975
PositionRelativeAlt=24.0
```

Fig. 8 DJI_RECORD .info file content

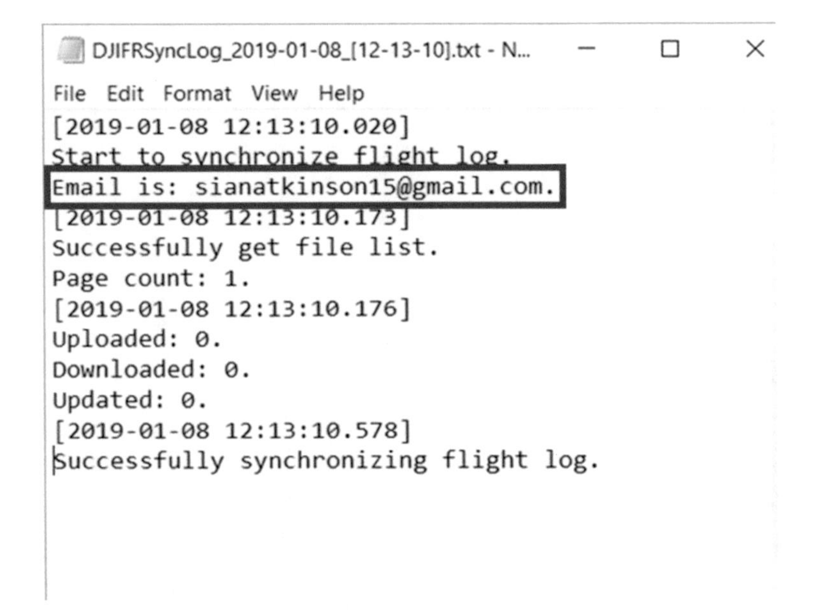

```
DJIFRSyncLog_2019-01-08_[12-13-10].txt - N...
File Edit Format View Help
[2019-01-08 12:13:10.020]
Start to synchronize flight log.
Email is: sianatkinson15@gmail.com.
[2019-01-08 12:13:10.173]
Successfully get file list.
Page count: 1.
[2019-01-08 12:13:10.176]
Uploaded: 0.
Downloaded: 0.
Updated: 0.
[2019-01-08 12:13:10.578]
Successfully synchronizing flight log.
```

Fig. 9 DJI ‘FlightRecord’ > ‘SyncResults’ file content

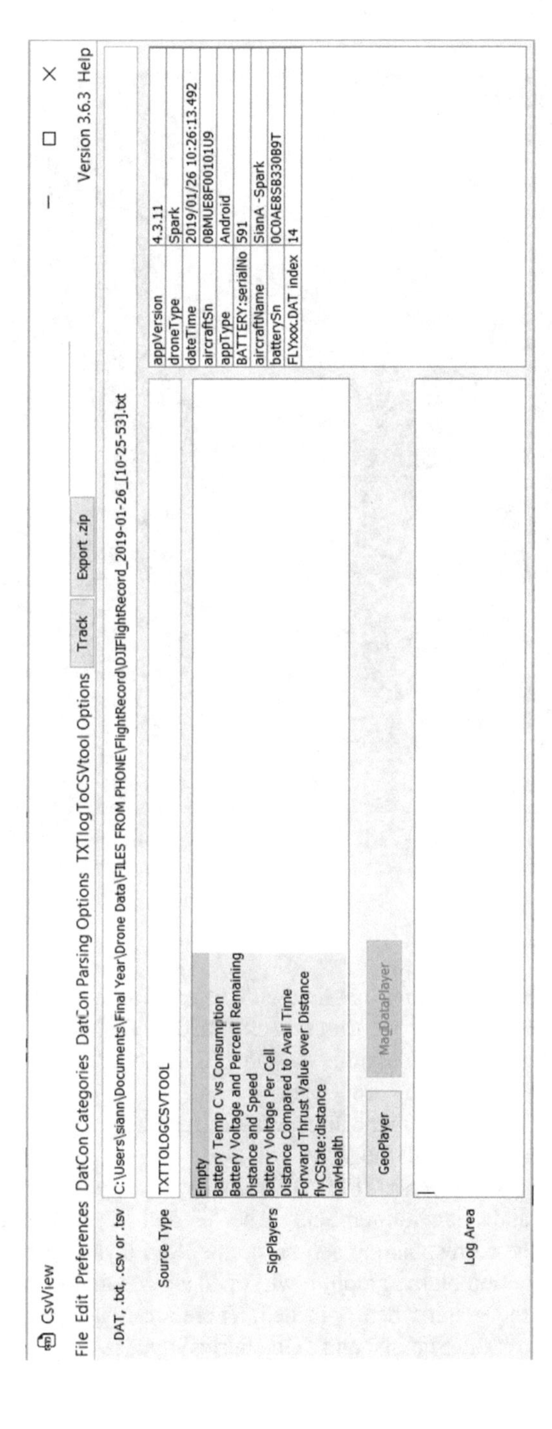

Fig. 10 CsvView .txt flight log upload

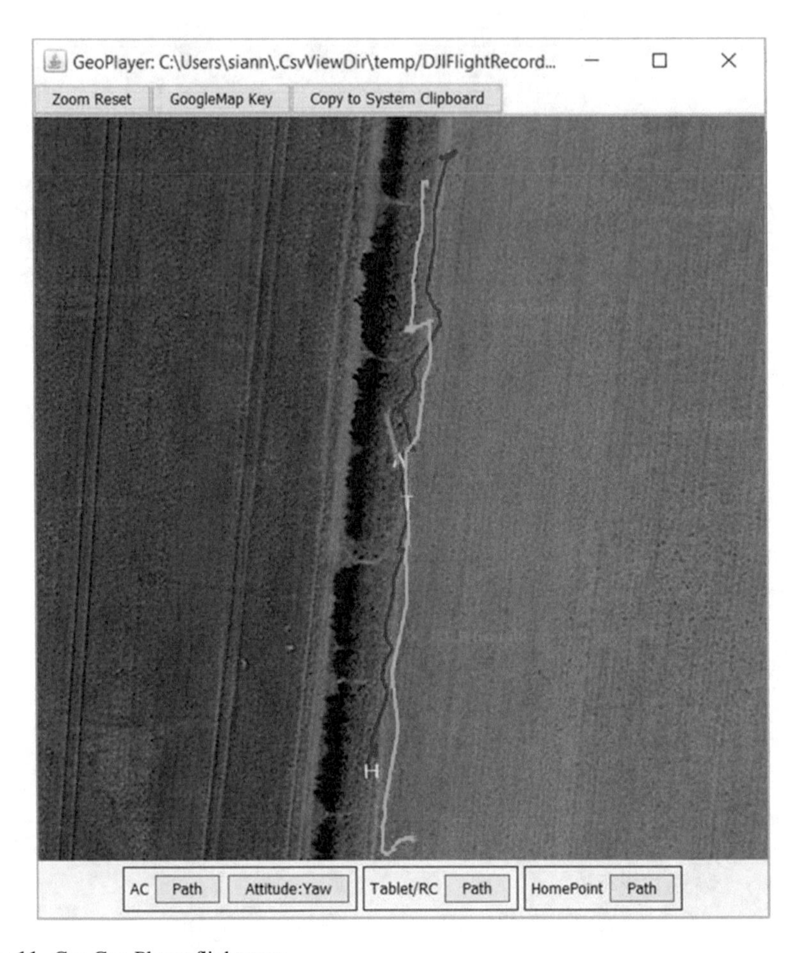

Fig. 11 Csv Geo Player flight map

The map has highlighted options that link with different aspects on the map. 'AC Path' refers to the path taken by the drone itself (RED). 'Tablet/RC' refers to the path taken by the RC and we can assume the pilot (GREEN). 'HomePoint' is used to identify the take-off point of the drone (H).

The 'AC Attitude: Yaw' refers to the direction the drone is facing; this is displayed by the small green beam linked to the 'A'.

The section for 'SigPlayers' (Fig. 12) displays a graph of the flight 'General:navHealth' and 'General:numSats'. The 'T' and 'A' on the 'GeoPlayer' map will move with the corresponding actions on the 'SigPlayers' graph. By going to the 'Pick Signals' option on the graph, it will open a new window of signals that can be displayed on the graph when uploaded. There are two types of signals to choose from; these are 'StateSignals' and 'TimeSeriesSignals'.

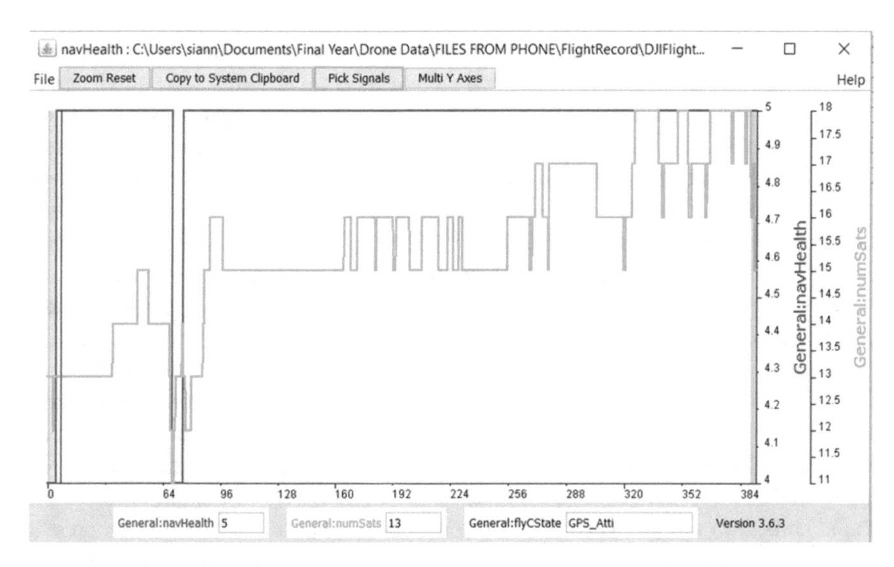

Fig. 12 Csv Sig Players default graph

4 Results and Discussion

This section focuses on the results from both the questionnaire and experimental work and what can be interpreted from the results.

4.1 Questionnaire Result

In order to gain a good and relevant understanding of how investigators examine UAV devices and the current protocols associated with such examinations, conducting an interview with a qualified forensic investigator from a police force is a worthy way to gain an insight into the workings of a forensic investigation unit. A mixture of qualitative and quantitative approaches will be used during the interview. By answering the questions, the interviewee will be able to provide a clear and concise impression of the depth of their knowledge on the subject and how useful their input will be to this research.

4.1.1 Interview with Digital Forensic Analyst

In order to gain a clear understanding of the current protocols, procedures and knowledge surrounding UAV devices in a forensic investigation environment, an interview was sought from a Digital Forensics Analyst working for a regional police force. By coincidence, as well as being employed by a regional police force, the interviewee is an associate lecturer at an institution teaching a module called 'Investigative Forensics'. By attending their lectures, a good idea of their capabilities and knowledge of the subject of digital forensics was gained. It was clear from these sessions that the interviewee was extremely well informed and was more than capable of providing quality responses to the questions provided. Although the digital forensics analyst stated that they did not house a vast amount of knowledge on UAV devices, it was clear that their input would be extremely valuable to this project. The interview took place on 27th February 2019 and took approximately one hour to complete.

In the interview, a range of questions surrounding UAV/Drone devices were asked to the analyst using both open and closed question methodology. The analyst has not had a terribly long career in digital forensics, having only worked in the sector for a total of approximately four years, but in that time has gained a lot of knowledge and experience in the field. However, it is stated in the interview that they are not an expert in the field of drone forensics. In fact, it is so rare that the department get such a device in for analysis, it is an exciting occurrence for all members of the digital forensics team. Thus, it is not an everyday occurrence. As stated, the analyst has worked in a Digital Forensics environment for approximately four years and is currently only aware of around three devices ever coming in for data extraction/analysis. On average, maybe one per year.

When asked about the types of crime that UAV devices are involved in, it is stated that every one of the cases they are aware of have been involved in drug related crimes. An example of such a case would be the investigation of a drone device that was found crashed in the vicinity of a prison. Officials wanted to ascertain where the device was flying from, when it did so and whether or not a payload was attached. In essence, whether or not the device was being used in a malicious manner i.e. delivering contraband into prisons, or just simply happened to be flying within the area of the prison. Both of which are illegal acts. The result being the latter. A second case example is also drug related, but the analyst did not work closely on the case, but simply knew about it; as stated previously it is a culture within the department to be up to date regarding the process of extraction and analysis of drone devices.

In addition to these questions, queries regarding knowledge of current legislation were also asked. As stated previously, there are vast amounts of laws and legislation being implemented surrounding UAV devices and as of the time of writing, the lack of appropriate laws surrounding drones is a very current issue. Therefore, it was a surprise to learn that the analyst did not have a lot of knowledge on these devices and even less on legislation. When asked about the area of legislation, the analyst was not able to provide distinctive answers regarding the legislation currently in place or due to be implemented. This was a surprise as it was thought when constructing the questions to ask that it would be a requirement by Digital Forensics department for all personnel to keep up to date with current legislation of digital devices to ensure they are aware of them. Nevertheless, this was proved to be a false assumption. It was deduced that investigation personnel are not required to keep up to date on laws and legislation but 'learn them as they go'. Although they are aware of legislation such as RIPA (Regulation of Investigatory Powers Act) [22], it is not something that is used every day.

However, it is also stated that in a lot of cases that do, or in the future, will involve drone devices, they will be dispatched straight away to the CAA for investigation at their specialist forensic investigation department. In addition to this, the analyst stated that by the time devices get to their department, whichever type of device it may be, issues relating to legislation are generally already dealt at the beginning of the evidence chain. Therefore, it is understandable why the force may not be willing to spend public funds sending its personnel on training courses when it is not viable with the amount of tasks they are asked to carry out which include legislation.

Although it seems forensic investigators/analysts are not required to know legislation relating to their devices, they are required to adhere to the four ACPO principles. It is suggested by [38] that these principles are not up to scratch for examination of modern devices and are not applicable in the slightest to UAV devices. Consequently, this was put to the analyst during the interview to gain their personal views on the theory and recommendation. The outcome of this question was a contradiction to the paper. The analyst's personal view is that the ACPO guidelines are still very relevant to modern investigations. One argument that was agreed was that on modern mobile devices, data is changed as soon as the device is powered on; therefore, breaking the first rule in the set of guidelines. Nevertheless, it was reasoned that the second rule of 'if original data must be changed, the investigator must be competent to do so', and that if the investigator is not competent, they should not be carrying out an examination in the first place. The third rule of 'make notes' is also still a very current and necessary task to undertake. However, they stated that they would remove or amend the final rule of 'the officer in charge is in charge' as it does not make much sense. A very well-made argument in favour of the ACPO guidelines was made by the analyst during the interview. However, if this rule were to be removed or altered drastically, it could result in the investigating personnel becoming solely responsible for the law and principles being followed. Although investigators should always be anally retentive in ensuring this, the removal of the final ACPO principle may lead to investigators becoming lacks in enforcing them.

In relation to the section of the said conference paper where it is recommended that UAV devices gain their own set of principles/guidelines; the analyst disagreed with the recommendation. Stating that it was a 'slippery slope' as almost every device is different, in particular mobile devices, and could lead to having hundreds of different sets of ACPO principles. This is a valid point as mobile devices can be seen to 'break' the guidelines just as much as UAVs. Therefore, it could lead to a set of guidelines for mobile devices, of which many operate differently which could lead to a set of principles for every model of device which would be unsustainable by DFUs.

Finally, the analyst was questioned about Anti-Forensics techniques. According to the interviewee, Anti-Forensics techniques are encountered at regular intervals when investigating devices, especially PCs. Popular techniques used by end users are apparently tools such as 'CCleaner', 'BitLocker' and 'VeraCrypt'. These tools have the ability to forensically wipe everything or encrypt the data from storage to make extracting data from them near to impossible for investigators. Although it is possible for investigators to gain a RIPA 49 order (essentially forces the suspect to reveal their password or face a two year custodial sentence), it is not an easy task as the order has to be signed off by a judge in court who has to be satisfied that everything has been done by investigators to attempt to gain the data from the device. The consequences of not complying with a RIPA 49 order, although could act as a deterrent and 'scare' suspects to reveal their credentials, could also act as a means of lowering the suspect's sentence. The reasoning behind this being that if the entire prosecution case relied upon the content of the 'locked' device, the suspect could simply take the two years' imprisonment instead of revealing the content of their device and risk gaining a longer custodial sentence.

In relation to Anti-Forensics in UAV devices, the analyst felt that it was unlikely that users would be able to implement Anti-Forensics techniques on such a device, unless the manufacturers installed one as default. A feature such as encrypting all data on a UAV device, which would mean the data could not be extracted and analysed by the investigations team. Nonetheless, a lot of the data captured and analysed when investigating a UAV device is extracted from the mobile app used to control the device. Flight paths for example are stored there. It is possible to download a scheduler on Android devices that wipes the data from apps at designated times. Thus, it could be possible for a user to install such an app and set it to delete everything from the drone app if it has not been opened in 'x' amount of days. This would count as an Anti-Forensic technique and is something that could be carried out by an end user. Nevertheless, it is unlikely that a standard or computer illiterate user will know about advanced features such as this; it would take an advanced 'tech savvy' user to carry out the technique. However, this can be said for almost any Anti-Forensic technique. The user would have to be aware of the presence of data in order to hide/delete it.

4.2 Experimental Work

As well as analysing drone data, multiple experiments were conducted to create a link with 2 main drone crimes. This is to show how the drone data can support or refute claims of a crime. For these experiments, smuggling and spying were the drone crimes chosen. Below explains the outcome of the experiments.

4.2.1 Smuggling Contraband

The flight logs provide data showing changes to the battery and motor data. The most significant changes to that data come from the 146 g payload flight. This outcome is to be expected due to the fact the payload was just under half the weight of the DJI Spark drone itself, therefore, to fly with the excess weight, the drone motors needed to use more power. The expectation for this flight was that the payload would be too much extra weight meaning the drone would not be able to take off.

However, this was not the case. Looking at the data for the motor speeds it's clear to see how the drone used the motors to balance out the added weight to gain altitude. Some of the data didn't show any considerable differences from the no payload flight and the 18 and 10 g payload flights, such as the 'motor voltage_output', which remained in the range of 9.3–9.6. The 146 g payload 'motors voltage_output' ranged from 31.3 to 87, showing significant difference to the amount of power needed to adapt to the added weight of the payload.

4.2.2 Spying Claims

By examining the flight log in CsvView, it's clear to see the flight path, as well as the path taken by the RC/pilot. For the spark drone, the maximum transmission distance from the drone to the controller is 1.2 miles. For this experiment, the pilot isn't in the vicinity of the drone when the flight ends, meaning there is a possibility the accused would remain a suspect of the crime.

The DJI Spark drone has a 2-axis mechanical stabilisation system (pitch and roll), as well as a controllable gimbal range of −85° to 0° (pitch). Pitch, yaw and roll are based on the drone's rotor movements and can be useful to determine the location the drone's camera was facing at this time of the crime. The yaw axis shows the investigator the direction the drone was facing during the flight, which in turn indicates the direction the camera was facing due to the camera on the Spark drone being located at the front.

Using CsvView, the investigators can determine which direction the camera was facing at the time of the reported crime. In this experiment the yaw axis is facing in a different direction to where the drone is being flown.

4.3 Data Analysis

As mentioned previously, a mixture of both Primary and Secondary data was gained in relation to UAV devices. This includes flight data, logs, contents of on board storage etc. This data has been processed and analysed in order to ascertain what types of data is available on the devices and how useful this data could be to an investigation. Also, as the data includes varying device models of UAV, it is interesting to see whether the amount of data exported from these devices differs by manufacturer and/or model.

4.3.1 Drone DJI Go 4

Below are screenshots of location logs and maps that have been extracted from the DJI GO 4 app of the mobile device used to fly the device. Cellebrite Physical Analyser is used to extract the data from the mobile device. This is the data relating to the UAV device that was flown personally and was not gained from the NIST dataset.

The above screenshots (Figs. 13 and 14) show the 'Extraction Summary' of the mobile device used to carry out the test flight of the obtained UAV device. The extraction summary tab itself contains a lot of useful information regarding the related

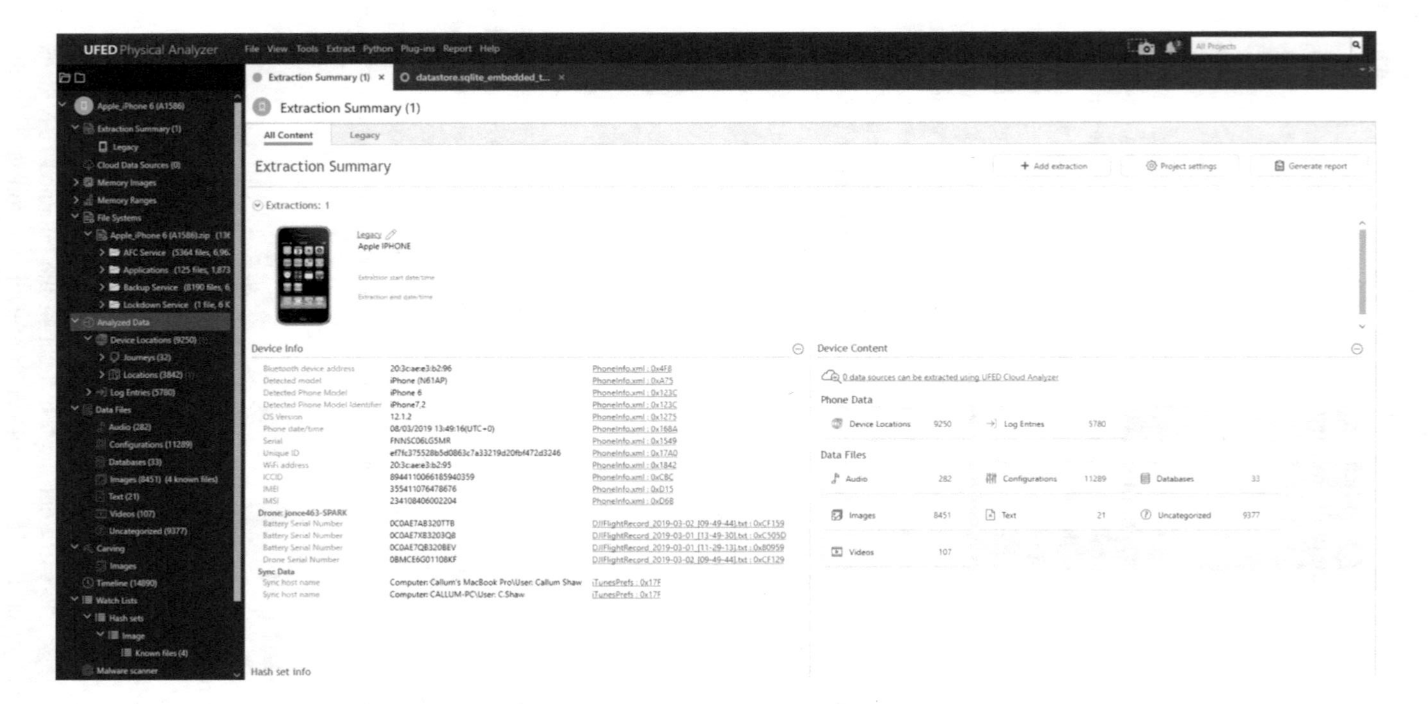

Fig. 13 STYLEREF 1 \s 5. SEQ Figure * ARABIC \s 1 1- Extraction summary of mobile device

Fig. 14 Extraction Summary of mobile device used during test flight

devices. For example, it is easy to ascertain uniquely identifiable information such as the mobile device's Serial Number and IMEI number. Also visible is the name(s) of the computer(s) used to sync data. However, most importantly, information relating to the UAV device used is displayed. The Serial Number of the UAV device and the batteries used are displayed. Very useful information for investigators. Also present are the files that the data was found in, more useful information for investigators.

In addition to these useful artefacts, Physical Analyser also has the ability to examine the device's file system (Fig. 15). After utilising this feature, it is found that the DJI GO 4 app's folders within the file system house some extremely useful items. For example, it seems that the app caches video files recorded by the device and stores them in the 'videoCache' folder of the file system. Present also are an array of video files (.mp4) which it is assumed were recorded on the UAV device test flights, due to the file name being a date. Also present in the same folder are files with the extensions '.mapv2′ and '.infoV2′. Although the description of these extensions is unknown, it is assumed that the content would contain data regarding the flight path of that particular flight. The various different information objects that could be gained from the file system could assist investigators to match the mobile device to a specific UAV device.

The screenshots below display the contents of the analysed data of the DJI Go 4 app. The user is able to view logs made by the app during use.

These logs are in .DAT format (Fig. 17) which allows the Physical Analyser software to generate a map and flight path (Fig. 16). The user is then able to view the flight path by clicking on the 'Play/Stop' toggle button, seen in Fig. 18. Once initiated, this would simulate the flight path made by the device (Fig. 16).

In addition to Cellebrite's Physical Analyser tool, Magnet's AXIOM Process/Examine software was also utilised in order to analyse the mobile device data (Fig. 19). Alike Physical Analyser, this software also allowed the user to examine the file system. Additionally, the software displays useful metadata about the current file,

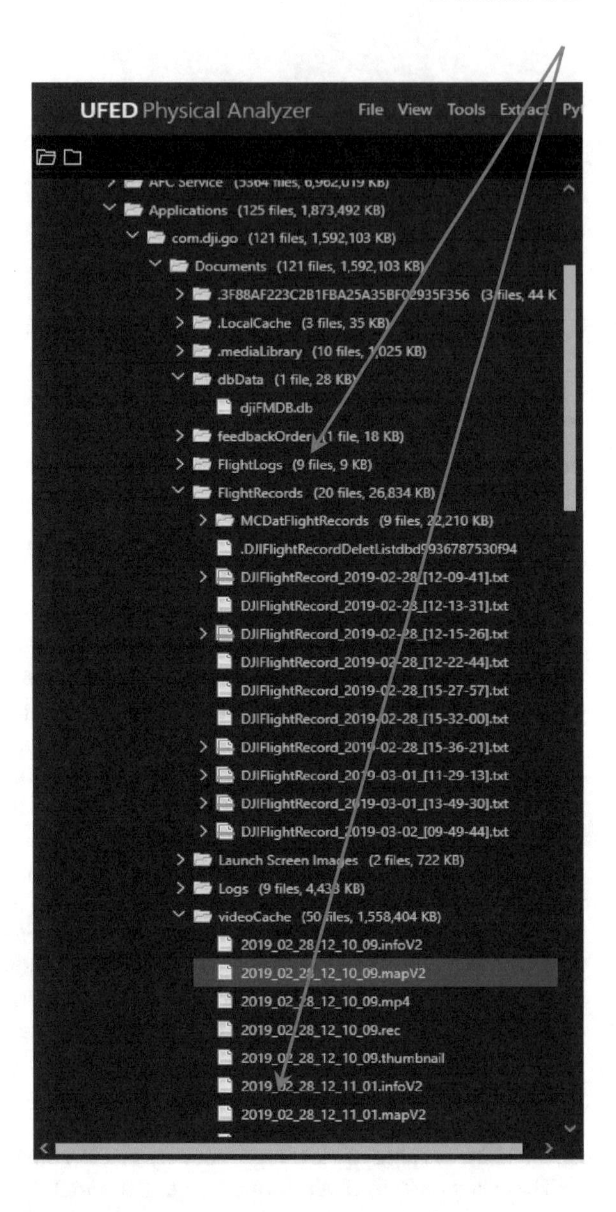

Fig. 15 Extract from the mobile device's file system

such as 'File Name', 'Creation Time', 'Logical Size' and 'Last Accessed'. All these examples, albeit present in the Cellebrite software, are very useful to investigators.

After analysing the evidence extractions in two different software, it seems that no crucial evidence/data has been missed by either. The device used to carry out the test flights is a DJI Spark device. The reasoning behind choosing this particular device is purely due to the fact that it is the most assessable. DJI is the largest UAV manufacturer (Unmanned Aircraft Systems (UAS) [9]: Commercial Outlook for a New Industry,

Fig. 16 Physical Analyser map of test flight

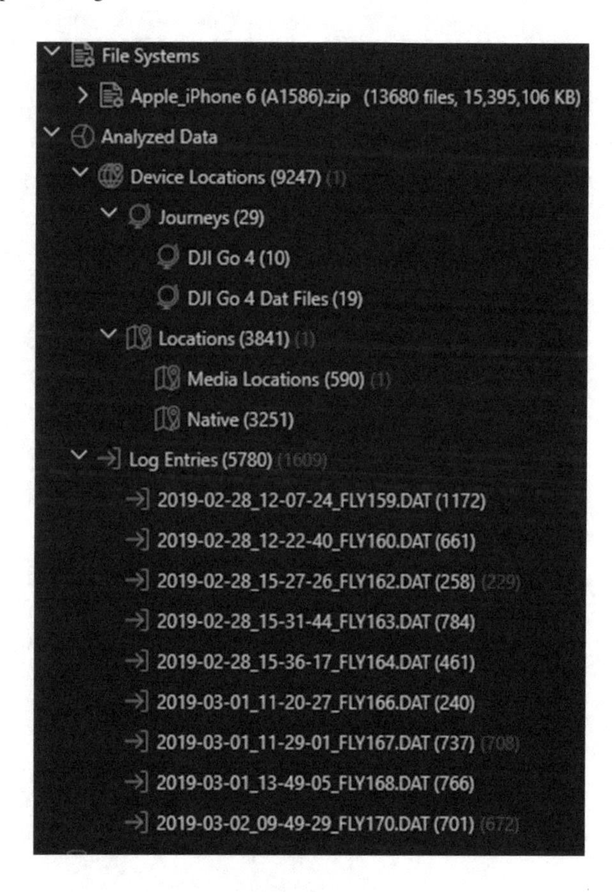

Fig. 17 Extract of the DJI GO 4 app's file system

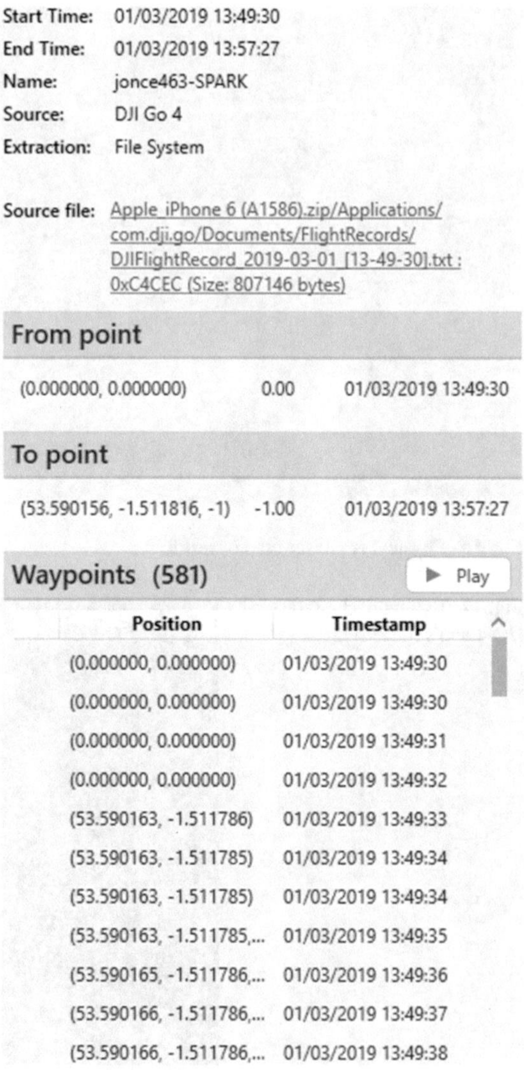

Fig. 18 Waypoints and timestamps of the GPS coordinates

Start Time: 01/03/2019 13:49:30
End Time: 01/03/2019 13:57:27
Name: jonce463-SPARK
Source: DJI Go 4
Extraction: File System

Source file: Apple iPhone 6 (A1586).zip/Applications/com.dji.go/Documents/FlightRecords/DJIFlightRecord_2019-03-01_[13-49-30].txt : 0xC4CEC (Size: 807146 bytes)

From point

(0.000000, 0.000000) 0.00 01/03/2019 13:49:30

To point

(53.590156, -1.511816, -1) -1.00 01/03/2019 13:57:27

Waypoints (581) ▶ Play

Position	Timestamp
(0.000000, 0.000000)	01/03/2019 13:49:30
(0.000000, 0.000000)	01/03/2019 13:49:30
(0.000000, 0.000000)	01/03/2019 13:49:31
(0.000000, 0.000000)	01/03/2019 13:49:32
(53.590163, -1.511786)	01/03/2019 13:49:33
(53.590163, -1.511785)	01/03/2019 13:49:34
(53.590163, -1.511785)	01/03/2019 13:49:34
(53.590163, -1.511785,...	01/03/2019 13:49:35
(53.590165, -1.511786,...	01/03/2019 13:49:36
(53.590166, -1.511786,...	01/03/2019 13:49:37
(53.590166, -1.511786,...	01/03/2019 13:49:38

2015) and the Spark model being relatively low priced at approximately £449 (DJI, 2019), it is assumed that this is the model most assessable to the general public and would be more likely to be involved in criminal activities due to this.

4.3.2 DJI SPARK SD Card

Due to unforeseen issues regarding the Cellebrite UFED 4PC software see Appendix 2, FTK Imager had to be used to create an E01 evidence file of the internal SD card

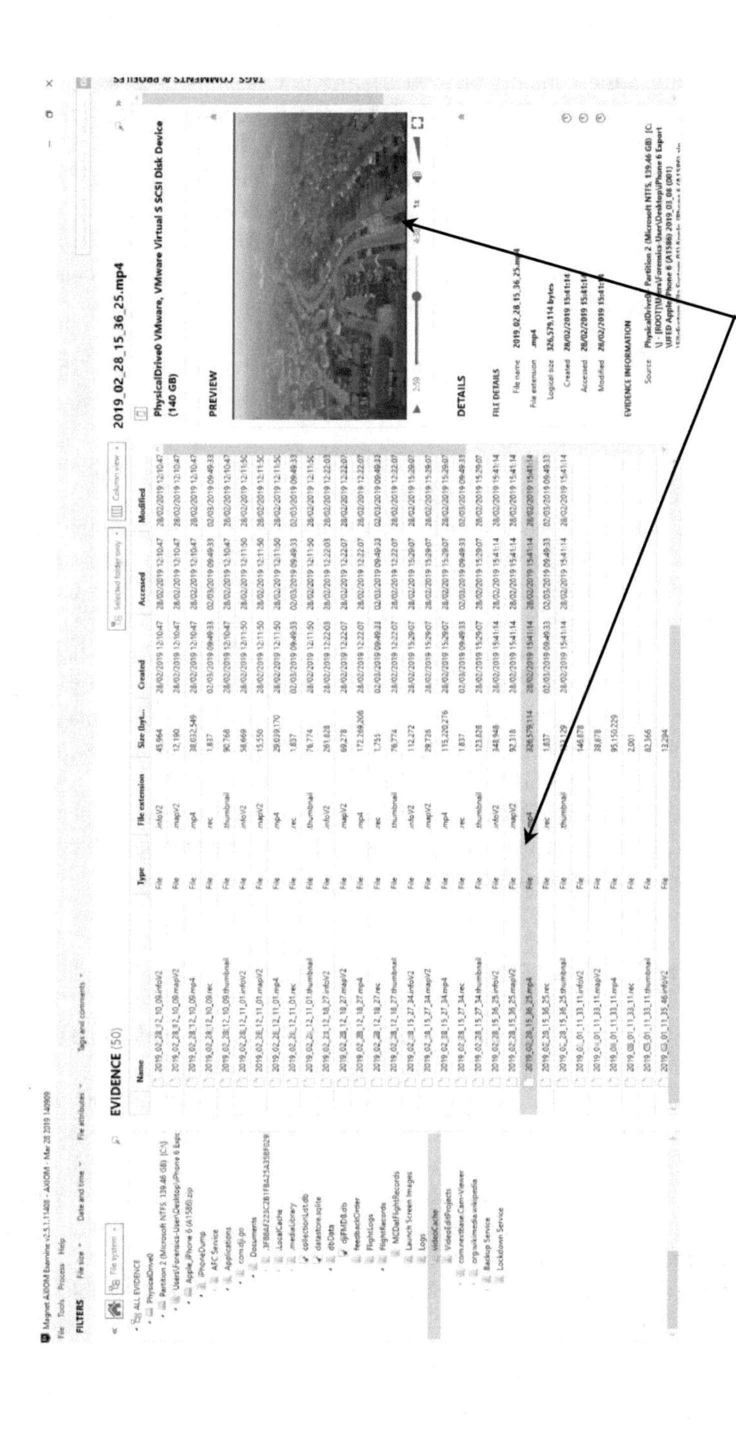

Fig. 19 Magnet AXIOM Examine software displaying the mobile device's file system

of the UAV device. The E01 file format is widely supported within forensic toolkits, which make it an ideal candidate.

In order to analyse the contents of the E01 file, EnCase was used. Below is a screenshot of the contents of the SD card (Fig. 20). A high volume of both video and image files are present on the card which can be viewed by the user.

Also, the user is able to view an array of metadata regarding the artefact (Fig. 21). Metadata such as the item path, creation date and last accessed. This is very useful information that could be pertinent in a case. In addition to this, the physical location of the file within the file system is also displayed. By using this data to locate the physical location on the drive it may make the case more solid.

4.3.3 DJI Phantom 4

Whilst investigating the Internal SD card a total of 11 flight of logs were located, alongside a text document titled 'PARM.LOG' and SYS.DJI (Fig. 22).

By using the CsvView application, data was able to be obtained from the flight logs and data was able to be extracted that could otherwise not have been seen by opening the .dat file manually. FLY008.DAT was inputted into CsvView and the following information was presented as shown in Fig. 23.

Using the feature GeoPlayer within CsvView allows for the user to see both the flight path which was taken by the drone and the home-point which was set at the start of the flight by the user. This home point is represented by the 'H' on the maps, whilst the flight path is shown as a red continuous line. Figure 24 shows the specific flight path of the file 'FLY008.DAT'.

Upon further examination of flight log 'FLY008.DAT', a number of graphs were able to be created showing a range of signals which are recorded by the drone at the point of flight. By extracting specific signals (Appendix 3).

4.3.4 External and Internal SD Card

Whilst examining the external SD card a number of files were located inside both a folder titled 'DCIM' and a hidden folder titled 'MISC' (Appendix 3).

Within the DCIM folder there was another folder titled '100Media' and within this there was a number of .JPG files and .mov files. In total 17 images were stored on the external SD card and a further 6 videos were also stored (Appendix 3).

Upon further investigation into the image files via the image properties key points of interest were located. The date of creation, modified date and accessed date were all available for the user to view (Fig. 25).

When viewing the details tab within the properties, further data was able to be acquired. The date taken was located as 29/06/2017 12:39. Details regarding the size of the image, resolution and dimension can also be viewed (Fig. 26).

Information regarding the camera which has taken this image can also be located in the details tab. Furthermore, GPS co-ordinates of the image can also be viewed,

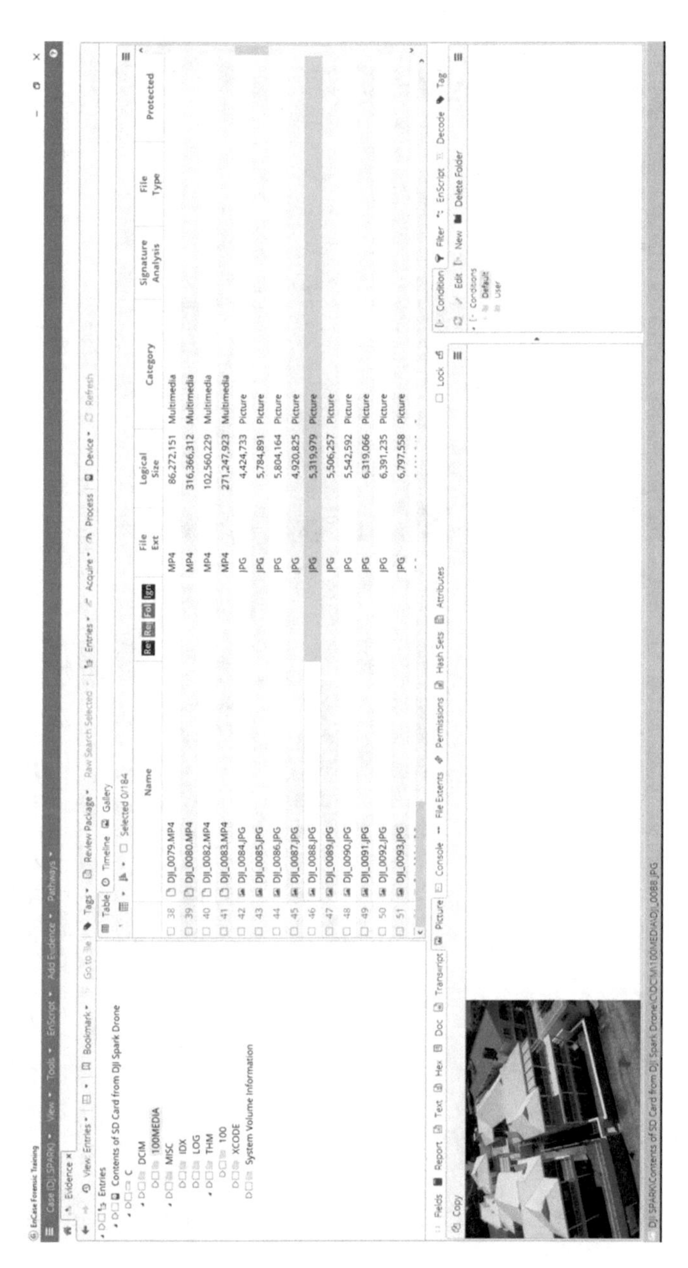

Fig. 20 EnCase loading the contents of the UAVs on board SD Card

Fig. 21 EnCase extraction showing the true path of file

Name	Date modified	Type	Size
FLY000.DAT	29/05/2017 10:03	DAT File	1,364 KB
FLY001.DAT	29/05/2017 10:10	DAT File	75,884 KB
FLY002.DAT	28/06/2017 12:54	DAT File	85,920 KB
FLY003.DAT	28/06/2017 12:59	DAT File	23,540 KB
FLY004.DAT	28/06/2017 12:59	DAT File	24,916 KB
FLY005.DAT	29/06/2017 12:06	DAT File	36,156 KB
FLY006.DAT	29/06/2017 13:30	DAT File	33,748 KB
FLY007.DAT	29/06/2017 13:51	DAT File	89,936 KB
FLY008.DAT	29/06/2017 15:49	DAT File	160,748 KB
FLY009.DAT	29/06/2017 15:27	DAT File	1,112 KB
FLY010.DAT	29/06/2017 15:33	DAT File	40,820 KB
PARM	29/06/2017 15:27	Text Document	21 KB
SYS.DJI	29/06/2017 15:26	DJI File	1 KB

Fig. 22 File's located within internal SD card directory

providing an insight as to where the drone was located when the image was taken (Figs. 27 and 28).

Through viewing the images that have been found on the external SD the user can view what the drone has been taking pictures of. This can provide evidence of specific crimes regarding spying and reconnaissance. A selection of the pictures

Firmware Date	May 25 2017
ACType	P4P
mcID(SN)	07JDE580020139
mcVer	v3.2.35.5
BatterySN	1133||
dateTime	2017-6-29 20:26:54 GMT
geoDeclination	8.91 degrees
geoInclination	66.26 degrees
geoIntensity	52038.92 nanoTesla

Fig. 23 Flight details of FLY008.DAT from CsvView

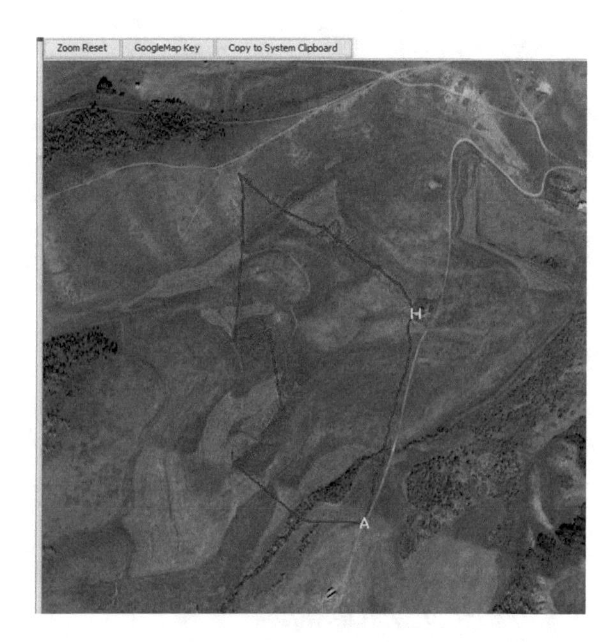

Fig. 24 Flight path obtained from FLY008.DAT

extracted from the external SD provide an insight as to what the drone operator was taking photos of and whether they were legal or illegally taken (Figs. 29 and 30).

Videos located as .mov files within the 100Media folder contain less information in the properties when compared to the still images acquired. The created, modified and accessed dates and time appear, however the accessed time appears to be before the created time (Fig. 31). Within the details tab, the media created date and time is obtained, this time is shown as 29/06/2017 13:46 which is just before the file creation time (Fig. 32). The video's properties do not include information regarding the location through GPS contrasting with the images that were obtained.

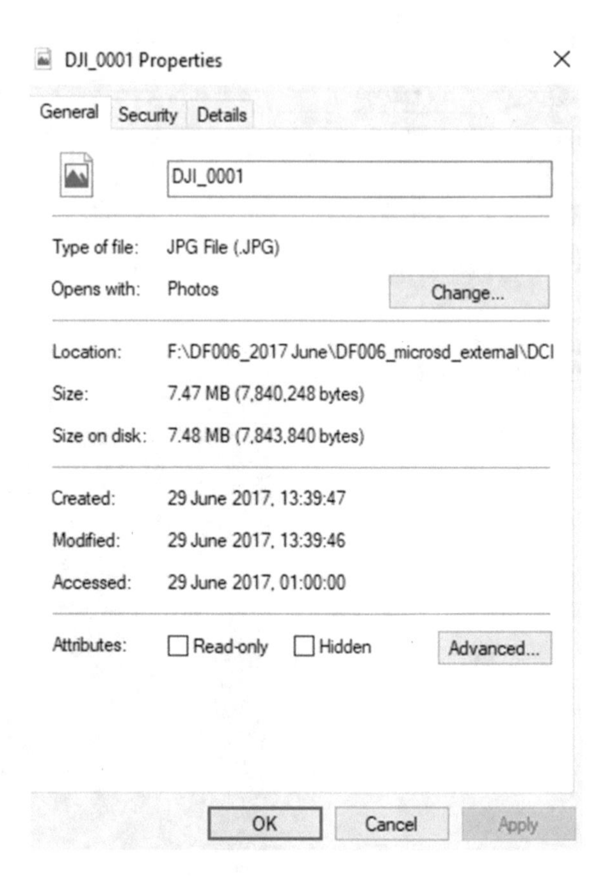

Fig. 25 Properties of DJI_0001 image

4.4 Further Discussion

In this section, any further considerations will be discussed regarding each component of this study. A lot of results have been gained from carrying out the experiments stated in this chapter; the outcome of which have fuelled various recommendations regarding the examination of UAV devices as well as thoughts on future work in this area.

In regard to the experiments, the scenarios and data interpretation are formed from the use of the DJI Spark. The outcomes are likely to vary depending on the model and brand of the drone used for criminal activity. In the case of the payloads, there would need to be further experimentation with drones that have a more intelligent stability system, to determine how the data output is affected by the additional weight.

With regards to the rotor movements and gimbal, it is important to note that at the time of experimentation, there was no data to determine the degree of the gimbal. This data would be an essential piece of evidence as the investigators would be able to determine if the camera was facing a specific direction, such as into the victim's property, rather than just facing in the general direction. This type of evidence may

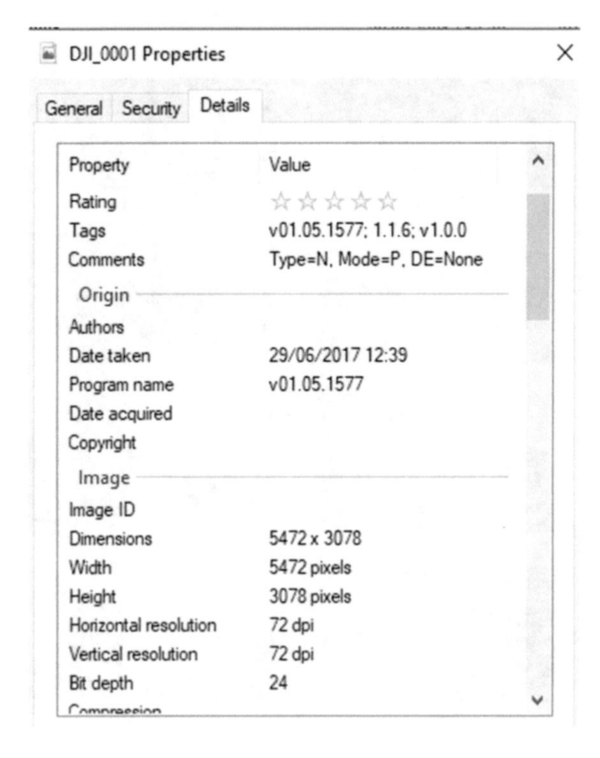

Fig. 26 Image details in properties of DJI_0001 image

Camera

Camera maker	DJI
Camera model	FC6310
F-stop	f/4.5
Exposure time	1/200 sec.
ISO speed	ISO-100
Exposure bias	0 step
Focal length	9 mm
Max aperture	2.97
Metering mode	Centre Weighted Average
Subject distance	0 mm
Flash mode	No flash function
Flash energy	
35mm focal length	24

Fig. 27 Camera details in properties of DJI_0001 image

GPS

Latitude	39; 57; 40.2909999999975...
Longitude	106; 12; 59.355099999986...
Altitude	2541.56

Fig. 28 GPS details in properties of DJI_0001 image

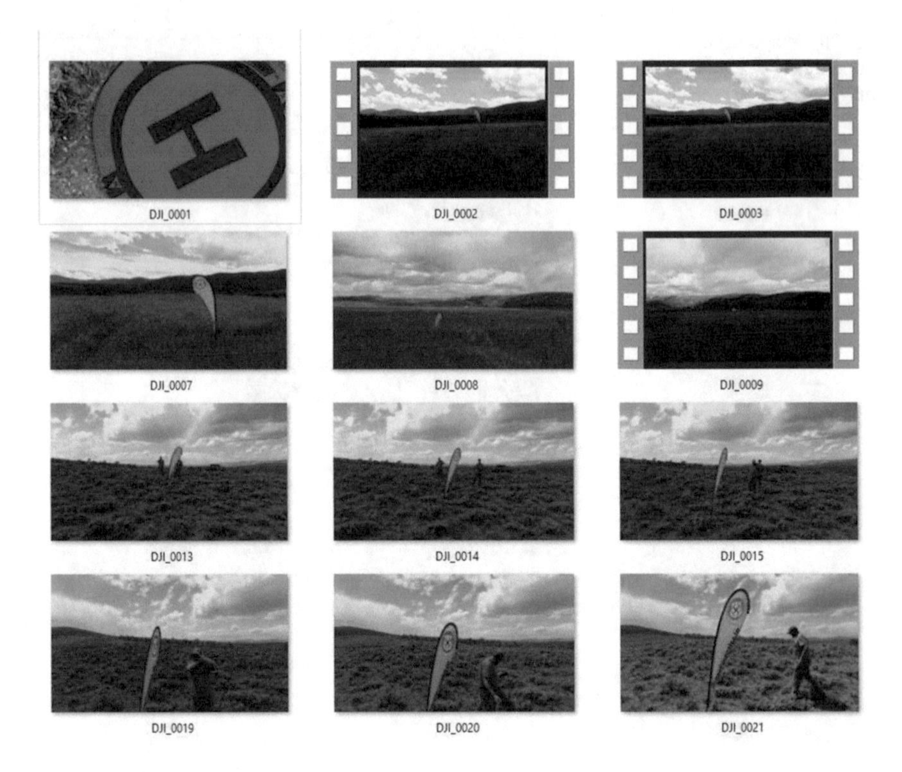

Fig. 29 Images and video's found on external SD

also aid in creating a timeline, to determine when the suspect first started observing the victim during their drone flight.

5 Conclusion and Future Work

5.1 Introduction

This chapter will provide the overall conclusion of the results which have been obtained through the experiment and will incorporate parts of the previous chapter discussing those results. Furthermore, this chapter will look at to what extent this project has added to the existing literature, whether the objectives and aims of the project were met and recommendations for future work within this area of research.

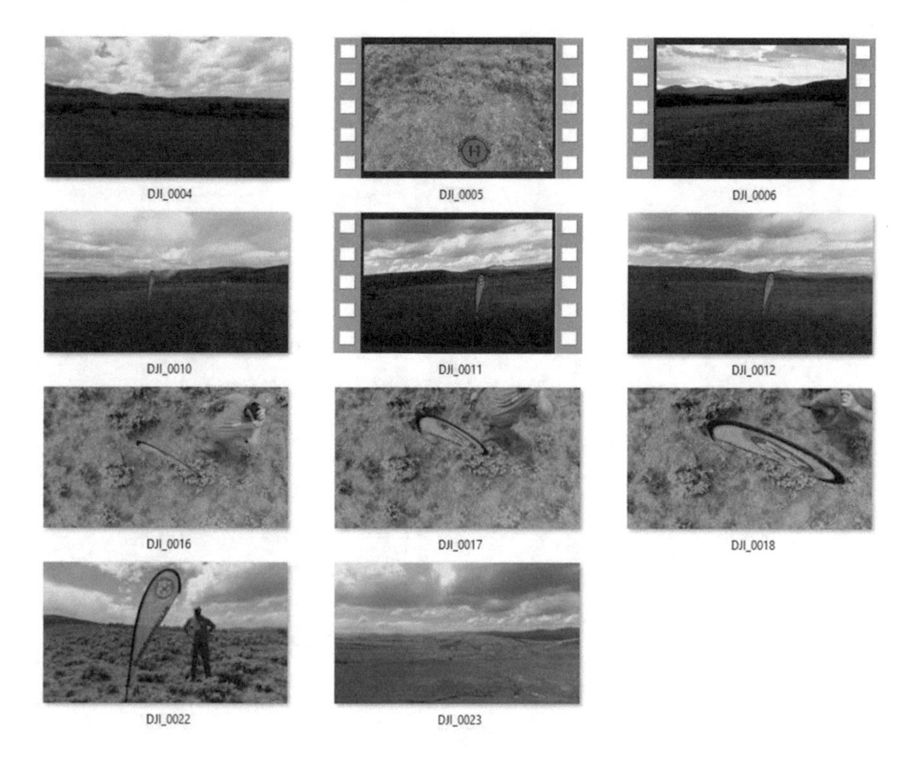

Fig. 30 Further image's and video found on external SD

5.2 Summary of Findings

Based on the findings, it is shown that with the right tools, not necessarily specialist software, a drone forensics team would be able to find appropriate evidence to support their investigations. However, it is important to note that there can be challenges during the extraction and analysis, that depending on the specific case, would need to be examined to retain the integrity of the data.

Through completion of the experiment a number of key items of interest were discovered that can help forensic examiners with criminal investigations. Flight data stored on the internal SD was analysed and provided a vast range of information relating to the signals which are stored on the device itself. This analysis of results was more in-depth than previous research papers detailed in the literature review, this study provides a more detailed analysis and highlighted signals such as airport limits, emergency brakes information and whether the controller was connected to the device at all times during the flight.

Media stored within the external SD and mobile application data was also found through the experiment. This data was congruent with previous research [3, 21, 33, 39] on different models of drones, by exploring the metadata the GPS co-ordinates of where the image was taken were located as was the date and time of creation.

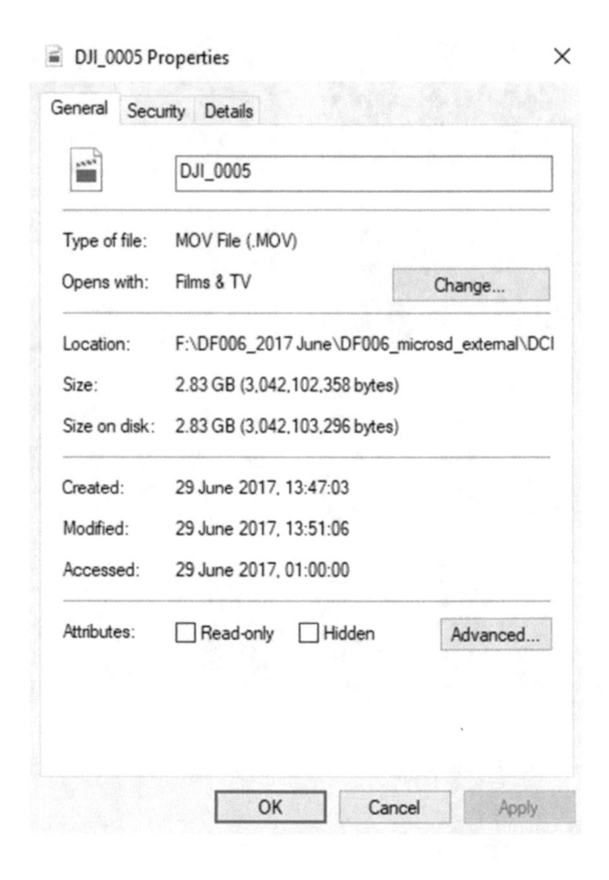

Fig. 31 DJI_005.mov properties

Ownership between the drone and the controller was explored, however due to limitations with the data which had been acquired form VTO it was not possible to provide concrete evidence that the two can be linked together. However, serial numbers of the drone were located on the internal SD flight logs and also discovered on the flight log which had been stored on the mobile application.

5.3 Limitation

It is important to note that all drones have different components and have different capabilities, depending on the range or brand, therefore each individual drone will hold unique limitations. As such, the experiments carried out with the DJI Spark drone presented a limitation of the internal memory size only being 4 GB. The issue with a smaller internal memory means that the existing data, and possible evidence, is overwritten when the capacity is reached.

There is currently limited research in the field of drone forensics as found in the literature review within the chapter. This means that researchers only have a limited

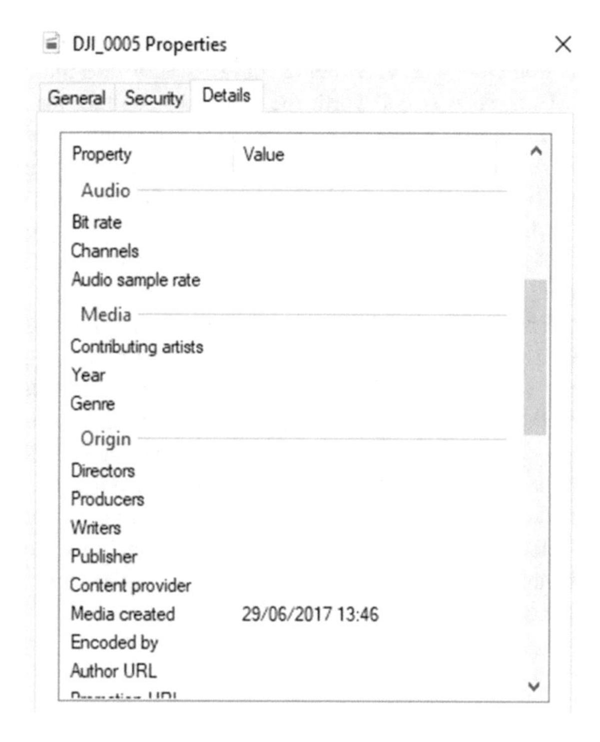

Fig. 32 DJI_0005.mov details

basis of understanding of this topic and there is a lack of unique research. Whilst this work will quickly become outdated due to the fast-growing nature of drones and their constant upgrading, this research provides a further understanding of drone forensics. However, more research needs to be carried out on a range of drones to establish common themes of data extraction and data locations. Only when many drones have been researched can a general framework for drone extraction be developed, as has been done with mobile forensics.

5.4 Recommendation and Future Work

Whilst carrying out this project, a number of obstacles have been encountered. Therefore, there are a few recommendations that this study would like to make in order to make the industry more 'drone/UAV friendly'.

Firstly, when attempting to extract data from a mobile device using EnCase software, the evidence was unable to be acquired. This is something that EnCase must work on to ensure that almost all devices are supported by their software.

Secondly it was discovered during the interview with the Digital Forensics Analyst that there is not a great amount of knowledge of UAV devices within DFUs. Although the analyst only works with one Police Force, they are in contact with people from other forces and states that views on UAV devices are very similar across all forces nationwide. The view being that it is very rare that a UAV device is brought into a Unit, so it is not worth the resources sending personnel on training courses. In addition to the lack of UAV Forensics training, DFU personnel are not required to have knowledge on current laws and legislations of any devices. Although standards such as ISO 27001 and ISO 17025 must be adhered, as well as the ACPO guidelines [2], knowledge of other legislation is not required and generally analysts/investigators 'learn as they go' in relation to laws and legislation. This is something that this study finds astounding as having up to date knowledge of the current legalities relating to the device(s) would be very beneficial to an investigation. For example, if the use of a device were an illegal act in a specific area e.g. the use of a UAV device in the vicinity of a school and it was found that this occurred, it should be included in the final report. Nonetheless it is likely that an investigator/analyst would look up such laws and legislation on an ad-hoc basis, but this would waste time during an investigation. Therefore, it is a recommendation of this study that DFUs at least make all of their personnel aware of relevant laws and legislation and send them on 'refresher' courses on a regular basis e.g. annually. Refresher courses would be very beneficial also for standards such as ISO 27001 and ISO 17025 to ensure that they are being continuously obeyed.

Appendix 1

Freedom Of Information
PO BOX 9
Laburnum Road
Wakefield
WF1 3QP

Tel: 01924 296006
Fax: 01924 292726
Email: foi@westyorkshire.pnn.police.uk
Website: www.westyorkshire.police.uk

Our ref: 04779/18

Date: 29/10/2018

Dear Mr Carr

Thank you for your request for information, received by West Yorkshire Police on 01/10/18.

You requested the following information:

The number of incidents including unmanned drones, split by incident type since data began being recorded, split by calendar year.
The number of these incidents that resulted in arrest/charged per calendar year.

Please see the table below showing the number of incidents reporting the use of Unmanned Aerial Vehicles (UAVs) between 01/04/2009 and 30/09/2018.

Please note that incident (report) data is separate to crime (arrest/charge) data. Arrests and charges are not held within incident data.

	2009	2010	2011	2012	2013	2014	2015	2016	2017	2018
Transport Related	0	0	0	0	0	0	2	1	2	1
Crime	0	0	0	0	0	0	5	10	17	21
Public Safety/Welfare	0	0	0	0	0	6	45	90	82	59
Anti-Social Behaviour	0	0	0	0	0	2	13	40	32	24
Crime	0	0	0	0	0	0	0	0	0	0
Admin/Other	0	0	0	0	2	1	5	10	38	43
Total	**0**	**0**	**0**	**0**	**2**	**9**	**70**	**151**	**171**	**148**

Figures show the number of incidents which:
- were recorded during the calendar year shown (NB: 2009 covers 01/04/2009 - 31/12/2009, 2018 covers 01/01/2018 - 30/09/2018)
- contained one or more of the search terms %UNMANNED AERIAL VEHICLE%, %UAV%, %DRONE% within the incident log text
- were, following a manual assessment of all matching records, identified as an incident referring to the use of an Unmanned Aerial Vehicle / drone

Appendix 2

When using EnCase to acquire the data from the mobile device used to control the UAV device, issues were encountered. The data had previously been extracted using Cellebrite and a '.UFD' dump file created. EnCase is able to read the dump file in order to acquire the data. However, it was found that the software was unable to read the data; the error message stating that the task was not possible whilst being run in a Virtual Machine. As the EnCase software is not available outside of the designated Virtual Machine, it is not known whether or not EnCase is able to read and acquire the dump data from the iOS device used to control the UAV device.

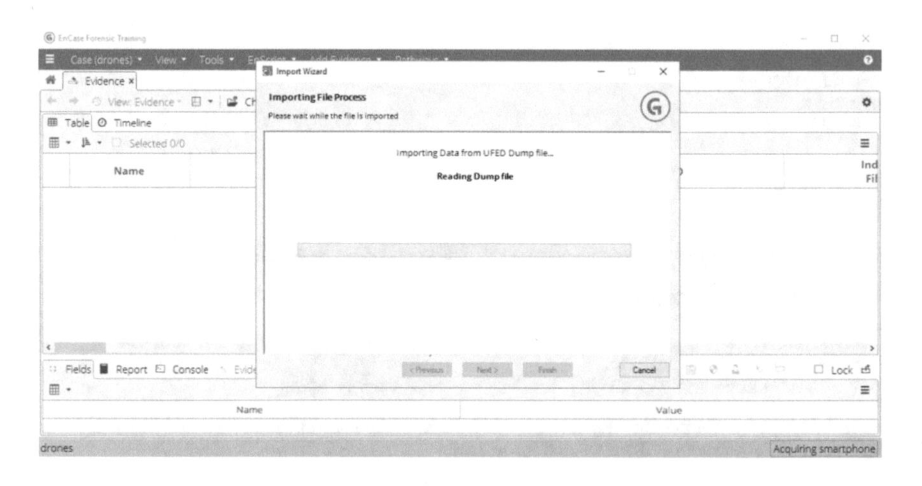

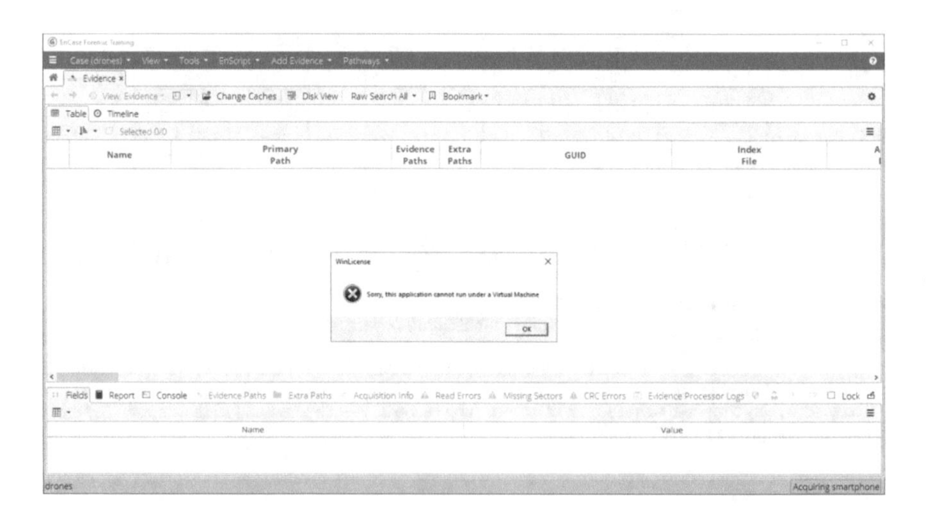

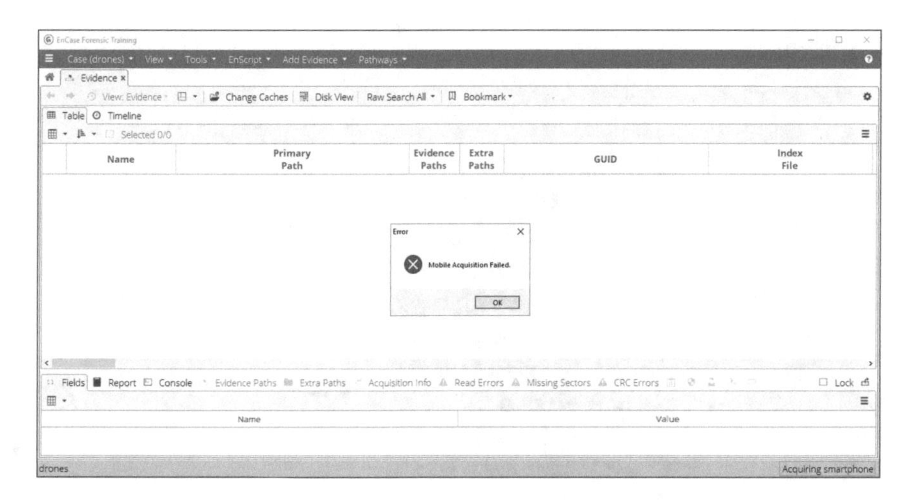

The following screenshots note the process and error messages displayed when attempting to carry out this task.

Appendix 3

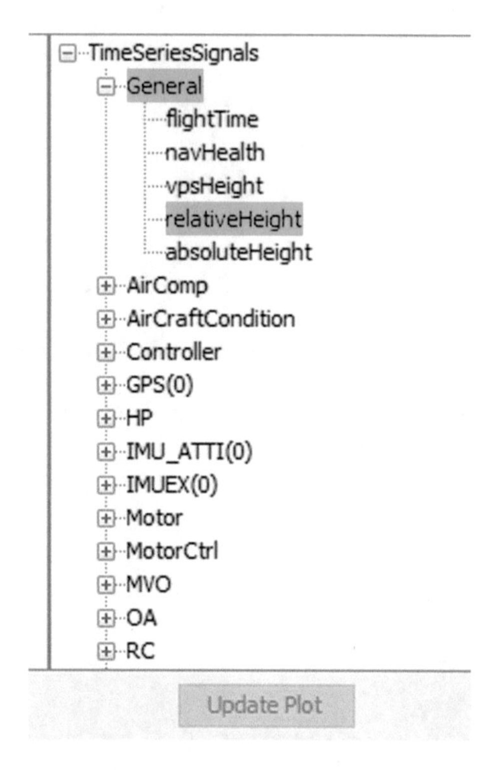

Signals used to present controller connection

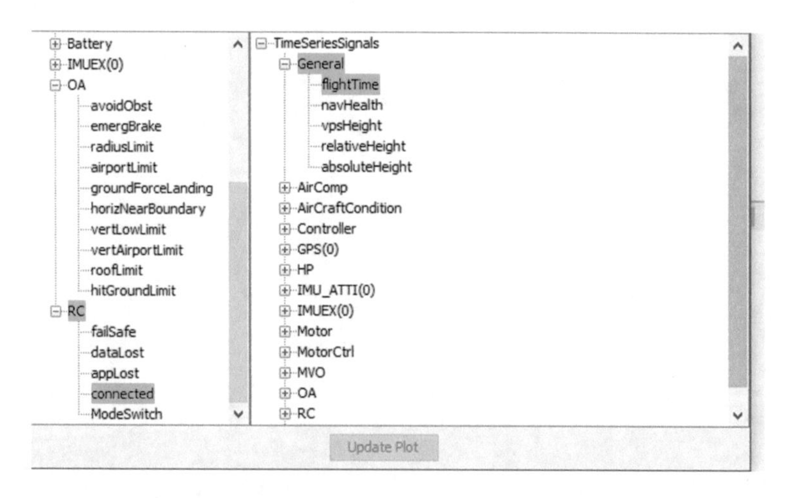

Signals used to present altitude

Drone Data (F:) › DF006_2017 June › DF006_microsd_external ›

Name	Date modified	Type	Size
DCIM	01/06/2017 10:21	File folder	
MISC	01/06/2017 10:21	File folder	

Root folder of external SD Card

Drone Data (F:) › DF006_2017 June › DF006_microsd_external › DCIM ›

Name	Date modified	Type	Size
100MEDIA	29/06/2017 13:39	File folder	

100Media Folder stored in DCIM Folder

Drone Data (F:) › DF006_2017 June › DF006_microsd_external › DCIM › 100MEDIA

Name	Date modified	Type	Size
DJI_0001	29/06/2017 13:39	JPG File	7,657 KB
DJI_0004	29/06/2017 13:42	JPG File	1,958 KB
DJI_0007	29/06/2017 15:37	JPG File	2,641 KB
DJI_0008	29/06/2017 15:39	JPG File	2,445 KB
DJI_0010	29/06/2017 15:42	JPG File	2,417 KB
DJI_0012	29/06/2017 15:43	JPG File	2,635 KB
DJI_0013	29/06/2017 15:44	JPG File	2,930 KB
DJI_0014	29/06/2017 15:44	JPG File	2,947 KB
DJI_0015	29/06/2017 15:44	JPG File	2,890 KB
DJI_0016	29/06/2017 15:44	JPG File	5,492 KB
DJI_0017	29/06/2017 15:45	JPG File	5,277 KB
DJI_0018	29/06/2017 15:45	JPG File	5,089 KB
DJI_0019	29/06/2017 15:45	JPG File	2,981 KB
DJI_0020	29/06/2017 15:45	JPG File	2,980 KB
DJI_0021	29/06/2017 15:45	JPG File	3,336 KB
DJI_0022	29/06/2017 15:45	JPG File	3,380 KB
DJI_0023	29/06/2017 15:46	JPG File	2,686 KB
DJI_0002	29/06/2017 13:41	MOV File	50,523 KB
DJI_0003	29/06/2017 13:47	MOV File	3,997,788 KB
DJI_0005	29/06/2017 13:51	MOV File	2,970,804 KB
DJI_0006	29/06/2017 15:42	MOV File	3,996,674 KB
DJI_0009	29/06/2017 15:42	MOV File	691,736 KB
DJI_0011	29/06/2017 15:46	MOV File	2,642,475 KB

Images and videos stored within 100Media Folder

Appendix 4

Question	Response
• Could you please state your official job title?	• Digital Forensic Analyst
• How long have you been working in forensics and within your current role?	• Placement year in Wales and then worked there over the summer in the last two years of university • Then worked there for 1 year before moving to SYP • Total of around 2 ½ years at SYP
• Do you have an area of forensics that you find more interesting than the rest, or a specialism?	• Computers more than phones are his specialism, centring around the operating system, Internet artefacts etc. • Phones are very interesting but change too often. Skills gained may change two years after you've learned them. Also there are skills/methods that are specific for one particular phone model • Not disinterested in any part of forensics, but the file system and operating system is where the interest lies
• How often would you say that you get Drone devices in your forensic department? – What are the types of crimes associated with these cases?	• Very rarely get a device in. Maybe one a year on average • The very few they have had have been drug related cases flying drugs into prisons • One example was a drone that had crashed in the vicinity of a prison and it had to be examined to see whether or not it had been used for the drug delivery purpose or had the drone just happen to have been flying around the prison. The outcome was the latter • The second case was also drug related and the suspects had been caught and the drone device seized. The investigation wanted to know where it had been used. Didn't directly work on the case but was aware of the process and outcome as the department personnel get 'excited' whenever a drone comes in. Drone was traced back to Manchester where it was being used to drop drugs to prisons
• You worked at a Welsh constabulary before SYP, were Drone devices more popular there?	• The Welsh constabulary cases were drug related. Wanting to know the flight path etc.
• Do you see the field of Drone forensics expanding?	• Yes. Criminal cases involving drones are rare. Mainly civil cases. See it staying that way for now, staying fairly rare • Foresees new laws surrounding drones coming in regarding videos and images with the new laws coming in regarding upskirting. Thinks in the future that flying a drone over a beach for example and capturing images of people sunbathing will be an illegal act • Thought there would be a massive increase in the use of drone devices when they were commercially available. However, there wasn't really any surge in the use of them in criminal acts. Thinks that it comes down to people still delivering/exchanging drugs by hand and using technology such as burner phones to suddenly buy and start flying drones to deliver goods

(continued)

(continued)

Question	Response
• What are the current protocols that must be followed when extracting evidence from a Drone device? Are they similar to Mobile protocols?	• Pretty similar to mobile devices. Essentially just remove the SD Card and forensically image it, following the same procedures as dealing with other mobile devices and removable media. Write blocker, make an E01 etc. • If it a more serious case, a chip off could be called for. But that takes a long time to carry out and all drones are different. It will probably land to someone with expertise in mobile phones to try and work out which pins to utilise etc. • Drones are at the minute a bit of an unknown • Normal procedure is to analyse the SD card and if the officer is satisfied with the evidence located, it probably wouldn't go any further. Especially if nothing can be gained from the SD Card. The more data that can be gained from the SD Card, the more likely it is that the officer will request further analysis of the device, i.e. chip off and apps on phones. As drones can be expensive, you have to justify taking it apart and risk damaging the device • Tend to get more information off the phone app • Some drone controllers store data also. colleagues in Derbyshire informed me, they have a specialist drone unit • Not getting enough devices in at the minute to warrant a specialised set of procedures • Could outsource the extraction of data if required e.g. chip offs. Never outsourced any though
• Do you think that the ACPO principles need to be updated to be more inclusive of Drone devices as well as mobile devices? – Do you think there should be a separate set of principles solely for Drone devices?	• Don't think they are as outdated as they could be considering how old they are. The first principle of 'don't change data' can be an issue, especially with phones as you change data as soon as it is powered on. However, the second rule of 'if you are competent to change data, you can' is a good cover as if you are working in forensics you will generally be competent • If I were to change ACPO principles, it wouldn't be any of the first three. It would be point four. Maybe re-word it or something similar • Generally points 1, 2 and 3 are still very valid in current times. They have aged very well • Even relevant for drone work • Don't think drones justify getting their own set of principles, as then there would be a different set of principles for a lot of other types of devices and would end up with ACPO Principle 99, 125 etc.
• Do you know anything about the current legislation on Drone devices?	• Do not know very much about current drone legislation, only that it is an illegal act to fly them in the vicinity of the airport • It isn't a requirement of the job role to be aware of current legislation • Generally if crimes involving breaking legislation rules were committed, the CAA would take the investigation not the police. It is thought that they will have their own team of forensic experts/investigators to analyse devices

(continued)

(continued)

Question	Response
• Does your forensics department send its staff on regular training courses regarding new methods of extracting data and changes/additions to legislation?	• Not required to attend any courses or read new legislation on a regular basis • Investigators get to know a lot of the legislation whilst they are working on devices • Legislation is mentioned within other courses on forensics • Normally the process of dealing with legislation is dealt with before the artefact gets to us. We usually only have to worry about working within RIPA
• Have you encountered any Anti-Forensics techniques when examining devices? • Do you think that Anti-Forensics techniques could be used on a Drone device? – If so, which ones? – How difficult do you think it could be to do this, i.e. could a novice user do it or would it have to be a highly skilled technically minded user?	• Yes. It is a big issue that can cover a lot. Tools such as BitLocker and CCleaner are classed as Anti-Forensics tools. CCleaner forensically wipes (i.e. 0 s everything) and encrypts and can do so on schedule. BitBleach, Eraser are also classed as Anti-Forensic tools. Tend to comment of their instillation within the final forensic report. Encryption is also seen as Anti-Forensics. BitLocker, VeraCrypt etc. Section 49 of RIPA allows officers to force suspects to disclose their password depending on the case. It is very difficult to get a RIPA 49 order. It has to go through a judge to be signed off and has to be proved that you have tried to get into the evidence • It depends on whether the user or the manufacturer implemented them. For example, fairly easy for the manufacturer to make every bit of data encrypted on the board. More difficult for a user of the drone to implement. Maybe set up a schedule on an Android to wipe the content of the app if it hasn't been opened in x amount of days. Standard users would find it difficult to carry out anti-forensics techniques on drone devices

References

1. Admin (2017) Drone Components_Quick list of it's parts. https://grinddrone.com/drone-features/drone-components
2. Association of Chief Police Officers (2012) ACPO good practice guide for digital evidence. https://www.digital-detective.net/digital-forensics-documents/ACPO_Good_Practice_Guide_for_Digital_Evidence_v5.pdf
3. Barton TEA, Hannan Bin Azhar MA (2017) Forensic analysis of popular UAV systems. IEEE. https://doi.org/10.1109/EST.2017.8090405
4. BBC (2019). Drone no-fly zone to be widened after Gatwick chaos. www.bbc.co.uk: https://www.bbc.co.uk/news/business-47299805, 20 Feb 2019
5. Bouafif H, Kamoun F, Iqbal F, Marrington A (2018) Drone forensics: challenges and new insights. In: 2018 9th IFIP international conference on new technologies, mobility and security (NTMS). IEEE, Paris, France
6. Brown R (2018) Fears burglars are using drones to case homes—as drone reports to police rocket. The Cambridgeshire Live. https://www.cambridge-news.co.uk/news/cambridge-news/burglars-drones-homes-reports-rocket-14785783
7. CAA, NATS (2019) The drone code. Drone Safe UK: https://dronesafe.uk/wp-content/uploads/2019/02/Drone-Code_March19.pdf, 19 Feb 2019

8. CAA, NATS (n.d) Drone Safe UK: https://dronesafe.uk/
9. Canis B (2015) Unmanned aircraft systems (UAS): commercial outlook for a new industry. Congressional Research Service. https://goodtimesweb.org/industrial-policy/2015/R44192.pdf
10. Conti M, Dehghantanha A, Franke K, Watson S (2018) Internet of things security and forensics: Challenges and opportunities. Future Gener Comput Syst 78:544–546. https://doi.org/10.1016/j.future.2017.07.060
11. CopterSafe (n.d) NFZ mod for Phantom 4 (not for PRO). CopterSafe: https://www.coptersafe.com/product/nfz-mod-phantom-4/
12. Crawford J (2018). 10 crimes committed using a drone. https://listverse.com/2018/07/26/10-crimes-committed-using-a-drone/
13. DJI (n.d.) FlySafe. DJI website: https://www.dji.com/uk/flysafe
14. DJI (2019) Matric 600. https://www.dji.com/uk/matrice600
15. DJI (2018) Phantom 4. https://www.dji.com/uk/phantom-4/info
16. Dormehl L (2018) The history of drones in 10 milestones. https://www.digitaltrends.com/cool-tech/history-of-drones/
17. Flynt J (2017) 21 types of drones. https://3dinsider.com/types-of-drones/
18. Fussell S (2018) Who will police drones? https://gizmodo.com/who-will-police-police-drones-1826891119
19. Haylen A (2019) Civilian drones. (Briefing Paper No. CBP 7734). www.parliament.uk/commons-library
20. Hegarty R, Lamb DJ, Attwood A (2014) Digital evidence challenges in the internet of things. Paper presented at the INC, pp 163–172
21. Horsman G (2016) Unmanned aerial vehicles: a preliminary analysis of forensic challenges. Digit Investig 16:1–11. https://doi.org/10.1016/j.diin.2015.11.002
22. HM Government (2000) Regulation of investigatory. Legislation.gov: https://www.legislation.gov.uk/ukpga/2000/23/pdfs/ukpga_20000023_en.pdf, 1 Aug 2000
23. HM Government (1978) Protection of Children Act 1978. Legislation.gov: https://www.legislation.gov.uk/ukpga/1978/37/pdfs/ukpga_19780037_en.pdf
24. HM Government (2003) Sexual Offences Act 2003. Legislation.gov: https://www.legislation.gov.uk/ukpga/2003/42/pdfs/ukpga_20030042_en.pdf
25. House of Commons—Science and Technology Committee (2019) Commercial and recreational drone use in the UK. https://publications.parliament.uk/pa/cm201719/cmselect/cmsctech/2021/2021.pdf
26. Jain A, Chhabra G (2014) Anti-forensics techniques: an analytical review. In: 2014 seventh international conference on contemporary computing (IC3). IEEE, India, p 7
27. James H (n.d.) No Fly Drones: https://www.noflydrones.co.uk/
28. James H (n.d.) Contact. No Fly Drones: https://www.noflydrones.co.uk/contact
29. Liao S (2017) DJI drones can get past no-fly zones thanks to this Russian software company. The Verge: https://www.theverge.com/2017/6/21/15818344/drones-russian-software-hack-dji-jailbreak, June 2017
30. Kessler GC (2007) Anti-forensics and the digital investigator. In: Proceedings of the 5th Australian digital forensics conference. Edith Cowan University, Perth Western Australia, p 8
31. Kovar D (2016) UVA (aka drone) forensics. [Slide Presentation] Cyber Security Summit. https://www.sans.org/cyber-security-summit/archives/file/summit-archive-1492184184.pdf
32. Kovar D, Bollo J (2018) Drone forensics. Digit Forensics Mag 34:14–19
33. Maarse M, Sangers L, van Ginkel J, Pouw M (2016) Digital forensics on a DJI phantom 2 vision UAV. University of Amsterdam
34. Mercer D (2019) Revealed: drones used for stalking and filming cash machines in the UK. https://news.sky.com/story/police-warn-drone-users-after-incidents-soar-by-40-in-two-years-11637695
35. NIST (2018) Drone forensics gets a boost with new data on NIST website. NIST: https://www.nist.gov/news-events/news/2018/06/drone-forensics-gets-boost-new-data-nist-website, 6 June 2018

36. NLD (n.d) NLD MOD client license key. No Limit Dronez: https://nolimitdronez.com/activation-key-for-nld-mod-client
37. PWC (2018) Skies without limits. https://www.pwc.co.uk/intelligent-digital/drones/Drones-impact-on-the-UK-economy-FINAL.pdf
38. Roder A, Choo K-K, Le-Khac N-A (n.d) Unmanned aerial vehicle forensic investigation, p 14
39. Roder A, Choo KR, Le-Khac N (2018) Unmanned aerial vehicle forensic investigation process: Dji phantom 3 drone as A case study
40. Rouse M (2018) Drone (unmanned aerial vehicle, UAV). https://internetofthingsagenda.techtarget.com/definition/drone
41. Rubens T (2018) Drug-smuggling drones: how prisons are responding to the airborne security threat. https://www.ifsecglobal.com/drones/drug-smuggling-drones-prisons-airborne-security-threat/
42. SmashingDrones.com (2019) Best camera drones for sale UK 2019. https://smashingdrones.com/
43. The Civil Aviation Authority (n.d) Safety apps. DroneSafeUK: https://dronesafe.uk/safety-apps/
44. The Daily Mail (2017) Ten drone crimes a day: surge in popularity sees police report for 12-fold jump in offences linked to the gadgets. https://www.dailymail.co.uk/news/article-4373806/Police-report-12-fold-jump-drone-offences.html
45. The Office of the General Counsel (2016) The air navigation order 2016 and regulations. The Civil Aviation Authority. https://publicapps.caa.co.uk/docs/33/CAP393_Fifth_edition_Amendment_13_March_2019.pdf, Aug 2016
46. UK Civil Aviation Authority (2019) Drone code. https://dronesafe.uk/drone-code/
47. Uleski M (2017) The top 6 reasons police UAV programs fail [Blog]. https://www.dartdrones.com/blog/top-police-uav-fails/
48. Waddell K (2017) The invisible fence that keeps drones away from the President. The Atlantic: https://www.theatlantic.com/technology/archive/2017/03/drones-invisible-fence-president/518361/, 2 Mar 2017
49. Watson A (2019) 5 ways commercial drones are pushing the boundaries of crime [Blog]. https://www.cellebrite.com/en/blog/5-ways-commercial-drones-are-pushing-the-boundaries-of-crime/

Intrusion Detection and CAN Vehicle Networks

Ashraf Saber, Fabio Di Troia, and Mark Stamp

Abstract In this chapter, we consider intrusion detection systems (IDS) in the context of an automotive controller area network (CAN), which is also known as the CAN bus. We provide a discussion of various IDS topics, including masquerade detection, and we include a selective survey of previous research involving IDS in a CAN network. We also discuss background topics and relevant practical issues, such as data collection on the CAN bus. Finally, we present experimental results where we have applied a variety of machine learning techniques to CAN data. We use both real and simulated data, and we conduct experiments to determine the status of a vehicle from its network packets, as well as to detect masquerading behavior on a CAN network.

1 Introduction

Research in automotive security is of increasing importance due to cars being more networked and interconnected than ever before. Providing security to consumers and maintaining their safety requires a considerable focus on automotive security as well as from users regarding security related issues [25]. In recent years, hackers and security researchers have demonstrated the ability to remotely breach vehicle security systems and gain unauthorized access. In one costly example, the successful hacking of a Jeep Cherokee led to the recall of 1.4 million vehicles in 2015 [25].

Intrusion detection systems (IDS) have been widely studied in the information security research literature. Such systems also have relevance to vehicle security,

A. Saber · F. Di Troia · M. Stamp (✉)
San Jose State University, San Jose, CA, USA
e-mail: mark.stamp@sjsu.edu

A. Saber
e-mail: ashraf.saber@sjsu.edu

F. Di Troia
e-mail: fabioditroia@msn.com

R. Montasari et al. (eds.), *Digital Forensic Investigation of Internet of Things (IoT) Devices*, Advanced Sciences and Technologies for Security Applications,
https://doi.org/10.1007/978-3-030-60425-7_5

as effective IDS would enable us to identify malicious users or activities on a CAN network. Yet, in comparison to IDS in a more general setting, relatively little research has been conducted on IDS in CAN networks.

In this paper, we apply machine learning techniques to CAN traffic data for both user activity analysis and masquerade behavior detection. In addition, we include an extensive survey of relevant IDS and masquerade detection topics. The remainder of this paper is organized as follows. In Sect. 2, we discuss relevant background topics, including a brief introduction to CAN networks and a similarly brief discussion of IDS in a general setting. We provide a selective survey of IDS in CAN networks and we discuss some related topics in Sect. 3. Then in Sect. 4 we outline research in the area of masquerade detection and discuss why this is likely a fertile area of research for CAN networks. In Sect. 5 we consider collection issues related to CAN data and, finally, Sect. 6 gives our conclusion and points to directions for future work.

2 Background

In this section, we introduce several relevant background topics. Specifically, we discuss the CAN bus and basic notions of intrusion detection.

2.1 CAN Bus

Automotive vehicles manufactured in the US after 2008 have a standard internal controller area network (CAN), which is commonly referred to as the CAN bus. Each vehicle has several small electronic control units (ECUs) that are responsible for controlling different car components. These ECUs communicate over the CAN bus, sending and receiving packets during vehicle operation. The CAN bus serves to replace a complex wiring harness that would otherwise be required in modern vehicles.

CAN is a message broadcast system—a sending node broadcasts its packet, the receiving node takes the packet, and other nodes should drop the packet. While CAN is conceptually similar to Ethernet, CAN is slower but offers reliable service, in the sense that the most important data will be transmitted with a higher priority. This makes CAN suitable for the challenging and safety-critical environment found in an automobile.

Another important component of vehicle networks is the on-board diagnostic (OBD-II) port. In the majority of vehicles, this port is to the left and below the steering wheel, and in some cars it is visible to the driver, while in others it is hidden. The OBD-II port enables a user or mechanic to check various engine conditions. For our purposes, the most significant use of the OBD-II port is to sniff traffic on the CAN bus [41].

Automotive security researchers have primarily focused on two aspects of the CAN bus. First, the possibility of breaching the vehicle network has been widely considered [30, 31], and second, attacks based on packet injection have been studied [17].

Next, we provide a high level discussion of IDS. Then we turn our attention to a more detailed discussion of IDS, with an emphasis on CAN networks.

2.2 Intrusion Detection Basics

Intrusion detection systems (IDS) are a fundamental tool in the field of information security. The purpose of an IDS is to notify users when their systems are compromised. IDS is typically considered to be distinct from an intrusion prevention system (IPS). As the names indicate, IPS is designed to prevent attacks, whereas IDS is designed to detect attacks once they have occurred—an IDS would be useful when, for example, an IPS fails to prevent an attack.

IDS can operate at the host level or the network level, or some combination thereof. Whether at the host or network level, there are many approaches to detecting a breach. From a high level perspective, anomaly detection and signature detection are the main techniques used by IDS [44]. IDS methods analogous to those used in general networks can be applied to automotive vehicle systems to detect malicious behavior.

2.3 Host Based IDS

A host-based IDS attempts to detect intrusions using information available at the host, without taking network behavior into consideration. That is, host based IDS monitors behavior on a specific host or set of hosts to detect malicious behavior [44]. This method of intrusion detection relies on data stored in logs, audit trails, checksum values, characteristics of user behavior, and so on. One potential advantage of host based IDS is that it may be able to determine the individuals behind the malicious behavior, since logs can reflect the actions of individual users [38].

Host based IDS has some disadvantages, depending on the specific implementation. For example, host based IDS might require large storage to maintain the necessary data that the IDS relies on [37]. And typically, multiple hosts need to each have their own host based IDS, which makes setup, configuration, and maintenance challenging [38].

2.4 Network Based IDS

Network based IDS attempts to detect intrusions at the network level. Such an IDS monitors network traffic and typically inspects the header and possibly other aspects of packets passing through the network [38]. An advantage of network based IDS is that it can detect scans (e.g., Nmap scans), DoS attacks, and other network based attacks [44]. This type of IDS is also easier to deploy and maintain, as compared to a host based system. One drawback to a network based IDS is that it does not have a clear view of host behavior.

In practice, host based and network based IDS are typically both used, at least to some extent. This provides layered security and allows for defense in depth.

2.5 Anomaly and Signature Detection

Anomaly detection and signature detection can be considered as two broad categories for classifying IDS systems—whether host based or signature based. In anomaly detection, we attempt to model characteristics of the system and when the behavior of the system diverges sufficiently from the model, we flag it as a possible attack. Although challenging, anomaly detection can potentially enable us to detect zero-day attacks [40]. In contrast, signature detection is a form of pattern matching. In such an approach, we extract a pattern or signature from a known attack, then when this pattern is detected, the corresponding attack may have occurred [60]. While relatively accurate and precise, signature scanning can only detect known attacks for which a signature has been previously extracted.

3 Selective Survey of IDS

In this section, our primary focus is to survey selected work on intrusion detection, with an emphasis on research that is most relevant to CAN networks. We have organized the material among several subtopics that are not mutually exclusive.

3.1 Anomaly Detection

As previously mentioned, a strength of anomaly detection is that it holds out the possibility of detecting new attacks based on zero-day vulnerabilities [60]. Generically, in anomaly detection, we train a model on normal behavior and significant deviation from the norm is considered a potential attack [11].

A major issue with anomaly detection is the inherent challenge in trying to model normal benign activities [40]. It is possible for benign activity to occur that was not modeled during the training phase, and it is likely that normal behavior will change over time. These and other similar issues can lead to an excessive number false positives. However, even with the drawback of false alarms, anomaly detection is popular in security research because of the potential to detect zero-day attacks. Detection of zero-day attacks can be viewed as the holy grail of security research [51].

The work of Ye et al. [60] relies on using anomaly detection methods for cyber-attack identification. The authors of [60] discuss the differences between several anomaly detection techniques. First, they consider a process that they refer to as specification-based anomaly detection. In this approach, the benign network events are well defined and properly described. The ordering of network events is also important to establishing a benign condition.

A second technique discussed by Ye et al. [60], is statistical-based anomaly-detection. In such an approach, the ordering of events is not important. The model learns benign behavior from historical data and, hence reliable historical data is essential. These authors argue that including the ordering of events guarantees a reduction in false alarms. In their work, they use network data and audit-trail data to train a Markov-chain model for the purpose of detecting intrusions. The model is tested for robustness and accuracy by altering the test data and checking the percentage of false alarms. This approach was shown to be reliable throughout the conducted tests.

Similar to Ye et al. [60], the work of Feng et al. [11] as well as that of Shon and Moon [40] relies on anomaly detection. However, these authors do not use Markov-chain models. They instead rely on the well-known support vector machine (SVM) algorithm. In both of these papers, the authors considered SVM as their starting point and make several modifications to the algorithm in an effort to improve their detection results.

Shon and Moon [40] use an enhanced SVM algorithm that is applicable to both supervised and unsupervised learning. As for Feng et al. [11], their research combines SVM and clustering, based on a self-organized ant colony network (SOACN) algorithm [11]. Both models were tested for accuracy and yielded low rates of false alarms. These papers indicate that customizing standard machine learning algorithms for the specific task at hand can sometimes yield better results than simply using the baseline algorithm.

The work by Tsai et al. [51] compares different machine learning algorithms for network intrusion detection. The authors of this paper consider 55 research papers on machine learning and intrusion detection in the period between 2000 and 2007. They categorize these papers according to the machine learning algorithm used, as well as their effectiveness in detecting intrusions. On average, hybrid methods—where baseline algorithms are modified—yielded the best results, both in terms of effectiveness and popularity among researchers [51]. The work of Javaid et al. [14] applies methods of deep learning to anomaly detection.

These authors use self-taught learning (STL) on the NSL-KDD dataset, a dataset that was provided by the Canadian Institute for Cybersecurity. These researchers use both unlabeled and labeled data in their models. The first phase of their work is referred to as unsupervised feature learning (UFL), and in this phase unlabeled data is used to train a model based on a sparse auto-encoder. In the second phase, the authors consider learning based on labeled data with a softmax function used for classification.

It is worth mentioning that the applications of machine learning in intrusion detection go well beyond computer networks. That is, the same machine learning and deep learning techniques used to detect intrusions in a computer network can be applied to other networks, such as CAN bus networks in vehicles. The type of data used for training will differ, but the underlying mathematical model can remain the same. For example, Naduri and Sherry [29] propose an anomaly detection model for aircraft that relies on recurrent neural networks (RNNs). In their paper, data from X-Plane simulation software is used, along with the X-Plane Software Development Kit (XSDK). Later in this section, we will discuss other research that uses RNNs for anomaly detection. A key point here is that machine learning algorithms have a wide variety of applications and are easily adapted to different problem domains.

In the CAN bus IDS papers that we are aware of, researchers focus on a certain feature to use in training their models. That is, researches extract a specific feature based on what they believe will provide the best representation of "normal behavior." Different features tend to prove more effective against specific types of attacks.

There are several limitations that researchers face when studying CAN networks. For one, high quality data is difficult to obtain—this is a topic that we discuss in more detail in Sect. 5. Another factor is that it may be challenging to form a complete understanding of the meaning of all CAN packets. This is due to the fact that car manufacturers tend to keep such information confidential. In many cases, researchers need to sniff traffic or simulate traffic and observe the behavior of the system. This diagnostic work might include replaying a packet in question repeatedly to observe its effect on the vehicle.

Anomalous behavior is usually detected by observing the data values in CAN messages, the sequence of the messages, the IDs associated with each packet, the timing of packets, and the frequency of the packets. Below, when we discuss various CAN based IDS systems, the relevance of these various attributes should become clear.

3.2 Sequence Anomalies

In this section, we discuss the specific topic of CAN network anomaly detection research that relies on sequences of data and packet IDs. Wang et al. [54] propose a live anomaly detection system for CAN networks that is based on hierarchical temporal memory (HTM). The goal of this model is to alert the user of an attack, based on sequential anomalies. Since each data packet is associated with a particular

ID, the model processes the data section of each ID. Then, the model predicts the next data packet and the actual packet is compared to the predicted packet to generate a score. If the score is below a certain threshold, the systems identifies it as an anomaly.

To collect their dataset, Wang et al. [54] sniffed 20 h of data packets from a Subaru Impreza vehicle. The data gathered was divided into sections, with the first 70% used for training and 10% for validation. Finally, the remaining 20% of the data was split into normal data and anomalies. The anomalies consisted of altered normal data, based on the authors' expectations of anomalous CAN traffic. The authors show that anomalies can be discerned from the frequency of a specific packet. That is, a packet that appears at a rate different than what is expected can provide a strong indication of anomalous behavior. In addition, anomalies can be present in the data fields of the packets, either in the form of extra data or truncated data. Wang et al. [54] compare their HTM model to a hidden Markov model (HMM) and a recurrent neural network (RNN). Their HTM model is shown to be superior to both of these other popular models.

The work of Marchetti and Stabili [20] presents an anomaly detection algorithm that is based on the analysis of ID sequences. In their work, these authors collected more than 10 h of CAN bus data. Then they analyzed the possible ID transitions throughout the collected CAN packets and created a transition matrix. The transition matrix includes a Boolean value to indicate whether a transition is possible from ID_i to ID_j, for each possible i and j. This transition matrix can be viewed as a representation of the normal behavior of the CAN network. Thus, the authors are able to detect possible anomalies based on anomalous ID transitions.

An advantage of the system proposed by Marchetti and Stabili in [20] is that the intrusion detection algorithm can operate in either a centralized or distributed mode. The authors emphasize that their algorithm could be implemented inside any of the gateway ECUs that have full visibility of the CAN network, and they note that their algorithm can operate in a distributed mode by implementing the algorithm in one ECU in each subnetwork.

To evaluate their algorithm, Marchetti and Stabili [20] conduct two main sets of experiments. In the first test, they inject realistic CAN traffic with IDs that have different frequencies. Injected packets alter the expected frequencies. A single message injection was detected with a high rate when the injected packet was from the pool of IDs corresponding to low frequency packets. In contrast, when IDs corresponding to high frequency packets are introduced, they usually go undetected. In other words, high frequency IDs have a low rate of detection, which is intuitively clear. These results improve when the number of packets used in the attack are increased. A strength found in these experiments is that no false positives were generated.

As a second set of experiments, Marchetti and Stabili [20] conducted replay, "bad injection," and "mixed injection" attacks. In the replay attack, a set of previous CAN messages is repeated, while for bad injection, a new set of CAN messages is introduced, that is, the injected packets are new to the CAN network under consideration. Finally, the mixed injection attack consists of injecting several random messages. In all cases, these messages are crafted by the attacker without taking transitions into consideration—only the function of the messages is considered. The detection of replay attacks was sporadic and did not follow a clear trend. However, for the bad injections, the detection rate was 100%, and for the mixed injections, the detection percentage increases with the number of messages. For a one message mixed injection, detection was 40%; when this number increased to two CAN messages, the detection rate improved dramatically, reaching virtually 100%.

Malhotra et al. [19] propose an anomaly detection technique based on long short term memory (LSTM) neural networks. The datasets used in this research consist of long patterns of data of variable length. LSTM was chosen due to the long term memory capabilities it provides, which enables such a model to take advantage of long sequences of data. Four datasets were used in the experiments. The first was electrocardiogram (ECG) data, which contained only one anomaly. The second was space shuttle valve data, which contains three anomalous regions. The third was a power demand dataset, containing actual power consumption data. The fourth and final dataset was multi-sensor engine dataset. This latter dataset includes the behavior of 12 engine sensors.

Malhotra et al. [19] used stacked LSTMs in their models. They showed that no prior exposure or knowledge of pattern duration is required when such an LSTM is used. In these experiments, LSTMs yielded results that were either better than or equivalent to those obtained with standard RNNs.

Taylor et al. [48] also consider an LSTM based anomaly detection method. These authors focused on detecting anomalies in CAN data sequences. Their work demonstrated that LSTM does not necessarily need to understand the target protocol. However, they highlight that LSTM networks have some drawbacks, including the fact that LSTMs deal with each CAN message ID sequence independently. These authors conjecture that if all IDs were considered at the same time—and hence the relationships between ID sequences could be taken into account—then the model would yield significantly higher accuracies. But, such an approach would likely be computationally expensive.

3.3 Physical Anomalies

Integrating the physical environment into an IDS can yield better detection of security breaches. For example, Wasicek et al. [57] proposes a proof of concept anomaly based IDS that they call context aware intrusion detection (CAID). In this paper, the authors build a model that can detect alterations to the physical systems. They rely on sensors to detect changes in the physical medium. The data they measure includes speed, rpm, fuel rate, pedal position, temperature, and fuel-to-air ratio. For anomaly detection, the CAID system relies on an artificial neural network (ANN).

The CAID framework consists of three main modules. The first modules are monitors that are used to collect raw data from the vehicle network. The second modules are known as detectors, which are responsible for analysis. The third and final modules are reporters, which, not surprisingly, communicate the detector result to the user. The CAID framework was tested on a 2015 vehicle—to test the model, a modified chip (used for vehicle tuning) with enhanced parameters was introduced to the vehicle. The CAID framework in [57] successfully detected the deviations in behavior.

3.4 Time Windows

Malicious activity on CAN networks would typically impact the timing of packets, as well as the frequency with which certain packets appear. As a result, several papers focus on detecting anomalies in packet timing. In this section, we discuss some examples of this type of research.

Taylor et al. [47] proposed to detect CAN attacks based on packet timing. The authors claim that packets usually arrive at a certain frequency and timing, and thus they consider an anomaly detection system based on historical timing behavior. In their algorithm, the authors measured inter-packet timings over a sliding window. An anomalous behavior is detected whenever a sufficient deviation from historical behavior occurs. Note that the time sequence of CAN messages is essential for this analysis.

To collect their data, Taylor et al. [47] logged the CAN packets of a 2011 Ford Explorer. They made five trips, each lasting five minutes. During those trips the driver did not operate any user controls and they maintained a low speed and came to a complete stop. The first three trips were used in training and the last two trips were used for attack simulation. Attack simulation was conducted by inserting new packets at different timings. The result showed that inter-packet timing yielded strong detection results.

Tomlinson et al. [49] also investigated anomalies in time windows and proposed an IDS that detects intrusions based on deviations in these timings. They utilized three statistical methods, namely, autoregressive integrated moving average (ARIMA), the well known Z score, and a supervised threshold. In their work, they consider non-overlapping windows. They preprocess the data in each window to reduce the necessity of recalculating their various metrics. Each metric was used to classify all broadcasts within the same window.

For data collection, Tomlinson et al. [49] logged 127 min of driving data from an unspecified target vehicle. Then they analyzed the frequency of broadcast packets. They found that packets with a higher priority and low IDs (priority and ID numbers are inversely related) had the least variation in timing, and these were also broadcast with the highest frequency. The analysis of the data showed that the majority of ECUs would broadcast at a consistent rate of at least 100 times per second, while other ECUs broadcast at least 10 times per second. After this analysis was complete, the authors simulated several attacks on the CAN network. To do so, they altered the sniffed CAN packet data to create two simulated malicious datasets. In the first set, they dropped several packets from a normal broadcast, while for the second set, they injected additional packets to existing broadcasts. For testing, the authors applied their three detection methods on a sample consisting of the five highest priority IDs.

They compared the broadcast interval against the mean of the normal window, which acted as a supervised threshold against which the Z score and ARIMA results were compared. Overall, the supervised threshold attained the best results, followed by the Z score, then ARIMA.

3.5 Entropy Based Anomalies

According to Marchetti et al. [21], entropy based anomaly detection models can be effective only against cyber attacks that cause a high rate of chaos or randomness. In other words, when the rate of malicious activity increases, entropy based IDS is more likely to be effective. These authors show that entropy based approaches enable the identification of malicious behavior without the necessity of disclosing manufacturer proprietary material related to the meaning of various CAN messages. From the manufacturer's point of view, this could be seen as a significant advantage.

The work of Müter and Asaj [28] also considers an entropy based IDS for CAN networks. These authors measure entropy in the context of coincidence in a given dataset—entropy and the proposed coincidence measure are directly proportional. Müter and Asaj make the point that in CAN networks, there is a low rate of coincidence between packets. Thus, when an attack occurs, their measure of coincidence (i.e., entropy) should increase.

To simulate attacks, Müter and Adaj [28] follow three approaches. In the first case, they slightly increase the frequency of a certain message, and in the second, they flood the network with a specific message. Finally, in their third approach, they consider the "plausibility of interrelated events," where the goal is to understand the correlation between certain events. The specific example they mention is that of a driver in the city who would reach a speed of 60 km/h then stop completely due to a stop sign. This higher level of understanding enables their model to detect sporadic single message injection.

3.6 Signature Detection

Of course, signature or pattern matching can be used to detect attacks. Each attack type, or related sequence of instructions, has a specific pattern and patterns collected from several known attacks can used to train a model. After training, such a model essentially acts as a meta-signature that is able to detect any of the patterns on which it was trained, and possibly other similar attacks. A disadvantage of such an approach is that it requires constant updating to the model as new attack patterns become available. Another disadvantage is that such a model is unlikely to be effective in defending against zero-day vulnerabilities [60].

3.7 *Language Theory Based Detection*

Studina et al. [46] proposed an intrusion detection method that relies on formal language theory. They derive attack signatures from different ECUs. Their method then detects malicious message sequences based on attack signatures, which depend on the fact that ECUs operate with consistent rules. Thus, the authors are able to leverage predictable ECU behavior to generate a language that characterizes certain types of attacks.

4 Masquerade Detection

The masquerade detection problem has been extensively studied in the literature. Specifically, masquerade detection based on UNIX commands has received considerable attention. The seminal work in this field is the Schonlau, et al., paper [13], which was published in 2001. There continues to be considerable interest in the topic, as evidenced by recent papers such as [7, 9, 12, 16, 18, 35, 53, 58, 59].

The authors are unaware of any masquerade detection research that has been specifically applied to CAN networks. Hence, the brief survey of masquerade detection that we provide in this section does not cite research that is directly related to vehicle networks. However, we believe that masquerade detection is highly relevant to the field of CAN networks, and in our experiments section, we return to this topic.

The survey article [4] cites approximately 40 relevant papers published prior to 2009, most of which use the Schonlau dataset. In [4], the authors identify the following general approaches to masquerade detection: information-theoretic, text mining, hidden Markov model (HMM), Naïve Bayes, sequences and bioinformatics, support vector machine (SVM), and other approaches

In the remainder of this section, we summarize some of the relevant work in each of the categories listed above, and we discuss a few examples of recent work.

4.1 *Information-Theoretic*

The original work by Schonlau et al. in [13] included analysis of a compression technique, based on the fact that commands issued by the same user tend to compress more than those involving an intruder. By the standards of subsequent work, the results are not particularly strong. More recently, related techniques have been pursued in [2, 3, 10], but the results have not improved dramatically.

4.2 Text Mining

In [16], a data mining approach is used to extract repetitive sequences of commands from training data. These sequences are then used for scoring. Other data mining approaches have been studied, including principle component analysis (PCA); for example, in [32], good results are obtained using PCA, although the computational cost is relatively high during training. Another example of a data mining technique being applied to this problem can be found in [6].

4.3 Hidden Markov Model

Hidden Markov model (HMM) techniques are considered in [33] and [34], for example. To date, HMMs have achieved some of the best detection results, and HMMs are often used as a baseline for measuring the effectiveness of proposed techniques.

4.4 Naïve Bayes

A Naïve Bayes classifier can be viewed as a static form of an HMM, in the sense that Naïve Bayes relies on frequencies, without using sequential information. Such an approach is applied to the masquerade detection problem in [22] and [23]. Although simple, Naïve Bayes often performs surprisingly well. Additional relevant work can be found in [61, 62], for example.

4.5 Sequences and Bioinformatics

Sequence-based and bioinformatics-like approaches are, in some sense, at the opposite extreme of Naïve Bayes. Recall that Naïve Bayes does not account for sequential information, while bioinformatics is focused on extracting sequence-related information.

In the Schonlau, et al., paper [13], a sequence-based analysis is considered. However, the only previous work on masquerade detection involving standard bioinformatics techniques appears to be [7], where the authors use the Smith-Waterman algorithm [52] to create local alignments of sequences. This alignment technique is analogous to a profile hidden Markov model (PHMM), as discussed, for example, in [45]. However, in [7], the resulting alignments are used directly for classification, whereas in a standard PHMM, we use these alignments to generate a model, which is then used for classification. Consequently, a PHMM based detection algorithm is considerably more efficient, while the training is no more costly.

4.6 Support Vector Machine

Support vector machines (SVM) are a class of machine learning algorithms that separate data points using a hyperplanes. The points in the original input space are typically mapped to a higher dimensional feature space, where separation is likely to be much easier. SVMs maximize the margin (i.e., the minimum separation between the sets of points), while keeping the computational cost low [8].

For example, in [55], an SVM-based masquerade detection system achieves results comparable to Naïve Bayes. Additional masquerade detection work involving SVMs can be found in [5, 15, 18], where the focus is primarily on improved efficiency, as compared to [55].

4.7 Other Approaches

Several other approaches that do not easily fit into any of the categories above have been considered. However, most of these other approaches have produced relatively poor results. For example, in [39] low frequency (i.e., not commonly used) commands form the basis for detection. In contrast, the paper [53] shows that relying on high frequency commands can yield comparable results.

Among other non-standard techniques, a "hybrid Bayes one step Markov" approach and a "hybrid multistep Markov" method (i.e., a Markov process of order greater than one) are considered in the paper [13]. Neither of these achieve particularly impressive results.

A non-negative matrix factorization (NMF) technique is developed and analyzed in [56]. These NMF results are improved upon in [24], where this approach is shown to achieve reasonable detection results

4.8 Discussion

The masquerade detection research discussed in this section highlights some important points that are relevant to IDS in CAN networks. First, the topic of masquerade detection seems particularly relevant to CAN networks. That is, an attacker that is aware that an IDS is in use, will likely try to masquerade as a normal user. Thus, a high degree of sensitivity will likely be needed to detect such attacks. Another important point to glean from the discussion above is that a standard dataset in invaluable in such research. The Schonlau dataset is far from perfect, but it has enabled researchers to directly compare their results, and hence the problem of masquerade detection based on UNIX commands has been thoroughly analyzed. A widely available standard dataset for masquerade detection on CAN networks would be an invaluable asset and would help focus research in this area.

5 Data Collection

Regardless of the IDS technique under consideration, researchers need access to data. In the case of CAN networks, according to Rajbahadur et al. [36] most researchers use simulated datasets, as opposed to data collected for real vehicles. Rajbahadur et al. studied 65 papers dealing with intrusion in vehicle networks and discovered that only 19 of these papers used real datasets. Here, we briefly discuss data related issues. In descending order from the most expensive to the least expensive, we consider the following three methods of obtaining data: real vehicles, ECU testbeds, and simulations.

5.1 Real Vehicles

The most expensive method for obtaining CAN data is using an actual vehicle. Note that a specific make and model would likely be needed to ensure consistency and so that the results could be easily reproduced. The advantage of such data is that it provides access to all in vehicle systems, including infotainment, air conditioning, GPS, door locks, etc. However, due to the cost, this option is not feasible for most research.

5.2 ECU Testbeds

Another option for data collection is an ECU testbed that includes actual vehicle components. Smith [42] explains how ECUs can be extracted from a vehicle and connected together to enable this type of data collection, and for automotive research in general. ECU benches could be relatively simple, including only a single ECU, or they could be very complex, including most of the components found in an actual vehicle. Such components would include the body control module, an engine control module, an instrument cluster, and so on.

Miller and Valasek [27] explain how ECUs can be connected together to build a test bench. They also show how data can be sniffed from an ECU and provide details needed to wire test bench components together.

Another option is the portable automotive security testbed with adaptability (PASTA), as developed by Toyota [50]. PASTA is the size of a suitcase, and significantly reduces the barrier to entry for researchers. This testbed contains several ECUs and displays simulated vehicle behavior on a screen. Thus, the data is generated from actual ECUs, and the result of the generated data is displayed to the user through simulated vehicle behavior.

5.3 Simulation

The least expensive option is to simulate the data in its entirety. Several tools are available to simulate CAN traffic. One of the best known simulation tools is ICSim [63], which is a vehicle simulator that runs on Ubuntu. ICSim allows users to generate CAN traffic and operate a virtual vehicle. It also includes sniffing, replay, and data injection capabilities.

6 Experimental Results

In this section, we discuss a variety of experiments that we have conducted that are relevant to the problem of IDS on CAN networks. We consider two main sets of experiments. The first set deals with classification, where our goal is to identify the behavior of a vehicle based on CAN packets. In this set of experiments, we rely on both real and simulated data. In our second set of experiments, our goal is to detect masquerade behavior. These experiments use only simulated data. Throughout all of the experiments, we employ a variety of machine learning techniques.

6.1 Datasets

For the classification experiments, we consider two sources of data. We use a dataset collected from a 2010 Ford Escape [26], and we also consider a simulated dataset that was generated using ICSim [63]. As mentioned in Sect. 5.3, ICSim enables CAN simulation, sniffing and injecting messages. The ICSim user interface is displayed in Fig. 1. In the classification experiments, we refer to the Ford Escape data as our real dataset, while the ICSim data is our simulated dataset.

Each CAN packet consists of eight bytes, where these byte can range over all possible values, that is, from `0x00` to `0xFF`. The real dataset contains CAN messages representing three different states, namely, idle, drive, and park. In contrast, the simulated dataset only contains CAN packets representing the idle and drive states.

For our masquerade detection experiments, we generated two simulated datasets. The first of these two simulated datasets represents the behavior of a specific user and

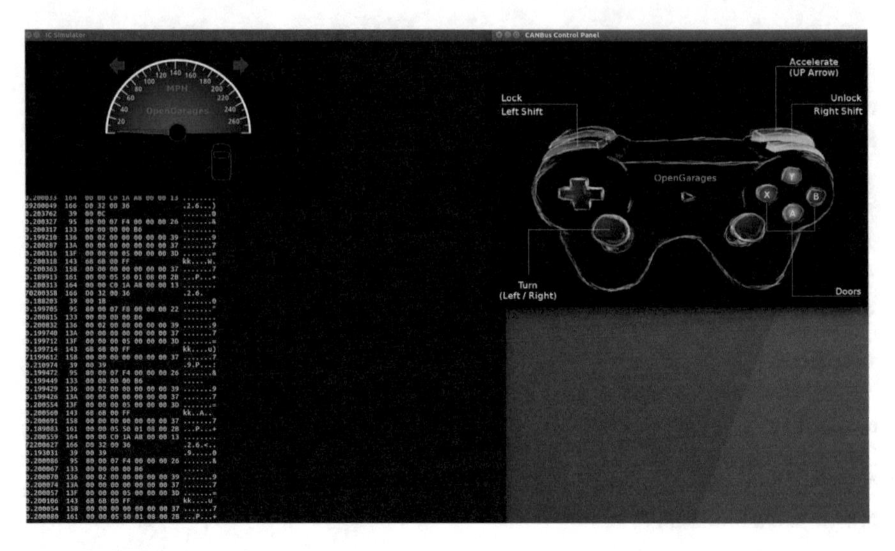

Fig. 1 ICSim display

Table 1 Datasets

Dataset	Source	Data	Experiments
1	Real	Idle/drive/park	Classification
2	Simulated	Idle/drive	Classification
3	Simulated	11 users	Masquerade detection
4	Simulated	7 actions	Masquerade detection

that of another masquerading users. One trip is the authenticated user, and another 10 trips are various masquerading users. Each trip consists of several actions. The 10 masquerading trips include five with large deviations from the authenticated trip as well as another five with small deviations. Our second masquerade dataset consists of seven simulations. Each of these simulations exclusively contains the packets related one specific action that the simulator allows. These seven actions are the following.

- Drive 20 mph
- Drive 40 mph
- Drive 60 mph
- Left turn
- Right turn
- Driver door open and close
- Right passenger door open and close.

A summary of our four datasets is provided in Table 1. Again, the first two of these datasets are used in our classification experiments, while the second two are used in our masquerade detection experiments.

6.2 *Feature Extraction*

For our initial experiment, each CAN message is converted to its decimal equivalent. We then use these decimal numbers as observations and train models based on sequences of numbers.

We also apply a word embedding technique, Word2Vec, to CAN packets (as numbers) and compare results obtained with and without the Word2Vec conversion.

Word2Vec is based on a shallow 2-layer neural network and is commonly used to find the context of words in natural language processing (NLP) [1]. In effect, Word2Vec groups common words together in a form that is suitable as input to other machine learning techniques. For our Word2Vec embedding, we consider CAN messages of length five, based on overlapping sliding windows. We train the Word2Vec model on these "words," with the output vectors serving as a feature set in some of the experiments discussed below.

6.3 *Classification Experiments*

As mentioned above, we apply various machine learning models to the CAN packets and also experiment with Word2Vec features. Specifically, we consider the following machine learning techniques: k-nearest neighbor (k-NN), hidden Markov models (HMM), a long short-term memory (LSTM) model, a deep neural network (DNN), support vector machines (SVM), and Naïve Bayes. In the classification experiments reported here, we trained on datasets 1 and 2. The goal is to identify the status of a vehicle from its network packets, without relying on any other information, such as a visual frame of reference.

6.3.1 k-NN Experiments

We first consider k-NN, which we apply to both the numerical CAN packets and the Word2Vec features. The real data and simulated data are treated as separate experiments. In these experiments, we measure how well we can distinguish "idle" CAN packets from "drive" packets.

Our k-NN results without Word2Vec are summarized in Fig. 2, while results for k-NN experiments based on the Word2Vec features are given in Fig. 3. From these results, we see that the Word2Vec features are far more informative, yielding much higher accuracies.

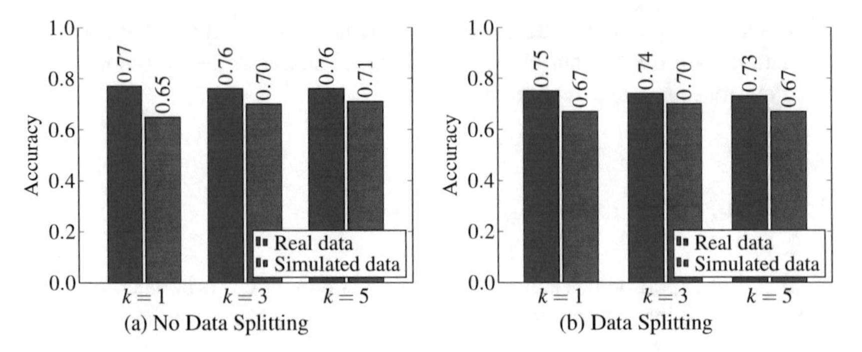

Fig. 2 k-NN results (CAN packets)

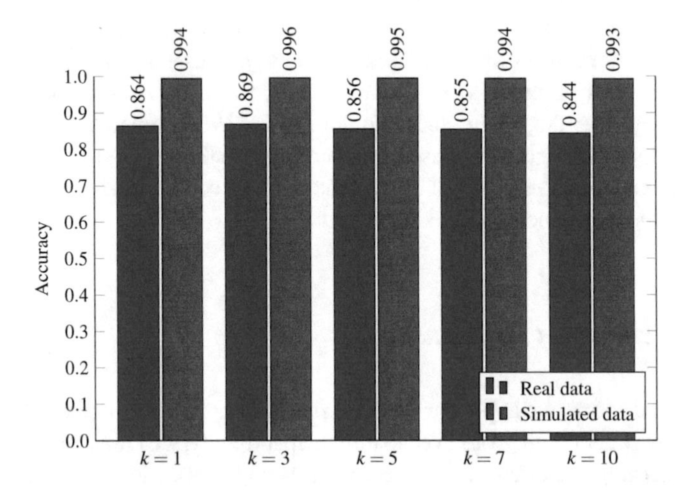

Fig. 3 k-NN with Word2Vec

6.3.2 HMM Experiments

A hidden Markov model (HMM) includes an underlying Markov process that is "hidden" in the sense that it is not directly observable. But, we do have access to an observation sequence that is probabilistically related to the hidden Markov process. In the standard terminology, as found in [43], for example, the A matrix drives the underlying (hidden) Markov process, while the B matrix relates the observations to the hidden states, and the π matrix contains the initial state distribution. All three of these matrices are row stochastic.

Note that for the experiments discussed in this section, we are operating in a data exploration mode. That is, we are training HMMs and we will then examine the models to see what they tell us about the data.

For our first HMM experiments, we train models treating the available CAN messages as observations, where we considered each byte as one observation.

Fig. 4 Converged B matrix (pairs of characters as observations)

121	0.00040	0.00017	0.00000
122	0.00000	0.00195	0.00000
123	0.00038	0.00181	0.00000
124	0.00000	0.00106	0.00000
125	0.00000	0.00408	0.00000
126	0.00000	0.01561	0.00000
127	0.00000	0.03051	0.00000
128	0.00000	0.01330	0.00000
129	0.00000	0.00798	1.00000
130	0.00000	0.00195	0.00000
131	0.00000	0.00248	0.00000
132	0.00063	0.00066	0.00000
133	0.00015	0.00079	0.00000
134	0.00000	0.00053	0.00000
135	0.00000	0.00302	0.00000
136	0.00000	0.00142	0.00000

Following the notation in [43], we have $M = 256$ distinct observations, and we train for 500 iterations of the Baum-Welch re-estimation algorithm. In addition, we have $T = 16800$ observations, and we consider a model with $N = 3$ hidden states. The idea here is that these three hidden states should correspond to idle, drive, and park.

A snippet of the final converged B matrix (which relates the hidden states to the observations) is given in Fig. 4. It is not immediately clear which column of this B matrix corresponds to which hidden state (idle, drive, park).

From Fig. 4, we see that observation 129 (hex representation of `0x81`), has probability 1.0 in the third state, while all other probabilities for the third state are, of course, zero. This signifies that a message with hex value of `0x81` is in the third state. However, we still do not know what this state actually represents.

To take this further, we trained another HMM with entire messages as observations. In this case, $M = 2830$, $T = 16800$, and we again choose $N = 3$ hidden states. A snippet of B matrix for this model is displayed in Fig. 5.

Similar results were observed for the third state in this case, where we find that an ID of 0 has a probability of 1. During the preprocessing of the data, each of the $T = 2830$ messages was assigned a unique ID. The ID 0 was assigned to the specific message `81 08 80 00 00 00 00 00`. This packet was from the autopark file, and hence the HMM results show that the model was able to correctly identify the packet that indicates when the car is in the park state. This was also in line with the results obtained from the first HMM model since we have `0x81` in the data section of this packet, and `0x81` does not appear elsewhere in the data. Again, this packet was only found in the autopark file and was absent from both the drive and idle files. Again, the HMM has associated the third state with the "park" state. This illustrates the strength of an HMM (and machine learning in general) for this data analysis problem.

```
final B^T =
0  0.00000  0.00151  1.00000
1  0.28850  0.09213  0.00000
2  0.00000  0.09400  0.00000
3  0.00000  0.00013  0.00000
4  0.04162  0.00000  0.00000
5  0.00000  0.00691  0.00000
6  0.00011  0.00000  0.00000
7  0.00000  0.00025  0.00000
8  0.00023  0.00000  0.00000
9  0.00034  0.00000  0.00000
10  0.00000  0.00025  0.00000
11  0.01410  0.02455  0.00000
12  0.02402  0.01201  0.00000
13  0.00011  0.00000  0.00000
14  0.00011  0.00025  0.00000
15  0.00011  0.00000  0.00000
16  0.00011  0.00000  0.00000
17  0.00000  0.04788  0.00000
18  0.01402  0.00000  0.00000
```

Fig. 5 Converged B matrix (packets as observations)

The A matrix obtained when the HMM was trained on CAN messages is displayed in Fig. 6. This matrix gives the transition probabilities between hidden states.

Based on the results discussed above, we observe that the third state is the park state. Further analysis shows that the first state is the idle state while the second is the drive state. As we can see from the A matrix in Fig. 6, from the park state, there is only one possible transition and that is to the idle state. This is entirely consistent with the way that a car is actually driven—a car cannot move directly from the park state to the drive state, as it must first pass through the idle state. Another key observation is

```
2827  0.00000  0.00000  0.00000
2828  0.00000  0.00000  0.00000
sum[0] = 1.000000 sum[1] = 1.000000 sum[2] = 1.000000
completed iteration = 499, log [P(observation | lambda)] = -81137.504826

T = 16800, N = 3, M = 2829, iterations = 500

final pi =
 0.00000  0.00000  1.00000 ,  sum = 1.000000

final A =
 0.10455  0.89545  0.00000 , sum = 1.000000
 0.99503  0.00497 -0.00000 , sum = 1.000000
 1.00000  0.00000  0.00000 , sum = 1.000000
```

Fig. 6 Converged A and π matrices

the fact that from the first two rows of the A matrix, we can see that there are frequent transitions from the idle to the drive state and vice versa. These results generated by our HMM nicely illustrate the "learning" aspect of machine learning models, since we never explicitly told the model anything about the states or about driving, yet the model was able to discern this information, which is completely consistent with real-life driving situations.

6.3.3 LSTM Experiments

Long short-term memory (LSTM) model is a type of recurrent neural network (RNN) that can be used to predict new information based the previous known information. Unlike other types of RNNs, LSTMs not only take into account recent past information, but also considers a much larger context to predict new information. The intuition behind using LSTMs in the CAN network context is that they are known to work well with time series and sequence data. We experimented with different number of LSTM layers and found that the best results were obtained with five LSTM layers, in which case we were able to obtain an accuracy of 100% for the problem of distinguishing idle and drive CAN packets.

6.3.4 DNN and SVM Experiments

Deep neural networks (DNN) are a type of complex artificial neural network (ANN) that includes multiple hidden layers between the input and output layer. We experimented with a DNN with two hidden layers, where each layer contains 128 neurons. We trained a DNN model on word embeddings based on various numbers of CAN messages. We found that models based on five CAN messages gave us the best results, yielding an accuracy of 99.46%. Again, these results are for the problem of distinguishing between idle and drive CAN packets.

Support vector machines (SVM) were also applied to the data. For our SVM experiments, we found that the linear kernel yielded the best results and hence we use this kernel function for all results reported here. Using SVMs, we obtained good results on both the real and simulated datasets. Our DNN and SVM results are summarized in Fig. 7.

6.4 Masquerade Detection Experiments

In this section we discuss the results of various basic masquerade detection experiments that we have performed. These experiments are based on datasets 3 and 4 and deal with different simulated users. The goal is to distinguish a (simulated) masquerader from a (simulated) authenticated user. Due to the superior results obtained above using Word2Vec, we employ the Word2Vec conversion in all experiments reported

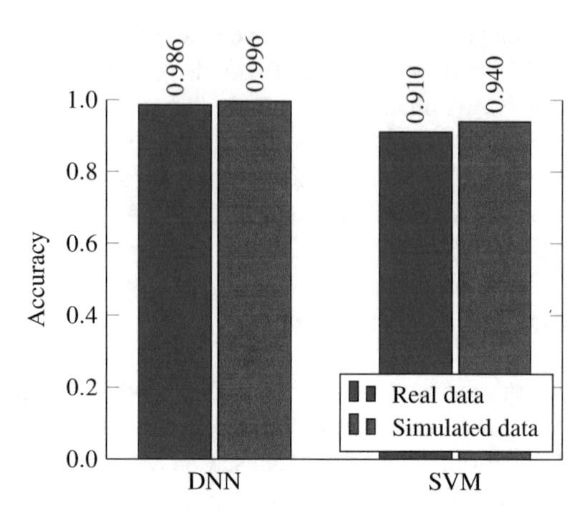

Fig. 7 DNN and SVM with Word2Vec

in this section. We experiment with three learning techniques, namely, *k*-NN, DNN, and Naïve Bayes.

6.5 k-*NN and DNN for Speed Detection*

For each simulated user, we generate two files, each at a different speed. For one file, the speed ranges from 20 to 40 mph and for the other, the speed ranges from 40 to 60. Then, the data from each file is converted with Word2Vec and labeled. Next, we apply the two machine learning methods under consideration, attaining high accuracy in both cases. These *k*-NN and DNN results are summarized in Fig. 8.

The use case here is to detect changed behavior of a user. For example, if a vehicle owner maintains a typical speed and that speed suddenly increases, this could be an indication of a theft.

6.5.1 *k*-NN for User and State Detection

In this section, we give the results of *k*-NN experiments based on the 11 simulated trips and the 7 actions found in datasets 3 and 4, respectively. First, we test a *k*-NN model on the simulated CAN traffic for the 11 trips, which we view as representing an authenticated user and 10 other trips. We split this data, with 70% used for training and 30% for testing. The results for this experiment are summarized in Fig. 9. As expected, these results are not as strong as the speed detection results in Fig. 8, but the results are still very good.

We also apply *k*-NN to differentiate between the 7 actions in dataset 4. The purpose of this experiment is to determine how well we can differentiate between specific actions, without assuming any specific knowledge of the details of the CAN data. In this case, the data was again split into 70% for training and 30% for testing. These

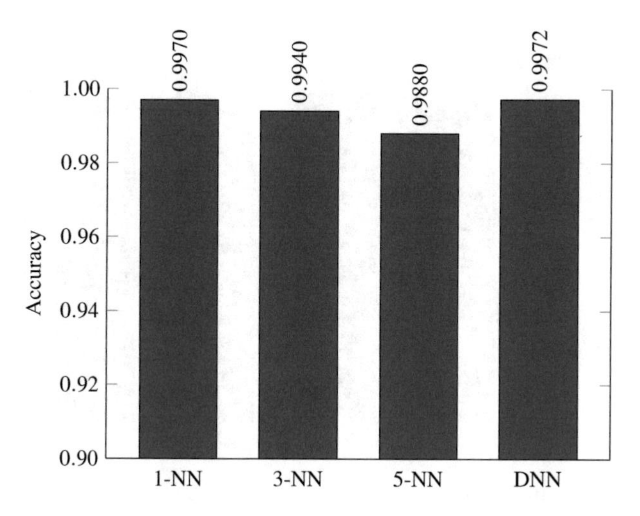

Fig. 8 *k*-NN and DNN for speed detection

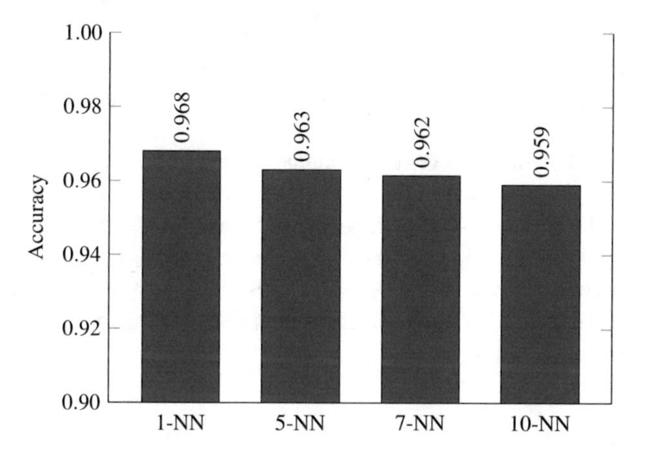

Fig. 9 *k*-NN user detection results

experimental results are given in Fig. 10. Note that we are able to distinguish between these actions with nearly 99% accuracy.

6.5.2 Naïve Bayes Experiments

In this section, we consider two sets of experiments based on Naïve Bayes. For our first set of experiments, we use Naïve Bayes to model the simulated CAN traffic for the 7 actions in dataset 4 and, independently, the 11 trips in dataset 3. In both of these cases, the data is split with 70% used for training and 30% reserved for testing. For

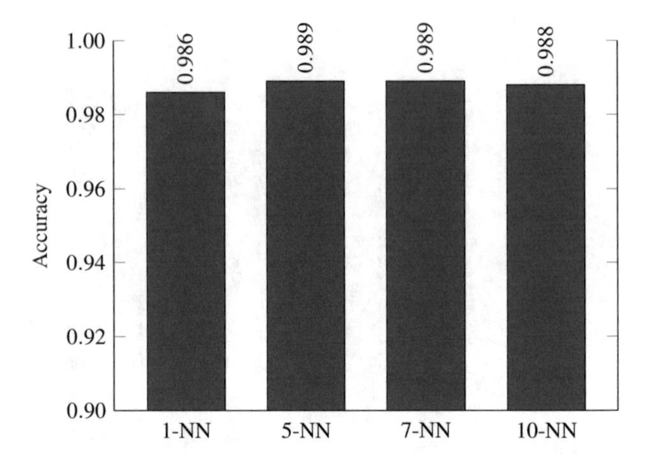

Fig. 10 *k*-NN state detection results

the 11 trips in dataset 3, we again view this data as an authenticated user plus 10 other trips. The results for these Naïve Bayes experiments are summarized in Fig. 11. In this graph, the "User" label represents the accuracies attained when we trained for the 11 trips in dataset 3, while the "States" label represents the accuracies for the 7 actions in dataset 4.

Our second set of Naïve Bayes experiments are more complex. We first use Naïve Bayes to model the simulated data for the 7 actions in dataset 4. Then we use the resulting model to predict what each packet represents for each of the 11 user trips in dataset 3. In this case, we obtain the accuracies in Fig. 12, where the bar for "User i" is the accuracy obtained for all actions corresponding to the ith user (or trip) in dataset 3. We see that Naïve Bayes fails to accurately predict the possible actions at the packet level.

Fig. 11 Naïve Bayes to distinguish users and states

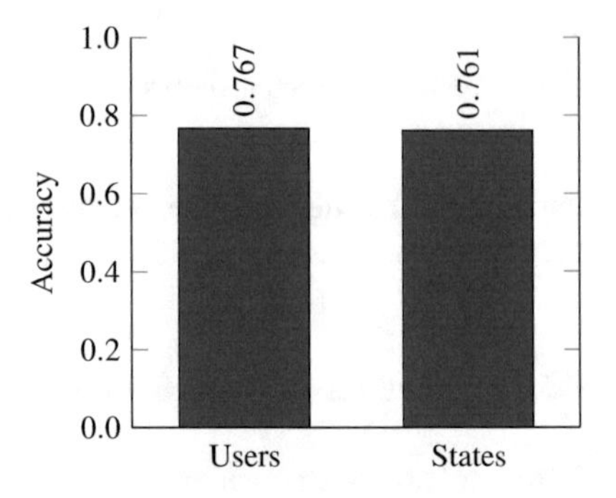

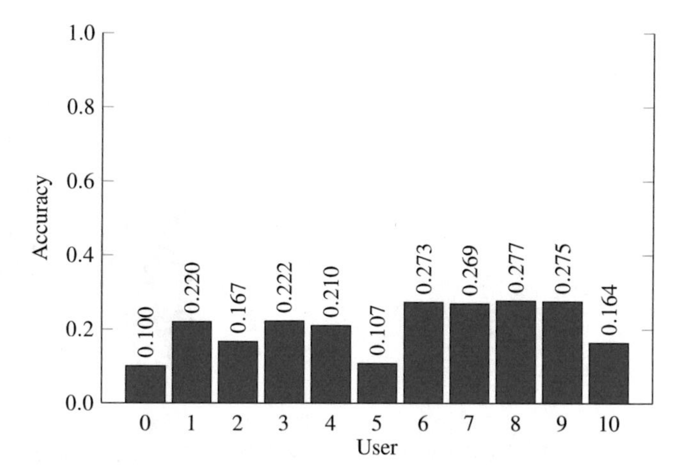

Fig. 12 Naïve Bayes trained on seven actions

6.6 Summary

The graph in Fig. 13 summarizes the accuracy of the various machine learning models considered for the problem of distinguishing the idle state from the drive state. From these results, we see that LSTM is the winner, giving perfect separation on both the simulated and real datasets. Our DNN model also performed well on both datasets. While k-NN does very well on the simulated dataset, it does poorly on the real data. Overall, it is clear that neural network based techniques have an advantage for this problem.

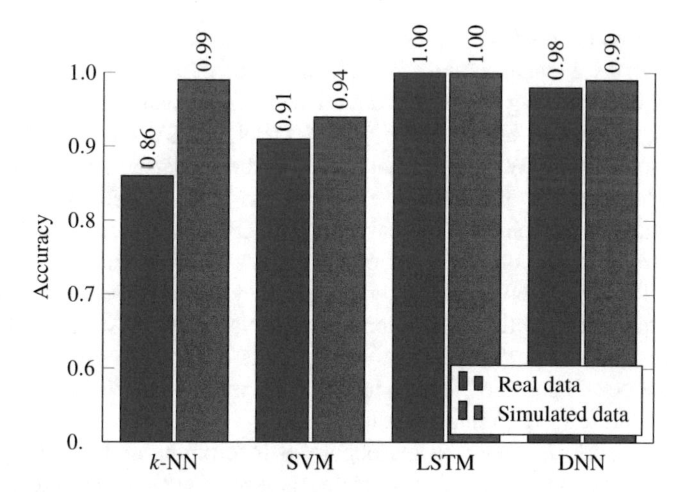

Fig. 13 Summary of CAN traffic analysis results

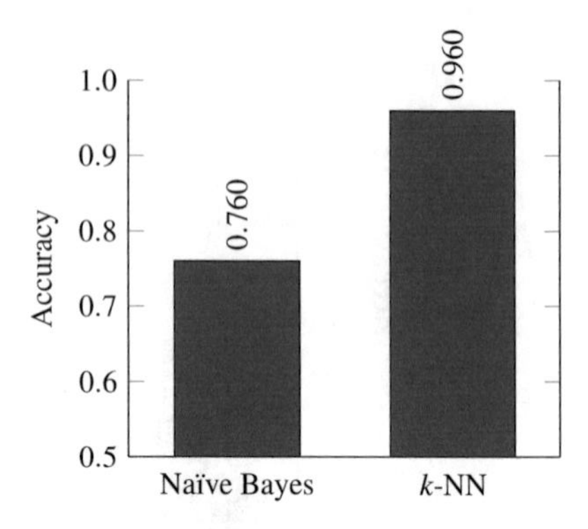

Fig. 14 Naïve Bayes versus *k*-NN for masquerade detection

In Fig. 14, we summarize the accuracies of both the Naïve Bayes and *k*-NN models for the masquerade detection problem considered here. Note that we have included the accuracies of the "user" detection experiments for Naïve Bayes, since the "state" detection results are nearly identical. We see that *k*-NN provides far better results than Naïve Bayes in this case.

7 Conclusion and Future Work

In this paper, we considered the problem of intrusion detection in CAN networks. We provided a selective survey of research in the field, organized around several major approaches. We also discussed masquerade detection research and suggested that this will likely be a useful path to follow for future research in CAN network security.

We also provided results for various sets of machine learning based experiments. We showed that we can accurately identify the status of a vehicle from its network packets without assuming any previous knowledge of the nature of the CAN traffic. Specifically, we applied *k*-NN, LSTM, HMM, DNN, and SVM to two datasets—on simulated, and one real. We showed that trained HMM models can be used to identify important characteristic of CAN packets. From the other models, we were able to accurately classify the vehicle status based only on CAN packets. Most of the models produced strong results on at least one of the datasets, with neural network based models (LSTM and DNN) giving us perfect accuracy (or nearly so) on both datasets.

Our masquerade detection experiments using *k*-NN produced strong results, while Naïve Bayes did not perform nearly as well. These particular experiments were limited in scope and designed to introduce the concepts and to illustrate some of the potential for future work in this domain.

For future work, it would be reasonable to incorporate additional data in the masquerade detection models. For example, the use of GPS location data together with CAN traffic data would seem to offer the potential for much stronger models.

A holistic masquerade detection system would rely on several steps. For example, we might initially authenticate a user based on a biometric feature—or a mix of several such features. Then by monitoring CAN traffic and other data, such as GPS location, we could detect significant deviations from the typical behavior of the authenticated user. When such deviations are detected, we suspect a masquerader, and in such cases we could require the user to re-authenticate.

References

1. A beginner's guide to Word2Vec and neural word embeddings. https://skymind.ai/wiki/word2vec, 2018
2. Bertacchini M, Benitez C (2007) NCD based masquerader detection using enriched command lines. In: Proceedings of the IV Congreso Iberoamericano de Seguridad Informatica, CIBSI 07, pp 329–338
3. Bertacchini M, Fierens PI (2007) Preliminary results on masquerader detection using compression based similarity metrics. Electron J SADIO 7(1):31–42
4. Bertacchini M, Fierens PI (2009) A survey on masquerader detection approaches. In: Proceedings of the IV Congreso Iberoamericano de Seguridad Informatica, CIBSI 09, 2009. http://www.criptored.upm.es/cibsi/cibsi2009/docs/Papers/CIBSI-Dia2-Sesion5(2).pdf
5. Chen L, Aritsugi M (2006) An SVM-based masquerade detection method with online update using co-occurrence matrix. In: Proceedings of the third international conference on detection of intrusions and malware & vulnerability assessment, DIMVA'06. Springer, Berlin, Heidelberg, pp 37–53
6. Chen L, Dong G (2006) Masquerader detection using OCLEP: one-class classification using length statistics of emerging patterns. In: Proceedings of the seventh international conference on web-age information management workshops, WAIM '06, p 5
7. Coull SE, Szymanski BK (2008) Sequence alignment for masquerade detection. Comput Stat Data Anal 52(8):4116–4131. https://www.cs.rpi.edu/~szymansk/theses/coull.ms.05.pdf
8. Cristianini N, Shawe-Taylor J (2000) An introduction to support vector machines: and other kernel-based learning methods. Cambridge University Press, New York, NY, USA
9. Dash SK, Reddy KS, Pujari AK (2011) Adaptive naïve Bayes method for masquerade detection. Secur Commun Netw 4(4):410–417
10. Evans S, Eiland E, Markham S, Impson J, Laczo A (2007) MDL compress for intrusion detection: signature inference and masquerade attack. In: Proceedings of 2007 IEEE military communications conference, MILCOM 2007, pp 1–7
11. Feng W, Zhang Q, Hu G, Huang X (2014) Mining network data for intrusion detection through combining SVMs with ant colony networks. Futur Gener Comput Syst 37:127–140
12. Geng D, Odaka T, Kuroiwa J, Ogura H (22011) An n-gram and STF-IDF model for masquerade detection in a UNIX environment. J Comput Virol 7(2):133–142. https://link.springer.com/article/10.1007/s11416-010-0143-3
13. Jakobsen T (1995) A fast method for the cryptanalysis of substitution ciphers. Cryptologia 19:265–274
14. Javaid AY, Niyaz Q, Sun W, Alam M (2015) A deep learning approach for network intrusion detection system. ICST Trans Secur Saf 3:1–6
15. Kim H-S, Cha S-D (2005) Empirical evaluation of SVM-based masquerade detection using unix commands. Comput Secur 24(2):160–168
16. Latendresse M (2005) Masquerade detection via customized grammars. In: Proceedings of second international conference on intrusion and malware detection and vulnerability assessment, DIMVA 2005, Vienna, Austria, July 2005. Springer

17. Lee H, Choi K, Chung K, Kim J, Yim K (2015) Fuzzing can packets into automobiles. In: International conference on advanced information networking and applications workshops, AINAW 2015, pp 817–821
18. Li Z, Li Z, Liu B (2006) Masquerade detection system based on correlation Eigen matrix and support vector machine. In: Proceedings 2006 international conference on computational intelligence and security, ICCIAS 2006, pp 625–628
19. Malhotra P, Vig L, Shroff G, Agarwal P (2015) Long short term memory networks for anomaly detection in time series. In: Proceedings of 23rd European symposium on artificial neural networks, ESANN 2015
20. Marchetti M, Stabili D (2017) Anomaly detection of CAN bus messages through analysis of ID sequences. In: IEEE intelligent vehicles symposium, IV 2017, pp 1577–1583
21. Marchetti M, Stabili D, Guido A, Colajanni M (2016) Evaluation of anomaly detection for in-vehicle networks through information-theoretic algorithms. In: 2nd international forum on research and technologies for society and industry, RTSI 2016. IEEE, pp 1–6
22. Maxion RA, Townsend TN (2002) Masquerade detection using truncated command lines. In: Proceedings of the 2002 international conference on dependable systems and networks, DSN 2002. IEEE Computer Society, pp 219–228
23. Maxion RA, Townsend TN (2004) Masquerade detection augmented with error analysis. IEEE Trans Reliab 53(1):124–147
24. Mex-Perera C, Posadas R, Nolazco JA, Monroy R, Soberanes A, Trejo LA (2006) An improved non-negative matrix factorization method for masquerade detection. In: Proceedings of the 1st Mexican international conference on informatics security, MCIS 2006
25. Miller C, Valasek C (2015) Remote exploitation of an unaltered passenger vehicle. https://www.ioactive.com/pdfs/IOActive_Remote_Car_Hacking.pdf. Presented at DEFCON 23, Las Vegas, Nevada, August 6–9
26. Miller C, Valasek C (2018) Car hacking. http://illmatics.com/carhacking.html
27. Miller C, Valasek C (2018) Car hacking: for poories. http://illmatics.com/car_hacking_poories.pdf
28. Müter M, Asaj N (2011) Entropy-based anomaly detection for in-vehicle networks. In: IEEE intelligent vehicles symposium, IV 2011, pp 1110–1115
29. Nanduri A, Sherry L (2016) Anomaly detection in aircraft data using recurrent neural networks (RNN). In: Proceedings of 2016 integrated communications navigation and surveillance, ICNS 2016, pp 5C2-1–5C2-8
30. Nie S, Liu L, Du Y (2017) Free-fall: hacking Tesla from wireless to CAN bus. https://www.blackhat.com/docs/us-17/thursday/us-17-Nie-Free-Fall-Hacking-Tesla-From-Wireless-To-CAN-Bus-wp.pdf
31. Nilsson DK, Larson UE, Picasso F, Jonsson E (2009) A first simulation of attacks in the automotive network communications protocol FlexRay. In: Corchado E, Zunino R, Gastaldo P, Herrero A (eds) Proceedings of the international workshop on computational intelligence in security for information systems, CISIS '08, pp 84–91
32. Oka M, Oyama Y, Abe H, Kato K (2004) Anomaly detection using layered networks based on Eigen co-occurrence matrix. In: Proceedings of RAID 2004. LNCS, vol 3224. Springer, pp 223–237
33. Okamoto T, Ishida Y (2007) Framework of an immunity-based anomaly detection system for user behavior. In: Apolloni B, Howlett RJ, Jain L (eds) Knowledge-based intelligent information and engineering systems. Springer, pp 821–829
34. Posadas R, Mex-Perera JC, Monroy R, Nolazco-Flores JA (2006) Hybrid method for detecting masqueraders using session folding and hidden Markov models. In: MICAI. Lecture notes in computer science, vol 4293. Springer, pp 622–631
35. Premaratne U, Nait-Abdallah A, Samarabandu J, Sidhu T (2010) A formal model for masquerade detection software based upon natural mimicry. In: Proceedings of the 2010 5th international conference on information and automation for sustainability, ICIAfS 2010, pp 14–19

36. Rajbahadur GK, Malton AJ, Walenstein A, Hassan AE (2018) A survey of anomaly detection for connected vehicle cyber security and safety. In: 2018 IEEE intelligent vehicles symposium, IV 2018, pp 421–426
37. Reilly M, Stillman M (1998) Open infrastructure for scalable intrusion detection. In: 1998 IEEE information technology conference, information environment for the future, pp 129–133
38. SANS Institute (2000) Host vs. network-based intrusion detection systems. https://cyber-defense.sans.org/resources/papers/gsec/host-vs-network-based-intrusion-detection-systems-102574
39. Schonlau M, Theus M (2000) Detecting masquerades in intrusion detection based on unpopular commands. Inf Process Lett 76(1–2):33–38
40. Shon T, Moon J (2007) A hybrid machine learning approach to network anomaly detection. Inf Sci 177(18):3799–3821
41. Smith C (2016) Bus protocols. In: The car hacker's handbook: a guide for the penetration tester. No Starch Press, San Francisco, CA, pp 15–35
42. Smith C (2016) The car hacker's handbook: a guide for the penetration tester, 1st edn. No Starch Press, San Francisco, CA, USA
43. Stamp M (2004) A revealing introduction to hidden Markov models. https://www.cs.sjsu.edu/˜stamp/RUA/HMM.pdf
44. Stamp M (2011) Information security: principles and practice, 2nd edn. Wiley
45. Stamp M (2017) Introduction to machine learning with applications in information security. Chapman and Hall/CRC, Boca Raton
46. Studnia I, Alata E, Nicomette V, Kaâniche M, Laarouchi Y (2014) A language-based intrusion detection approach for automotive embedded networks. In: Proceedings of the 21st IEEE Pacific Rim international symposium on dependable computing, PRDC 2015, Zhangjiajie, China
47. Taylor A, Japkowicz N, Leblanc S (2015) Frequency-based anomaly detection for the automotive CAN bus. In: Proceedings of 2015 world congress on industrial control systems security, WCICSS 2015, pp 45–49
48. Taylor A, Leblanc S, Japkowicz N (2016) Anomaly detection in automobile control network data with long short-term memory networks. In: 2016 IEEE international conference on data science and advanced analytics, DSAA 2016. IEEE, pp 130–139
49. Tomlinson A, Bryans J, Shaikh SA, Kalutarage HK (2018) Detection of automotive CAN cyber-attacks by identifying packet timing anomalies in time windows. In: 48th annual IEEE/IFIP international conference on dependable systems and networks workshops, DSN-W 2018, pp 231–238
50. Toyama T, Yoshida T, Oguma H, Matsumoto T (2018) Pasta: portable automotive security testbed with adaptability. https://i.blackhat.com/eu-18/Wed-Dec-5/eu-18-Toyama-PASTA-Portable-Automotive-Security-Testbed-with-Adaptability[1].pdf. Presented at BlackHat Europe
51. Tsai C-F, Hsu Y-F, Lin C-Y, Lin W-Y (2009) Intrusion detection by machine learning: a review. Expert Syst Appl 36:11994–12000
52. Wagner RA, Fischer MJ (1974) The string-to-string correction problem. J ACM 21(1):168–173
53. Wan MD, Wu H, Kuo Y, Marshall J, Huang SS (2008) Detecting masqueraders using high frequency commands as signatures. In: Proceedings of 22nd international conference on advanced information networking and applications, AINA, pp 596–601
54. Wang C, Zhao Z, Gong L, Zhu L, Liu Z, Cheng X (2018) A distributed anomaly detection system for in-vehicle network using HTM. IEEE Access 6:9091–9098
55. Wang K, Stolfo SJ (2003) One-class training for masquerade detection. In: 3rd IEEE conference data mining workshop on data mining for computer security, DMCS 2003
56. Wang W, Guan X, Zhang X (2004) Profiling program and user behaviors for anomaly intrusion detection based on non-negative matrix factorization. In: 43rd IEEE conference on decision and control. CDC 2004, vol 1, pp 99–104
57. Wasicek AR, Pesé MD, Weimerskirch A, Burakova Y, Singh K (2017) Context-aware intrusion detection system for autonomous cars. https://web.eecs.umich.edu/˜mpese/papers/Escar17_CAID_Paper.pdf

58. Wu H-C, Huang S-HS (2008) User behavior analysis in masquerade detection using principal component analysis. In: Proceedings 8th international conference on intelligent systems design and applications, pp 201–206
59. Wu H-C, Huang S-HS (2009) Masquerade detection using command prediction and association rules mining. In: proceedings of the 2009 international conference on advanced information networking and applications, AINA '09, pp 552–559
60. Ye N, Zhang Y, Borror CM (2004) Robustness of the Markov-chain model for cyber-attack detection. IEEE Trans Reliab 53(1):116–123
61. Yung KH (2003) Using feedback to improve masquerade detection. In: Zhou J, Yung M, Han Y (eds) Applied cryptography and network security, ACNS 2003. Springer, pp 48–62
62. Yung KH (2004) Using self-consistent Naïve-Bayes to detect masquerades. In: Dai H, Srikant R, Zhang C (eds) Advances in knowledge discovery and data mining, PAKDD 2004, pp 329–340. Springer
63. zombieCraig. https://github.com/zombieCraig/ICSim

Cloud Computing Security: Hardware-Based Attacks and Countermeasures

Reza Montasari, Alireza Daneshkhah, Hamid Jahankhani, and Amin Hosseinian-Far

Abstract Despite its many technological and economic benefits, Cloud Computing poses complex security threats resulting from the use of virtualisation technology. Compromising the security of any component in the cloud virtual infrastructure will negatively affect the security of other elements and so impact the overall system security. By characterising the diversity of cyber-attacks carried out in the Cloud, this paper aims to provide an analysis of both common and underexplored security threats associated with the cloud from a technical viewpoint. Accordingly, the paper will suggest emerging solutions that can help to address such threats. The paper also offers future research directions for cloud security that we hope can inspire the research community to develop more effective security solutions for cloud systems.

R. Montasari (✉)
Hillary Rodham Clinton School of Law, Swansea University, Richard Price Building, Sketty Ln, Sketty, Swansea SA2 8PP, UK
e-mail: Reza.Montasari@Swansea.ac.uk
URL: http://www.swansea.ac.uk

A. Daneshkhah
School of Computing, Electronics and Mathematics, Coventry University, Priory Street, Coventry CV1 5FB, UK
e-mail: ac5916@coventry.ac.uk
URL: https://www.coventry.ac.uk/

H. Jahankhani
Information Security and Cyber Criminology, Northumbria University, 110 Middlesex Street, London E1 7HT, UK
e-mail: hamid.jahankhani@northumbria.ac.uk
URL: http://www.london.northumbria.ac.uk

A. Hosseinian-Far
Faculty of Business and Law, University of Northampton, Waterside Campus, University Drive, Northampton NN1 5PH, UK
e-mail: Amin.Hosseinian-Far@northampton.ac.uk
URL: https://www.northampton.ac.uk/

R. Montasari et al. (eds.), *Digital Forensic Investigation of Internet of Things (IoT) Devices*, Advanced Sciences and Technologies for Security Applications,
https://doi.org/10.1007/978-3-030-60425-7_6

Keywords Cyber security · Cloud security · Virtualisation · Hypervisor attacks · Network attacks · Cyber physical systems · Digital forensics · Computer security · Network security

1 Introduction

Cloud Computing (CC), still an evolving paradigm, has become one of the most transformative computing technologies and a key business avenue, following in the footsteps of main-frames, minicomputers, personal computers, the World Wide Web and smartphones [35, 34, 41, 43]. CC is a shared collection of configurable networked resources (e.g., networks, servers, storage, applications and services) that can be reconfigured quickly with minimal effort [37, 30]. Its vital features have considerably reduced IT costs, contributing to its swift adoption by businesses and governments worldwide.

As a result, CC has drastically transformed the way in which Information Technology (IT) services are created, delivered, accessed and managed. Such a transformation, that offers many technological and economic benefits, has produced substantial interest in both academia and industry. However, despite all its benefits, CC poses numerous security threats with devastating consequences. As a result, many organisations do not move their business IT infrastructure completely to the cloud mainly due to the fears of cloud-related security threats. Some of these fears among others are due to the issues such as processing of sensitive data outside organisations, shared data and ineffectiveness of encryption, etc. [34, 21].

Considering its security requirements (confidentiality, integrity, availability, accountability, and privacy-preservability), this paper presents an analysis of security threats associated with CC. To this end, we identify both common and underexplored cyber-security attacks carried out in Cloud Computing environments (CCEs). Accordingly, we will also propose emerging solutions and lines of defence against each attack vector with a view to mitigating such threats. Our study will also provide insights into the future security perspectives related to the CC. This study only focuses on analysing the technical aspects of cyber-security threats in cloud. To this end, this analysis emphasises the complexity, intensity, duration and distribution of the attacks, outlining the major challenges in safeguarding against each attack. Investigating other security aspects such as organisational, compliance, physical security of data centers, and the way in which an enterprise can meet regulatory requirements is outside the scope of this paper. Similarly, providing an exhaustive list of attack vectors is outside the scope of this study. The remainder of the paper is structured as follow.

The remainder of the paper is structured as follow. Section 2 analyses attack vectors while Sect. 3 discusses countermeasures. The paper is concluded in Sect. 4.

2 Attack Vectors

There are certain vulnerabilities associated with computing hardware that attackers can exploit to launch destructive attacks which often go undetected by the existing software countermeasures against embedded device systems within cloud environments. One of the usual characteristics of a computer hardware component can be malfunction or some kinds of abnormal behaviour which can result in providing a backdoor access to a potential adversary. The following sub-sections analyse the cyber-attacks that can be carried out due to physical hardware flaws.

2.1 Side-Channel Attacks

Side Chanel Attacks (SCAs) are a type of hardware-targeted attacks that is almost impossible to detect. SCAs are based on Side-Channel information that leak through a medium that is not intended for communications. This medium is called a Side-Channel. Side-Channel information can be acquired from an encrypted digital device. SCAs could also stem from a leakage produced by electronic circuits as by-products that render it possible for an adversary without access to circuit, itself, to determine how the circuit operates and what type of data it is processing (heat and electromagnetic leakage are both feasible sources of information for an adversary). SCAs can be very detrimental if proper defence mechanisms are not implemented on a target device. The three major types of side-channel attacks can be categorised based on the leaked information, including: time, trace [25] and access-driven [53]. All the three types acquire sensitive information by observing the execution time or power consumption variations produced through cache hits and misses. However, they tend to vary on the details of the captured information. The time-driven attack observes the aggregate profile and the total number of cache hits and misses. It can be of two types: passive if adversaries have no direct access to the victims machine or active if there is a physical access to the machine [33, 36, 32, 39]. The following sub-sub sections provide an analysis of the variants of SCAs.

Prime + Probe Attacks Through a Prime + Probe Attack, a variation of CB-SCAs, the adversary could potentially attain co-residency and perform load measurement. The theft of sensitive information can be accomplished by exploiting three probable channels: pre-emptive scheduling, hyper-threading, and multi-core. In the first channel, the adversary exploits the context switch between his VM and the victims VM to observe the cache status as the victim had left it. In the second channel, the malicious operation is performed by breaching the CPU core sharing. In this case, the attacker exploits the multi-tenancy, realised with multiple threads running on a single processor. The third stage involves reading the L3 cache that is the only one shared when VMs are allocated to multiple cores instead of multiple threads [25]. Furthermore, by performing a Prime + Probe Attack, the attackers could extract an RSA secret key from a co-located instance in CCEs [22]. A Prime + Probe attack can also be launched against different processor caches such as the L1 data cache, L1 instruction cache and the branch prediction cache. Similarly, adversaries could also exploit Prime + Probe for LLC attacks by leveraging hardware elements that are beyond the control of the CSPs but activated in the VMM for operation reasons [29].

Time-Driven Attacks In a Time-Driven Attack, the attackers extract cipher keys by exploiting side-channel information vulnerabilities triggered by the execution of cryptographic algorithms and data-dependent behaviour of cache memory. A Time-Driven Attack determines the run times of victim processes by exploiting the connection between the secret key and the number of cache misses which in turn establishes the runtime to infer the key. Regardless of the differences between these approaches, the Cache-Based Timing Attacks rely upon the impacts that the number of cache misses have on the execution time of an encryption process. Furthermore, Time-Driven Attacks can be performed against AES in a virtualisation setting [20, 11, 3, 10, 40, 25]. An example of such an attack is the PikeOS Microkernel Virtualization Framework [12], which was can be mounted against AES on an actual CPS. Similarly, Address Space Layout Randomization (ASLR), a security technique used to prevent exploitation of memory corruption vulnerabilities, can be bypassed by applying the branch-target buffer.

It is also possible to locate the place in the kernel where codes were run based on the mapping from virtual addresses to the branch-target buffer cache [14]. In addition, a malicious operating system could potentially reverse-engineer the control flow of SGX enclaves via branch-prediction analysis or branch shadowing [28]. Likewise, adversaries might be able to mount Timing Attacks against secret-dependent data access patterns on the sliding-window modular exponentiation implementation [9, 29, 1]. The Scatter–Gather technique, a commonly-implemented method to stop Time-Based Attacks, can also be exploited through a variation of Timing Attack called CacheBleed [52], which takes advantage of cache-bank collisions [15] (Intel, 2016) to generate quantifiable timing differences [16]. In general, Time-Driven Attacks are simple to execute since they require less leaked information [7, 3].

Access-Driven Attacks In cases when multi tenancy is employed through Hyper-Threading method, information on cypher algorithms as RSA and AES could potentially be observed by the attackers. One of the most recent attacks is the Access-Driven Attack, via which the attacker can control the cache sets that the cipher process changes. For instance, the attacker is likely to be able to determine which aspects of the lookup tables have been accessed by the cipher. Another type of attack is Boot Integrity Attack, in which adversaries with either logical or physical access are likely to be able to damage boot integrity with bootkits or particular form of malware that exists outside the OS (e.g. within System Management Mode (SMM)).

Considering its location, this kind of malware is predominantly threatening as it can reinfect new OS installations. Susceptibilities have been discovered in the BIOS, UEFI, Master Boot Record (MBR), CPU Management Engines and PCI device option ROMs [50]. Access-driven attacks exploit the connection between the secret key and the cache use of a crypto process. Because the cache is divided between various processes, an adversary might be able to gain the cache usage of the victim process by monitoring a carefully created process, which executes together with the victim process [27].

2.2 *Cache-Based Attacks*

Confidential data can be safeguarded against unauthorized access by storing it in an encrypted form and transmitting it over encrypted channels. However, at some point, data need to be decrypted so as to perform the computation. Adversaries could potentially exploit the multi-tenant environment to gain access to physical resources such as memory bus, disk bus, and data and instruction caches in which they can locate decrypted data and the cryptographic keys of well-known algorithms (AES, DES, RSA) and of other VMs instances. An instance of this concerns the shared memory hierarchy of an Intel Pentium 4 with hyper-threading features. Both the L1 and L2 caches with the hyper-threading feature turned on can leak information from one process to the other [33, 36, 45]. This is called a Cache-Based Side-Channel Attack (CBSCA) and is part of a family known as Cross-VM Side-Channel Attacks. This attack evades the logical isolation provided by the hypervisor layer and can be launched by two types of malicious actors including: insider attackers (often cloud employees abusing their privileged position) and malicious customers (that in a first phase must land in the victim server and then initiate the attack).

2.3 *Flush + Reload Attacks*

By exploiting resource sharing features in virtual environments, adversaries will be able to carry out cross-VM Flush + Reload Attacks against VMs in a hypervisor such as VMware [23]. As a result, they could potentially extract an AES keys in OpenSSL 1.0.1 running inside a victims VM. Likewise, shared memory controllers are susceptible to Flush + Reload Attacks that could exploit memory interferences as timing channels [48]. Similarly, covert channels shared between processor resources could be exploited to facilitate secret communication between malign processes. Trojans and spies could be utilised to compromise Processor Branch Prediction Units [13]. By exploiting Branch Predictor conflicts, adversaries could establish covert channels enabling them to launch Flush + Reload attacks against Computing hardware devices.

2.4 Rowhammer Attacks

If a particular row of a Double Data Rate (DDR) memory bank is constantly activated (opened) and pre-charged (closed) within a Dynamic Random-Access Memory (DRAM) refresh interval, one or more-bit flips take place in physically adjacent DRAM rows to an incorrect value. Such disturbance is known as Rowhammer [26]. An advanced attacker can exploit the Rowhammer to compromise the DRAM of a computing device. This occurs by evading the defence mechanisms often deployed through traditional security software and features such as memory isolation to conduct the memory disturbance attack. Similarly, a Rowhammer Fault Injection, a recently discovered real-time Microarchitectural Attack, can be launched remotely to gain full access to the DRAM of a CC device. A Rowhammer Attack can pollute system memory, access and alter sensitive data and gain full control of the system.

2.5 Hardware Threading Attacks

Attackers can also exploit hardware threading to examine a competing threads L1 cache usage in real time [16, 40]. Simultaneous multithreading (the sharing of the operation resources of a superscalar processor between multiple execution threads) is a feature implemented into Intel Pentium 4 processors. Under this implementation, the sharing of processor resources between threads spreads beyond the operation units. This denotes that the threads also share access to the memory caches. Such shared access to memory caches can facilitate side channels and enable a malign thread with restricted privilege to scan the operation of another thread. In turn, this results in allowing the attackers to steal cryptographic keys [17, 29, 40]. Additionally, by exploiting side-channel information based on CPU delay, adversaries could potentially mount TBSCAs against the Data Encryption Standard (DES) implemented in some applications. Such cryptanalysis technique applies side-channel information on encryption processing to gather plaintexts for cryptanalysis and infers the information on the extended key from the acquired plaintexts [46]. Through this attack, the adversary will be able to break the cipher with plaintexts [33, 36, 32].

2.6 Data Loss and Data Breach

Data stored in the cloud could be lost because of the hard drive failure, its accidental deletion by CSPs or malicious modification by adversaries, etc. Data loss can have disastrous impacts on enterprises such as bankruptcy. A data breach occurs when a VM accesses data from another VM on the same physical host (when the tenants of the two VMs are different customers). For example, a data breach can be carried out through a Side-Channel Attack, in which adversaries could potentially access data from one VM through another by utilising their shared components such as processors cache.

3 Countermeasures

Shared Technology: Cloud Security Alliance (CSA) (Hubbard and Sutton 2010) recommends a defence in depth strategy that should include compute, storage, and network security enforcement and monitoring. According to recommendation, CSPs could deploy robust compartmentalization to ensure that individual customers do not affect the operations of other tenants running on the same CSP. This denotes that customers must not be able to have access to any other tenants actual or residual data, network traffic, etc. Data Loss and Data Breach: Thus, one of the most effective ways to safeguard against data loss is to have in place a proper data backup, which resolves data loss issues.

3.1 Side-Channel Countermeasures

The purpose of a SCA countermeasure must be to hide the leakage or reduce it so that it holds minor or no valuable information against which an adversary is motivated to launch an attack. One of the most widely used defence mechanisms against a SCA relies on rendering the security operation time delay constant or random irrespective of the microarchitecture components utilised [16]. However, implementing constant-time execution code is difficult because optimisations presented by the compiler must be circumvented. Therefore, dedicated constant-time libraries have been introduced to enable security developers to safeguard their applications against SCAs. Hardware partitioning can also be used as a countermeasure to safeguard against SCAs. This hardware partitioning must be based on inactivating hardware threading, page sharing, presenting Hardware Cache Partitions, quasi-partitioning, and migrating VMs within cloud services. Considering that SCAs exploit physical elements of a system, its countermeasures must take the approach of enhancing the security of the system design and development such as that of cache architectures. These cache mechanisms should be offered without vast performance costs.

Furthermore, efficient implementation of Advanced Encryption Standard (AES) algorithm in hardware could be utilised as a defence mechanism against SCAs. There are currently few manufacturers implementing better hardware support in the design of their processor technologies to offer better constant-time cryptography operations. For instance, Intel has introduced AES New Instructions (AES NI), which is a new encryption instruction set that enhances on the AES algorithm and speeds up the encryption of data in the two categories of Intel Xeon processor and the Intel Core processor. AES-NI provides an advantage in relation to speed over other implementations. Moreover, since AES-NI, which consists of seven new instructions, was specifically developed to be constant-time, it provides a better protection against SCAs over some other software implementations.

3.2 Cache-Based Countermeasures

Configurable Cache Architecture could be used as a countermeasure to provide hardware assisted defence against CBSCA [49]. The cache is dynamically divided into safeguarded regions and can be configured for an application. In partitioned caches, there is a section of the cache that is assigned exclusively to the safe-guarded process so as to avert information leakage. Therefore, partitioned cache mechanism can be deployed as a line of defence against CBSCAs. A partitioned cache must be included in devices that are susceptible to SCAs to separate the cache behaviour of one process to another. This will prevent process interference by providing adequate space to store the entire S-box in cache (It will be locked when it is pre-loaded). Segregation does not permit forcible flushing of the cache; furthermore, partitioned cache employs longer cache lines that render attacks more problematic. Similarly, a method called Partition-Locked Cache (PLcache) [49] can be used to deal with cache sharing issues. This method will rely on a fine-grained locking control to isolate only the cache lines that contain important data. By making private partitions only those cache lines that are of interest are locked.

McBits [4] and Bitslice implementation of the AES [5] is another constant time countermeasure. By not using any lookup tables, this implementation could essentially prevent information from leaking out via a side channel. Another countermeasure is to carry out Cache Warming or Pre-Fetching. Time-Driven and Trace-Driven SCAs distinguish cache-miss and cache-hits. Eliminating this distinction can be a robust countermeasure [38]. So as to prevent information from the leakage, one needs to warm up the cache into which the Lookup Tables must be loaded prior to the runtime being initiated. In this situation, no cache misses will occur on condition that data is loaded to cache prior to the runtime. As a result, there will be no leakage of data.

3.3 Rowhammer Countermeasures

To perform a successful Rowhammer attack, adversaries must undertake four steps consisting of identifying the target device and its specific memory architecture characteristics, activating rows in each bank in a swift manner to trigger the Rowhammer vulnerability, accessing the aggressor physical address from userland [44] and exploiting bit flips [26]. In order to counteract a Rowhammer Attack, Rowhammer-induced bit flips must be blocked by altering DRAM, memory controllers or the combination of both. It is important that specific rows not be repeatedly triggered during a specific refresh point if the adjacent rows are not simultaneously refreshed.

Targeted Row Refresh (TRR) mode and Maximum Activate Count (MAC) metadata field could also be used by a memory controller as countermeasures to safeguard against Rowhammer Attacks [24]. In the TRR, a memory controller would need the DRAM device to refresh a rows neighbours. In contrast, MAC metadata field specifies the number of activations that a given row can safely cope with before its neighbours require refreshing.

Another countermeasure against Rowhammer Attacks is to use the physical probing of the memory bus via a high-bandwidth oscilloscope. This can be achieved by determining the voltage on the pins at the DIMM slots [42]. Furthermore, time analysis based on the rowbuffer conflict can be used to determine address pair that is part of the same bank and then apply this address set to rebuild the precise map function automatically [42, 51]. Furthermore, a simple solution requires that DRAM vendors build Rowhammer mitigations internally within a DRAM device, which does not need special memory controller support.

There exist various other methods that can be used to mitigate Rowhammer Attacks. For instance, by constantly refreshing the entire rows, disturbance errors could be eliminated for sufficiently short refresh intervals (RI RIth) [26]. This is despite the fact that regular refreshing might diminish performance and energy-efficiency. Furthermore, a mechanism called Probabilistic Adjacent Row Activation (PARA) [26], which is implemented in the Memory Controller can also be utilised to prevent DRAM disturbance errors Manufacturers could also retire DRAM cells, identified as victim cells, and remap them to spare cells. The end-users, themselves, could also retire DRAM sells by assessing and utilising system-level techniques for deactivating faulty addresses or remapping defective addresses to reserved addresses [36, 26] (Montasari, hardware [26]).

Authors in [18] have also suggested a Run-Time Memory Hot Detector (ARMOR) to mitigate Rowhammer attacks. ARMOR is analogous to DRAM in that it is implemented at the memory level. According to the authors in [18], ARMOR is capable of detecting all the conceivable Row Hammer errors, screening the activation flow at the memory level and also identify hot rows (specific rows) that might be hammered at run-time.

One method of detecting a Rowhammer attack is to implement the last-level cache counter facility to generate an interrupt after N misses. This method involves monitoring the last-level cache misses on a refresh interval and row access with high temporal locality on certain processors such as Intel/AMD. In the event of missing cache surpassing a threshold, a selective refresh could be performed on the vulnerable row. Identifying the hot row (i.e. specific row or aggressor row) and refreshing its neighbouring rows is another technique to counteract a Rowhammer attack [2].

The CFLUSH command available on user space (userland) for x86 devices can also be used as a countermeasure to evict cache lines associated with the aggressor row addresses among its memory accesses [44, 26]. However, these countermeasures do not appear to be appropriate to deal with Rowhammer Attacks in CCEs. In the context of cloud, Rowhammer Attacks are executed on different attack interfaces (e.g. scripting language based attacks) by deploying web browsers that are activated remotely, a view supported by the authors in [19, 6].

Therefore, it is essential to develop new eviction methods to replace the existing flush instructions so that Rowhammer attacks within cloud environments can be addressed more effectively. These new methods must be able to identify an eviction set that would comprise of addresses which will be part of the same cache set of the aggressor rows. For instance, this can be accomplished by employing a Time Attack to identify the eviction set. Yet, another eviction method could be based on the reverse engineering analysis of the system that has come under attack [8]. However, this could be a complex task considering the modern Intel processors. Direct Memory Access methods could be utilised to bypass CPUs and their caches to address the Rowhammer attacks [47].

4 Conclusion

In this study, we identified and analysed both common and underexplored hardware-based attacks associated with CCEs. We then made several recommendations with a view to mitigating such attacks. This analysis was based on our own experience as well as various sources in the literature such as official documentations, white papers and existing research articles. As a future work, in order to realise the fullness of our recommended countermeasures, one could perform distinct studies related to the suggested methods. Practical assessments must be performed for each recommendation with a view to determining how effective the countermeasure against a given attack vector is and establish whether or not that mitigation mechanism can be bypassed. It is only by conducting these practical studies that we can truly provide adequate insight on the fullness of these countermeasures.

References

1. Acıiçmez O, Koç ÇK, Seifert J-P (2007) Predicting secret keys via branch prediction. In: Cryptographers track at the RSA conference. Springer, pp 225–242
2. Aweke ZB, Yitbarek SF, Qiao R, Das R, Hicks M, Oren Y, Austin T (2016) ANVIL: software-based protection against next-generation rowhammer attacks. ACM SIGPLAN Notices 51(4):743–755
3. Bernstein DJ (2005) Cache-timing attacks on AES
4. Bernstein DJ, Chou T, Schwabe P (2013) McBits: fast constant-time code-based cryptography. In: International workshop on cryptographic hardware and embedded systems. Springer, pp 250–272
5. Bernstein DJ, Schwabe P (2008) New AES software speed records. In: International conference on cryptology in India. Springer, pp 322–336
6. Bhattacharya S, Mukhopadhyay D (2016) Curious case of rowhammer: flipping secret exponent bits using timing analysis. In: International conference on cryptographic hardware and embedded systems. Springer, pp 602–624
7. Bonneau J, Mironov I (2006) Cache-collision timing attacks against AES. In: International workshop on cryptographic hardware and embedded systems. Springer, pp 201–215

8. Bosman E, Razavi K, Bos H, Giuffrida C (2016) Dedup Est Machina: memory deduplication as an advanced exploitation vector. In: 2016 IEEE symposium on security and privacy (SP). IEEE, pp 987–1004
9. Brumley D, Boneh D (2005) Remote timing attacks are practical. Comput Netw 48(5):701–716
10. Coppens B, Verbauwhede I, De Bosschere K, De Sutter B (2009) Practical mitigations for timing-based side-channel attacks on modern x86 processors. In: 2009 30th IEEE symposium on security and privacy. IEEE, pp 45–60
11. Crane S, Homescu A, Brunthaler S, Larsen P, Franz M (2015) Thwarting cache side-channel attacks through dynamic software diversity. In: NDSS, pp 8–11
12. Elphinstone KG (2013, November) From L3 to seL4 what have we learnt in 20 years of L4 microkernels?. In: Proceedings of the Twenty-Fourth ACM Symposium on Operating Systems Principles, pp 133–150
13. Evtyushkin D, Ponomarev D, Abu-Ghazaleh N (2015) Covert channels through branch predictors: a feasibility study. In: Proceedings of the fourth workshop on hardware and architectural support for security and privacy, pp 1–8
14. Evtyushkin D, Ponomarev D, Abu-Ghazaleh N (2016) Jump over ASLR: attacking branch predictors to bypass ASLR. In: 2016 49th annual IEEE/ACM international symposium on microarchitecture (MICRO). IEEE, pp 1–13
15. Fog A (2020) The microarchitecture of Intel, AMD and VIA CPUs: an optimization guide for assembly programmers and compiler makers. https://www.agner.org/optimize/microarchitecture.pdf. Accessed 16 May 2020
16. Ge Q, Yarom Y, Cock D, Heiser G (2018) A survey of microarchitectural timing attacks and countermeasures on contemporary hardware. J Cryptogr Eng 8(1):1–27
17. Genkin D, Pachmanov L, Pipman I, Tromer E (2015) Stealing keys from PCs using a radio: cheap electromagnetic attacks on windowed exponentiation. In: International workshop on cryptographic hardware and embedded systems. Springer, pp 207–228
18. Ghasempour M, Lujan M, Garside J (2015) Armor: a run-time memory hot-row detector
19. Gruss D, Maurice C, Wagner K, Mangard S (2016) Flush + flush: a fast and stealthy cache attack. In: International conference on detection of intrusions and malware, and vulnerability assessment. Springer, pp 279–299
20. Gullasch D, Bangerter E, Krenn S (2011) Cache games—bringing access-based cache attacks on AES to practice. In: 2011 IEEE symposium on security and privacy. IEEE, pp 490–505
21. Heiser J, Nicolett M (2008) Assessing the security risks of cloud computing. Gart Rep 27:29–52
22. Inci MS, Gulmezoglu B, Irazoqui G, Eisenbarth T, Sunar B (2016) Cache attacks enable bulk key recovery on the cloud. In: International conference on cryptographic hardware and embedded systems. Springer, pp 368–388
23. Irazoqui G, Inci MS, Eisenbarth T, Sunar B (2014) Wait a minute! a fast, cross-VM attack on AES. In: International workshop on recent advances in intrusion detection. Springer, pp 299–319
24. JEDEC (2014) Memory configurations: JESD21-C. JEDEC Global Standards for the Microelectronics Industry.
25. Kim T, Peinado M, Mainar-Ruiz G (2012) {STEALTHMEM}: system-level protection against cache-based side channel attacks in the cloud. In: Presented as part of the 21st {USENIX} security symposium ({USENIX} security 12), pp 189–204
26. Kim Y, Daly R, Kim J, Fallin C, Lee JH, Lee D, Wilkerson C, Lai K, Mutlu O (2014) Flipping bits in memory without accessing them: an experimental study of dram disturbance errors. ACM SIGARCH Comput Archit News 42(3):361–372
27. Kong J, Aciiçmez O, Seifert J-P, Zhou H (2009) Hardware-software integrated approaches to defend against software cache-based side channel attacks. In: 2009 IEEE 15th international symposium on high performance computer architecture. IEEE, pp 393–404
28. Lee S, Shih M-W, Gera P, Kim T, Kim H, Peinado M (2017) Inferring fine-grained control flow inside {SGX} enclaves with branch shadowing. In: 26th {USENIX} security symposium ({USENIX} security 17), pp 557–574

29. Liu F, Yarom Y, Ge Q, Heiser G, Lee RB (2015) Last-level cache side-channel attacks are practical. In: 2015 IEEE symposium on security and privacy. IEEE, pp 605–622
30. Mell P, Grance T et al (2011) The NIST definition of cloud computing. Computer Security Division, Information Technology Laboratory, National Institute of Justice. Accessed 15 May 2020
31. Mogull R, Arlen J, Gilbert F, Lane A, Mortman D, Peterson G, Rothman M (2017) SECURITY GUIDANCE for critical areas of focus in cloud computing, v4. 0., Cloud Security Alliance, Tokyo, Japan
32. Montasari R, Hill R, Hosseinian-Far A, Montaseri F (2019) Countermeasures for timing-based side-channel attacks against shared, modern computing hardware. Int J Electron Secur Digit Forensics 11(3):294–320
33. Montasari R, Hosseinian-Far A, Hill R, Montaseri F, Sharma M, Shabbir S (2018) Are timing-based side-channel attacks feasible in shared, modern computing hardware? Int J Organ Collect Intell (IJOCI) 8(2):32–59
34. Montasari R (2017) An overview of cloud forensics strategy: capabilities, challenges, and opportunities. In: Strategic engineering for cloud computing and big data analytics. Springer, pp 189–205
35. Montasari R, Hill R (2019) Next-generation digital forensics: challenges and future paradigms. In: 2019 IEEE 12th international conference on global security, safety and sustainability (ICGS3). IEEE, pp 205–212
36. Montasari R, Hill R, Parkinson S, Hosseinian-Far A, Daneshkhah A (2019) Hardware-based cyber threats: attack vectors and defence techniques. Int J Electron Secur Digit Forensics
37. Montasari R, Hosseinian-Far A, Hill R (2018) Policies, innovative self-adaptive techniques and understanding psychology of cybersecurity to counter adversarial attacks in network and cyber environments. In: Cyber criminology. Springer, pp 71–93
38. Page D (2002) Theoretical use of cache memory as a cryptanalytic side-channel. IACR Cryptol ePrint Arch 169:2002
39. Peltola P, Montasari R, Seco F, Jimenez AR, Hill (2019) Multi-platform architecture for cooperative pedestrian navigation applications. In: 2019 international conference on indoor positioning and indoor navigation (IPIN). IEEE, pp 1–8
40. Percival C (2005) Cache missing for fun and profit
41. Perry R, Hatcher E, Mahowald RP, Hendrick SD (2009) Force.com cloud platform drives huge time to market and cost savings. TechRepublic.com/White Papers. [en línea] Disponible en: https://whitepapers.techrepublic.com.com/abstract.aspx
42. Pessl P, Gruss D, Maurice C, Schwarz M, Mangard S (2016) {DRAMA}: exploiting {DRAM} addressing for cross-CPU attacks. In: 25th {USENIX} security symposium ({USENIX} security 16), pp 565–581
43. Ruan K, Carthy J, Kechadi T, Crosbie M (2011) Cloud forensics. In: IFIP international conference on digital forensics. Springer, pp 35–46
44. Seaborn M, Dullien T (2015) Exploiting the DRAM rowhammer bug to gain kernel privileges. Black Hat 15:71
45. Takahashi M, Nakatani M, Hiraoka S, Ushio H, Oshima A (2012) Component separation device. US Patent 8,273,302, 25 Sept 2012
46. Tsunoo Y, Saito T, Suzaki T, Shigeri M, Miyauchi H (2003) Cryptanalysis of des implemented on computers with cache. In: International workshop on cryptographic hardware and embedded systems. Springer, pp 62–76
47. Van Der Veen V, Fratantonio Y, Lindorfer M, Gruss D, Maurice C, Vigna G, Bos H, Razavi K, Giuffrida C (2016) Drammer: deterministic rowhammer attacks on mobile platforms. In: Proceedings of the 2016 ACM SIGSAC conference on computer and communications security, pp 1675–1689
48. Wang Y, Ferraiuolo A, Suh GE (2014) Timing channel protection for a shared memory controller. In: 2014 IEEE 20th international symposium on high performance computer architecture (HPCA). IEEE, pp 225–236

49. Wang Z, Lee RB (2007) New cache designs for thwarting software cache-based side channel attacks. In: Proceedings of the 34th annual international symposium on Computer architecture, pp 494–505
50. Weiß M, Weggenmann B, August M, Sigl G (2014) On cache timing attacks considering multi-core aspects in virtualized embedded systems. In: International conference on trusted systems. Springer, pp 151–167
51. Xiao Y, Zhang X, Zhang Y, Teodorescu R (2016) One bit flips, one cloud flops: cross-VM row hammer attacks and privilege escalation. In: 25th {USENIX} security symposium ({USENIX} security 16), pp 19–35
52. Yarom Y, Genkin D, Heninger N (2017) CacheBleed: a timing attack on OpenSSL constant-time RSA. J Cryptogr Eng 7(2):99–112
53. Zhang Y, Juels A, Reiter MK, Ristenpart T (2012) Cross-VM side channels and their use to extract private keys. In: Proceedings of the 2012 ACM conference on computer and communications security, pp 305–316

Aspects of Biometric Security in Internet of Things Devices

Bobby L. Tait

Abstract This chapter provides detailed insight into the general mechanisms utilized for biometric application in Internet of things devices. The mechanisms and internal working of these biometric technologies presented in this chapter are focused specifically on the applicability in IOT devices. IOT devices incorporates various scanners and sensors to allow the IOT device to biometrically interact with a human being. These scanners and sensors were primary designed to facilitate and ease user interaction with the IOT device in an effort to make the day to day usability of the IOT device faster and easier if you may. It must be noted that every biometric technology has certain strengths, but indeed, also certain noteworthy shortcomings. It is often these shortcomings that get exploited in a security subversion attempt. This chapter introduces and discusses the various biometric technologies used in IOT devices. Attention is given to the software and the hardware aspects of each biometric system. The generic working of these biometric technologies is presented. Attention is given to legacy biometric technology implemented on IOT devices, currently used biometric technology implemented on IOT devices, and finally, possible future biometric applications of biometric technology destined for IOT devices. In conclusion practical examples of biometric subversion on IOT devices such as fingerprint, facial and voice biometric subversion and hacking, will be investigated, discussed and evaluated.

Keywords Biometric security · Internet of Things · Biometric subversion · Fingerprint biometrics · Facial biometrics · Voice biometrics

B. L. Tait (✉)
University of South Africa, Pretoria, RSA, South Africa
e-mail: taitbl@unisa.ac.za

R. Montasari et al. (eds.), *Digital Forensic Investigation of Internet of Things (IoT) Devices*, Advanced Sciences and Technologies for Security Applications,
https://doi.org/10.1007/978-3-030-60425-7_7

1 Introduction

This chapter introduces and elaborates on various aspects relating to Biometrics with specific consideration of biometrics utilized in Internet of Things (IOT). The aim of this chapter is to introduce the reader to the fundamentals of biometric technology, and the status-quo of biometrics in use online and in IOT devices today. Strengths and weaknesses of this technology is discussed, possible solutions to mitigate some of the weaknesses are presented. This chapter aims to provide the reader with a deep understanding and appreciation of this technology. The chapter starts out with some background and historic information of biometric technology, followed by fundamental terminology and principals of biometrics. This is followed by a discussion on Internet of things (IOT) and the biometric technology found incorporated into IOT devices, as well as examples of how these biometric mechanism gets subverted. IOT devices and biometrics considerations are discussed in Sect. 6. Just before the chapter is concluded in Sects. 8 and 7 discusses a secure cloudbased biometric technology.

2 Background and History of Biometrics

The term "biometrics" is derived from the Greek words "bio" (life) and "metrics" (to measure). The use of Biometrics is by no means novel, Identification and authentication of a human by means of their physical traits can be traced back thousands of hears, One of the oldest and most basic examples of a biometric characteristic that is used for identification and authentication, is the human face. Since the very early times of civilization, humans have used faces to identify known and unknown individuals. This basic biometric recognition worked well in small communities and small villages, but as populations grew, and travel to distant places became common, recognition and authenticity of a individual became increasingly difficult. One must keep in mind that people were not just only recognized by their face, but also by means of their voice, they way they walked, and even smell. All of these factors contributed the the unique biometric characteristics that made up the uniqueness of a human being. in history many examples can be found of humans using biometric to authenticate an artifact or the be personally authenticated.

Caves are decorated with paintings, estimated to be 31,000 years old. These paintings found on the walls of the caves of ancient believed to be created by prehistoric men who lived there. Surrounding these paintings are numerous handprints that are felt to "have acted as an unforgettable signature of its originator [1].

As early as 500 B.C. "Babylonian business transactions are recorded in clay tablets that include fingerprints." [2].

Joao de Barros, a Spanish explorer and writer, wrote that early Chinese merchants used fingerprints to settle business transactions. Chinese parents also used fingerprints and footprints to differentiate children from one another [3].

In 1684 Dr. Nehemiah Grew published friction ridge skin observations in "Philosophical Transactions of the Royal Society of London" paper [4].

In 1788, German anatomist and doctor J. C. A. Mayer wrote "Anatomical Copperplates with Appropriate Explanations" containing drawings of friction ridge skin patterns, noting that "Although the arrangement of skin ridges is never duplicated in two persons, nevertheless the similarities are closer among some individuals. Mayer was the first to declare that friction ridge skin is unique" [5].

Around the later part of the 1800s, police departments formally started using fingerprints for identification of individuals. This process emerged in South America, Asia, and Europe. A method was developed to index fingerprints that provided the ability to retrieve records that was based on individualized metric- fingerprint patterns and ridges. The first robust system for indexing fingerprints was developed in India by Azizul Haque for Edward Henry, Inspector General of Police, Bengal, India. This system, called the Henry System, and variations on it are still in use for classifying fingerprints [6].

Automation of biometric recognition has only become available the past few decades.

True biometric systems began to emerge in the latter half of the twentieth century, coinciding with the emergence of computer systems. The nascent field experienced an explosion of activity in the 1990s and began to surface in everyday applications in the early 2000s.

The accuracy of automated biometric recognition is very reliant on various abilities provided to the automation process. The uniqueness in the biometric measurement if compared to the population, depends on the sensitivity (or measuring resolution) of the biometric process.

Each human consists of billions of atoms. The specific arrangement of these atoms are so unique, that the organization of these atoms for a given human, has never existed since the dawn of time, and will (considering the magnitude of atoms in a human), never exist again. If we could create a biometric sensor to measure the arrangement of each atom in a human being, the authentication of a human based such a technology would be irrefutable.

3 Biometrics Terminology and Principals

3.1 Terminology

A scholar of biometric technology will concur that there is quite a lot of confusing and contradicting terms used for the various aspects found in relation to biometric terms. Many research papers refer to biometric terminology in such a way that it is not entirely clear to what aspect of a biometric characteristic is being discussed. In various research papers [7–10], the authors use "biometric data" to describe biometric technology which has been digitized and translated into a binary representation of the biometric characteristic. In most cases, "biometric token" is used to describe the biometric characteristic to be digitized. However, this commonly used terminology, does not clearly define all aspects which form part if the complex biometric environment.

This section discusses the various aspects associated with biometrics, and proposes a terminology framework for the biometric environment. the proposed framework of the biometric environment is presented in an effort to expel the confusion that currently exists in the biometric technology community [11].

It must be noted that though the examples posed in the framework relates to fingerprint biometric characteristics, the basic principal of the framework is applicable for all biometrics that could be collected as a human interacts with his or her environment.

3.1.1 Biometric Characteristic.

As mentioned earlier, the word "biometric token" is often used to describe the physical biometric characteristic. However, the word token is associated with a man made article such as a RF-id Card, magnetic card or special key. Concatenation of the word biometric with token is ambiguous. In Fig. 1 the part of the individual presently to be offered to the biometric scanner, should be referred to as a biometric characteristic.

3.1.2 Fake Biometric Characteristic

Fake biometric characteristic is a biometric characteristic fraudulently manufactured [7] in order to be presented to the biometric scanner to masquerade as the real individual and is part of the undesired biometric path illustrated in Fig. 1.

3.1.3 Latent Biometric Image

Latent biometric image is an image left behind by an individual as he/she interacts with the environment, i.e. the imprints left on a glass, handled by an individual. This image can be used to manufacture (create) a fake biometric characteristic. In "Impact of artificial gummy fingers on fingerprint systems" [7] the author causes confusion by referring to these imprints as "biometric tokens"—clearly ambiguous. A latent biometric image is the starting point of the undesired biometric path, and can be used to create a fake biometric characteristic.

3.1.4 Biometric Data

Biometric data is customarily the term used to describe the digitized representation of a biometric characteristic. The term however does not distinguish between a legit-

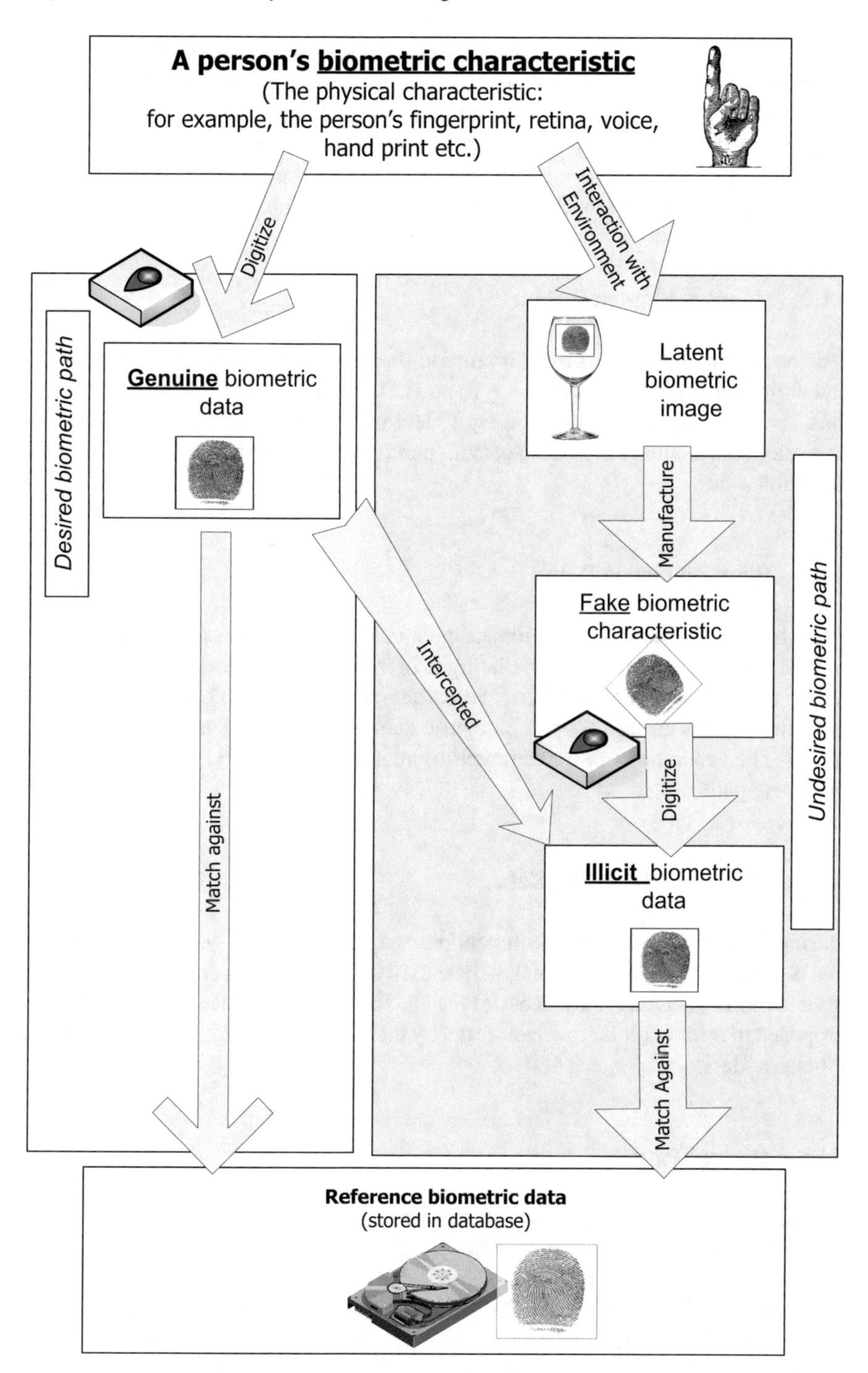

Fig. 1 Biometric terminology framework

imate and illegitimate (illicit) biometric data, both of which would be accepted as valid for authentication by the validation process. For example, if biometric data is intercepted for a replay attack [10], using the word "biometric data" is not accurate in the sense that, even though it is still this biometric data (digitized biometric characteristic), the current terminology does not allow for the fact that this biometric data is now in possession of a hacker, and is to be used for illicit purposes.

3.1.5 Genuine Biometric Data

This term defines the legitimate biometric data as indeed being generated by the authentic user (as illustrated in Fig. 1, as part of the desired path) "Genuine biometric data" is the result of generating a legit electronic representation of the biometric characteristic. Genuine biometric data is found if the biometric is digitized from the authentic user.

3.1.6 Illicit Biometric Data

Illicit biometric data refers to biometric data obtained by means of an illegitimate process and subsequently offered illicitly as biometric data of the authentic user. This term extends the understanding of the particular state of the biometric data. Illicit biometric data is all instances of biometric data that has not been acquired by legal means. The generation of illicit biometric data is illustrated in Fig. 1 in the undesired biometric path.

3.1.7 Reference Biometric Data

During the initial controlled enrollment process, the biometric system creates a special biometric template [12] which will be used by the matching algorithm to test any offered biometric data for a successful match. The term "reference biometric data" is proposed to refer to biometric data stored by the biometric system for testing offered biometric data.

3.1.8 Desired Pathway

The desired pathway is how the biometric authentication system is designed to function. An individual offers a biometric characteristic this action is intentional, and the individual wishes to be authenticated; the biometric characteristic is digitized and produces genuine biometric data. The genuine biometric data is compared to the reference biometric. This process followed the desired path, and in the instance that the genuine biometric data and reference biometric data match, the biometric characteristic, and thus also the individual, is authenticated.

3.1.9 The Undesired Pathway

A latent biometric image is the result of an individual interacting with the environment. It may be a fingerprint or even DNA in the saliva left on the rim of a glass that the individual drank from. The hacker filches [7] and use the latent biometric image to manufacture a fake biometric characteristic. If a fake biometric characteristic is created, it can be used to spoof the biometric digitizer. The illicit biometric data created from the fake biometric characteristic, will match the reference biometric data, and allow the hacker to be authenticated. In a similar fashion, the genuine biometric data, digitized along the desired pathway, can be intercepted, and illicitly offered for authentication. Considering that this intercepted biometric data is from the authentic individual, the system will match the biometric data to the reference biometric data, and authenticate the hacker along the undesired pathway, thus compromising the system when it is used for false authentication purposes.

3.2 Biometric Auth Process

Regardless of the biometric characteristic used for biometric authentication, a generic biometric process is described for all aspects involved in biometric matching. This generic process is illustrated Fig. 2.

Regardless of biometric characteristic used (Fingerprint, Iris, Retina, Gait, Palm print, gait, voice etc.) , or technology used (Personal computer or IOT), In all instances of automation of biometric authentication The biometric mechanism consist of four broad aspects [13]: (1) Data acquisition, (2) The Reference Data Store, (3) Signal Processing and finally (4) The decision policy.

During the *data acquisition* phase the biometric characteristic is digitized to a digital representation of the characteristic collected from the person, known as biometric data. This biometric data is then sent to the *signal processing* phase for feature extraction and segmentation, and the subsequently compared to the reference biometric data from the *data store*. This data store may be locally stored, stored in a cloud storage area, or even inside a Block-chain infrastructure [14]. A match & Quality score is calculated by the matching algorithm by comparing the reference

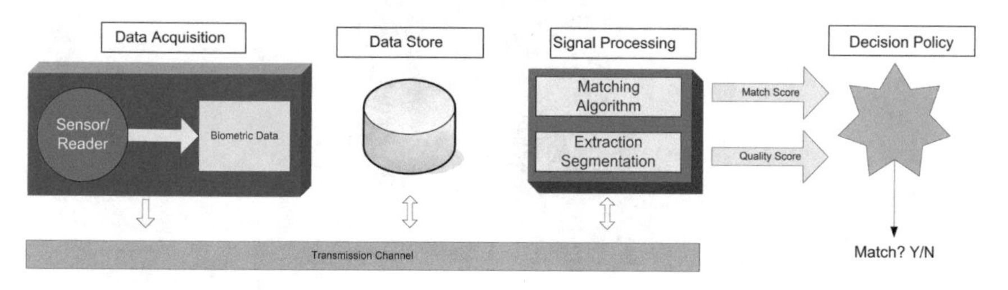

Fig. 2 Biometric mechanisms

biometric data with the newly acquired biometric data. The *decision policy* will, based on various parameters and constraint provide a match result (Yes or No).

4 Internet of Things

Generally speaking, IoT refers to the networked interconnection of everyday objects, which are often equipped with ubiquitous intelligence. IoT will increase the ubiquity of the Internet by integrating every object for interaction via embedded systems, which leads to a highly distributed network of devices communicating with human beings as well as other devices [15]. Thanks to rapid advances in underlying technologies, IoT is opening tremendous opportunities for a large number of novel applications that promise to improve the quality of our lives. In recent years, IoT has gained much attention from researchers and practitioners from around the world.

Current IOT devices incorporate biometric technology in varying degrees. Some IOT devices, such as the Apple watch series 5 as pictured in Fig. 3 has the ability to monitor a person's heartbeat through out the day [16]. However some IOT devices incorporates a number of biometric technologies. The modern cellular phone such as the Samsung Note series includes an Iris scanner [17], Fingerprint Scanner [18], facial recognition scanner [19] and hand writing recognition [20]. The Apple iPhone XS, includes a facial recognition mechanism as part of its authentication mechanisms [21]. The inclusion of biometric technology into IOT devices are clearly gaining momentum in the past number of years [22].

Considering that these biometric mechanisms give a user access to payment systems on the IOT device, it stands to good reason that a number of academic research and industry efforts demonstrated that these biometric mechanisms can be subverted. The current subversion efforts are discussed in more detail in Sect. 5. Subverting the biometric mechanism on a IOT device, makes identity theft and fraudulent transac-

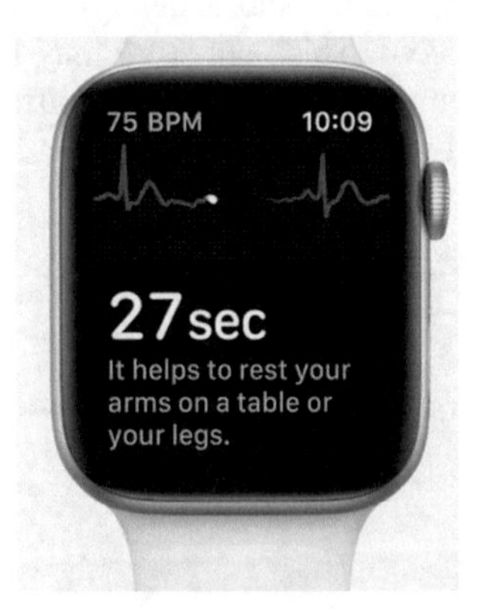

Fig. 3 Apple watch series 5

tions a reality. The Biometric mechanisms as found on IOT devices are clearly not safe from the methods used to spoof biometric devices as already demonstrated in earlier years on dedicated biometric systems [7, 23–25].

In conclusion, it is clear that IOT incorporates a number of biometric mechanisms, and it has been demonstrated that these devices are still vulnerable to biometric spoof attacks. In Sect. 7, a possible solution is presented in an effort to safeguard IOT devices. In the following section (Sect. 5), current spoof attempts on IOT devices is discussed.

5 Biometric Subversion of IOT

Manufacturers of IOT devices includes biometric technology in these IOT devices, and from a marketing point of view, sell their products based on the features of the device and the perceived security improvements of such a device. End users are seldom informed enough to weigh the advantages and disadvantages of such technology. it is stated at [26] that "Biometric scanners such as fingerprint and facial recognition systems have become increasingly popular on smart-phones because they are seen as less vulnerable and harder to discover or copy than passwords or number-based unlock codes", yet looking at the examples presented in Sect. 5.2, it is clear that the biometric technology built into these devices, are not safe at all. In Sect. 6, a number of these aspects is presented, and fundamental to IOT devices that incorporates biometric technology. This section illustrates two technical successful hacking attempts on current IOT barometric technology.

5.1 IPhone Facial Recognition Spoof

On the 7th of August 2019 at the Black Hat USA 2019 conference, researchers demonstrated an attack that allowed them to bypass a victim's iPhone FaceID and log into their phone simply by putting a pair of modified glasses on their face. During the demonstration the researchers merely placed tape carefully over the lenses of a pair of glasses and placing them on the victim's face. In doing so the researchers demonstrated that they could bypass Apple's FaceID in a specific scenario [27]. To launch the attack, researchers with Tencent tapped into a feature behind biometrics called "liveness" detection, which is part of the biometric authentication process that sifts through "real" versus "fake" features on people. One such biometrics tool that utilizes liveness detection is FaceID, which is designed and utilized by Apple for the iPhone and iPad Pro [27]. Liveness testing is an very important aspect of today's biometric authentication approach as this allows authentication algorithms to distinguish between real and fake biometric data. With the leakage of biometric data and the enhancement of AI fraud ability, liveness detection has become the Achilles' heel of biometric authentication. In previous spoof attempts, such as the research by

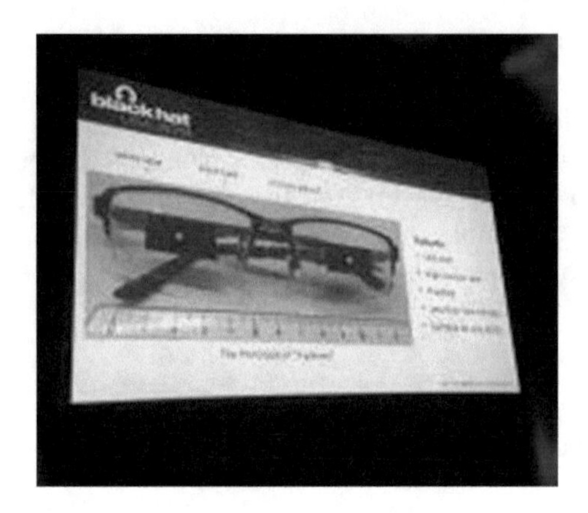

Fig. 4 Glasses made by researchers to bypass FaceID biometrics detection

[7] illustrated the focus was on creating fake biometric characteristics. in the attempt presented here, rather than attempting to generate a 3D model of a person's face, the researchers turned to the liveness aspect of the biometric authentication [28]. The researchers discovered that if a user is wearing glasses, the way that liveness detection scans the eyes changes. "After our research we found weak points in FaceID it allows users to unlock while wearing glasses if you are wearing glasses, it won't extract 3D information from the eye area when it recognizes the glasses." [27].

The researchers created modified glasses as illustrated in Fig. 4 with black tape on the lenses, and white tape inside the black tape. Using this trick they were then able to unlock a victim's iPhone phone and then transfer his money through mobile payment App to bypass the attention detection mechanism of both FaceID and other similar technologies [27]. Clearly, though the attack comes inherently with some shortcomings, it still managed to demonstrate that, given enough motivation, a spoof attempt is indeed possible, regardless of the promised made by the manufacturer of how secure the technology is.

5.2 Samsung Fingerprint Recognition Spoof

Mid April 2019, a researcher successfully illustrated that a Samung device van be unlocked, using a fake biometric characteristic printed on a 3D printer [29]. In illustrating the spoof attempt, the researcher took a picture of his fingerprint on a wineglass, processed it in Photoshop, and made a model using 3ds Max that allowed him to extrude the lines in the picture into a 3D version. Printing the fake biometric characteristic, took roughly 13-min, and three printing attempts with some tweaks to each repint, the researcher was able to print out a version of the fingerprint that fooled the Samsung S10's sensor. In Fig. 5, the 3D printed fake biometric characteristic is shown on the biometric scanner of the phone.

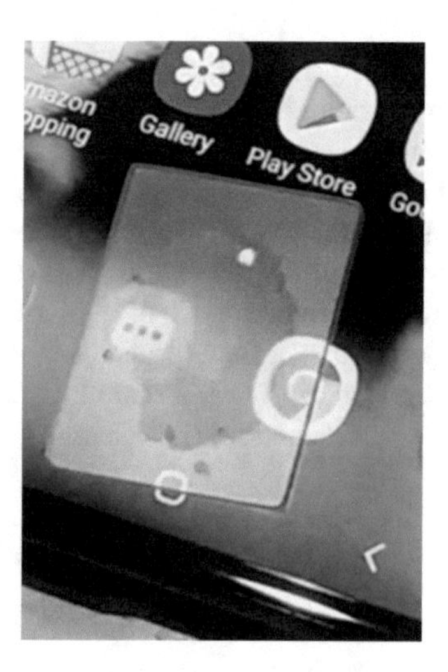

Fig. 5 3D printed fingerprint

In another example, at the end of October 2019 a serious problem with Samsung's in-screen fingerprint authentication mechanism was discovered. Researchers learned that if certain screen protectors are installed on the phone's screen, the fingerprint scanner, built into the screen stopped functioning correctly, resulting in the phone to unlock regardless of the fingerprint being presented [26, 30]. From this vulnerability is clear that anyone can unlock a this IOT device and gain access to all data and payment systems on the device by using biometric authentication mechanism on the device. It stands to good reason that if such a device is stolen, the thief can also easily gain access to the device.

These examples, are simply presented to illustrate that current, high-end IOT devices are indeed very vulnerable to Biometric spoof attacks. The general public has a false sense of security due to the marketing tactics of the companies manufacturing these devices, as also mentioned by [29]. The following section presents further Biometric considerations when used in IOT.

6 Biometrics and IOT Considerations

6.1 Touch Versus Non-touch Biometric Technology

Touch biometric devices are often used by a population for access control. If a hand palm biometric scanner is used for secure access to a specific area in a building, such as a biometric reader must touched in order to have the biometric characteristic digitized [31]. Users are often reluctant to touch biometric devices, which have been touched by other people [32]. Various hygienic implications arise whenever touch biometrics, such as fingerprint, palm print, hand shake biometrics et cetera are used. With the current Corona virus outbreak, it is advised by the World health organization that a person must sterilize as surfaces often touched by other persons [33]. Due to these concerns, often non-touch biometric technologies are suggested as a solution for the problems associated with touch type biometrics, it must be warned that this decision must not be taken without due consideration [11]. Aspects that must be considered, are presented in Sects. 6.2, 6.3 and 6.4. IOT devices incorporates touch biometric technology such as a fingerprint scanner and non-touch biometric technology, such as facial recognition or iris scanning. for this reason, the aspects listed below, are of extreme relevance to IOT devices.

6.2 User Intent

If a person needs to touch a biometric scanner to conclude a transaction, the user must willingly offer the biometric characteristic. This action of willingly offering the biometric to the biometric scanner can be considered as intent from the user's side to conclude the transaction. In comparison, if non-touch technology is used, such as facial recognition, a transaction might be concluded without the user even being aware that the facial recognition system authenticated him for the transaction. With non-touch biometric technology, it is a greater challenge for a person to withhold one's biometric identifiers. When non-touch biometric technology is used in IOT devices, A IOT device can simply be held in front of a user's face and then the system will unlock the device. This issue is also true for a person being forced to place his fingerprint on the fingerprint scanner. The authentication and unlock of the device can be forced on a person. If a different authentication mechanism is used, such as a pass phrase, the authenticator is not so easy to obtain from the user. Specifically in IOT devices a user is quite vulnerable when any form of biometric is used.

6.3 Privacy Considerations

The biometric identifier such as facial authentication is always visible and available to be scanned. This can happen without a person being aware that he was scanned. Privacy with non-touch technology is a major concern [34]. The ability of biometric facial recognition systems to track a person based on facial biometric technology poses many privacy concerns.

6.4 Replication Considerations

Lastly replication of biometric characteristics in order to create a fake biometric characteristic has been demonstrated in many research papers [7, 10] and even by industry researchers [23, 27, 29]. To replicate touch biometric such as a fingerprint, an object which the user has touched must be procured in order to replicate the biometric characteristic. For non-touch biometric technology, it is possible to obtain for example an individual's iris characteristic without the user being aware that the replication even occurred.

6.5 Strong Versus Weak Biometric Characteristics

A strong biometric characteristic requires the individual to be authenticated to be alive [35]. The particular biometric characteristic cease to exist the moment the individual dies. Examples of strong biometric characteristics are brainwave patterns, and electric currents generated by the heart, Hand writing recognition, Gait and voice recognition. On the contrary a weak biometric characteristic does not require the individual to be alive to be taken. These characteristics include finger-, hand-, and palm prints. It also includes biometric technology such as iris, retina, and facial scans [27]. In order to confirm that a user is indeed alive when the biometric is taken, vitality should then be established as an additional step during the scanning process [28].

If a biometric solution is considered for implementation in a IOT device, attention must be given to weak versus strong biometrics. Although a strong biometric system may be preferred, one should keep in mind that authentication can only be determined if the individual is alive. If, for example, a strong biometric such as facial thermography is used for authentication of a testament and will of an individual, once the person dies, the facial thermograph will not be detectable, which makes the authentication after the death impossible. If facial thermography is used for nationwide authentication, strong biometrics will preclude the authentication of any person after death.

In conclusion, if a biometric authentication system is proposed for the entire population, the most suitable authentication system should be contemplated. If post mortem authentication is needed, strong biometric systems will not be suitable.

6.6 Biometric as a Widely Accepted Authentication Standard

Biometrics is often considered as method for authentication. Heathrow airport to allows UK citizens access to the UK by means of iris biometrics instead of using a British passport [36]. Banking groups such as Capitec, have considered using fingerprint biometrics for authentication at their automated teller machines, replacing the need for pins [37]. If biometric technology is considered as a standard for authentication by homeland security or banking groups, subsequently to be used in IOT devices, the possibility must be kept in mind that the population for which biometric authentication is proposed, might not all possess the required biometric. If a bank group proposes that iris technology will be used to access their application on a IOT device, instead of using a pin, the possibility exists that not all people have an iris biometric characteristic. Fingerprint characteristics can only be scanned if the individual has fingers. It is often also difficult to obtain fingerprints from individuals who do a lot of manual labour (i.e. brick layers or mine worker) [38]. A Diamond mine in South Africa proposed to use fingerprint biometrics for all the mine workers, but soon found that many of the mine workers, due to the physical nature of their work, do not have fingerprints at all. These examples demonstrate that biometric characteristic required for a specific population group should be carefully selected.

To summarize, biometric characteristics required to authenticate a particular population group, may not be present in all the individuals of that group- they may lack limbs (making fingerprint, palm print, hand geometry, gait and even written signature not available for authentication for the group). It is also possible the users might not have eyes (making iris and retina unavailable for authentication). However, it can not be argued that facial thermography, facial recognition, DNA, body odour, skin luminescence and brainwaves are all examples of biometric characteristics which are found in all individuals in any population group.

In conclusion, whenever a biometric is chosen for a widely accepted standard for authentication, it is important to consider if all individuals in the population have this chosen biometric.

In Sect. 7, a framework is proposed as a possible solution to mitigate the inherent problems associated with biometric technology in general, but sacrificially developed for the use in IOT devices.

7 Secure Cloud-Based IOT Biometric Authentication

This section proposes an approach to allow a person to use a IOT device such as the latest iPhone, or latest Samsung smart phone for secure biometric authentication in a networked environment. It is argued in this section that a IOT device can be considered as a "IOT token", to address the security concerns associated with biometric technology discussed up to this point in this chapter.

During the research of biometric technology, it became clear that biometric data is asymmetrical in nature due to the fact that the biometric sensor does not collect all biometric markers in exactly the same way every time a biometric characteristic is scanned [10]. This aspect of biometric data can be used to uniquely identify every biometric characteristic ever presented for authentication by a given user of this system.

7.1 Biometric Asymmetry

The asymmetric nature of biometric data affords a unique benefit: All biometric data received from a user's biometric characteristic will almost always be unique. It is highly unlikely that, considering all the variables associated with the capturing and digitizing of a biometric characteristic, a 100% match will be found with any previously offered biometric data [39]. The fact that biometric data is uniquely identifiable is the first step towards a cloud-based biometric authentication system to prevent the possibility of replay of biometric data. Each instance of accepted biometric data can be linked to a given transaction performed by the user. In order to ensure that biometric data is not being replayed, and to link offered biometric data from the user to a specific transaction, a special biometric transaction log file must be used in the cloud environment. This log file is referred to as a cloud bio archive (CBA). The cloud-based biometric authentication system can detect any replay attempt, and log transactions by using a CBA. A second problem found with biometric technology, and mentioned in [29] relates to the possibility of sourcing a latent biometric image of a person's biometric characteristic.

7.2 Cloud-Based Biometric Authentication System

If biometric technology is to be used for secure authentication, a protocol should be used to ensure that problems as outlined earlier in this chapter can be concisely managed. Though attacks on the biometric system cannot be eliminated, the proposed system ensures that attacks on the system can be mitigated. To ensure that a hacker cannot use a fake biometric characteristic, a user side biometric archive (UBA) is introduced on a IOT device, and referred to as a IOT-token. The archive stored on the IOT-token, and contains finite (for example, 500) previously offered biometric data samples. The UBA is populated by the cloud-based authentication server, and gets updated under trusted situations. The system is outlined in Fig. 6.

In step 1, the user needs to conduct a transaction requiring authentication, and supplies a fresh biometric characteristic to the smart device's biometric scanner. In step 2, the smart device digitizes the biometric characteristic resulting in fresh biometric data. During previous communication with the cloud server, the authentication system sent a challenge to the IOT device. This process is outlined in detail in the journal paper [?]. For this discussion, it is stated that the authentication server requested specific biometric data stored in position 58 of the user bio archive (UBA). In step 3, the IOT device fetc.hes the biometric data stored in position 58 in the UBA. In step 4, the IOT device generates a biometric parcel (bio-parcel); The fresh biometric data is XOR'ed with the historic biometric data requested by the server from the UBA, resulting in a 'XOR bio-parcel'. In step 5, the XOR bio-parcel generated in step 4, is submitted to the cloud server for authentication. In step 6, the cloud-based server fetc.hes the historic data in the CBA corresponding with the biometric data

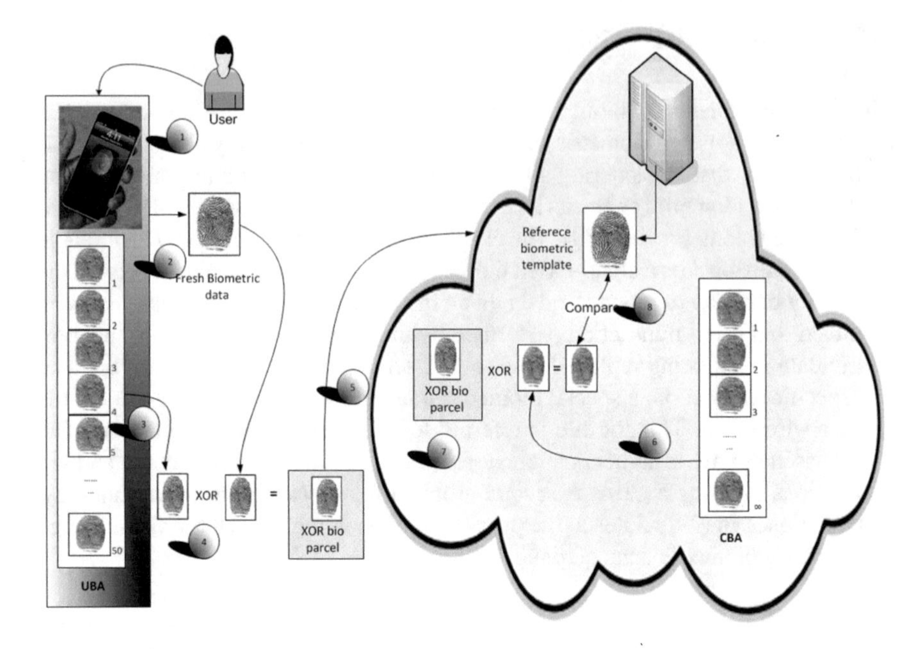

Fig. 6 Cloud-based biometric authentication system

challenged from the user's UBA. During step 7, the XOR bio parcel is unlocked in the cloud by applying the XOR calculation of the historic bio-data with the XOR bio parcel received from the user, yielding the fresh biometric data of the user recently scanned and digitized. In step 8, the fresh biometric data received from the user, is matched with the reference biometric template of this user on file by the cloud service. If the match falls withing the defined parameters of the matching algorithm, the user is considered authentic. The fresh biometric data is added to the CBA for this user's profile.

At this stage the user is successfully authenticated. The result of the authentication can at this stage be conveyed using various existing protocols to the user, or to institution requesting authentication.

For the protocol to function, a user must be able to supply fresh biometric data, and be in possession of a Registered IOT-token, which can supply historic biometric data from the UBA. Should the IOT-token, be stolen, an instruction can be sent to the IOT-token, to clear all UBA content. The IOT-token ensures that the UBA can only be used by the authentic user.

If a person at any stage feels that his or her identity is compromised in one or other way, this person can remove all existing biometric data by cleaning all UBA and CBA data, and removing the reference biometric template. A user can start with a clean slate, re-creating the UBA and CBA, with a fresh reference biometric template. The old "identity" is archived.

8 Conclusion

As long as a person uses the Cloud-based biometric authentication system, the person's biometric characteristics are safe, and such a person does not need to concern himself with the possibility that latent biometric data might be fraudulently used.

References

1. Mitra S, Gofman M (2016) Biometrics in a data driven world: trends, technologies, and challenges. Publisher, CRC Press, California
2. Hariharan R, Logeswari S (2019) Finger print as a personal identification tool in forensic science. Global J Res Anal 8.1
3. Oforka VO (2016) Evaluation of factors influencing biometrics adoption in Nigeria banks. Dissertation, Federal University of Technology, Owerri
4. Chauhan A, Jyoti S (2015) A review; timeline of palm prints since beginning till now. Int J Res (IJR) 2(06)
5. Cothron GR (2012) Fingerprint evidence Part I: tracing friction ridges through history. SSRN 2130808
6. Sodhi GS, Jasjeet K (2005) The forgotten Indian pioneers of fingerprint science. Curr Sci 88(1):185–191
7. Matsumoto T et al (2002) "Impact of artificial" gummy fingers on fingerprint systems. In: Optical security and counterfeit deterrence techniques IV, vol 4677. International Society for Optics and Photonics
8. Matyas Jr SM, Peyravian M (2003) Protection of biometric data via key-dependent sampling. U.S. Patent No 6,507,912. 14 January 2003
9. Borza SJ, Borza MA, Teitelbaum N (1999) Mouse adapted to scan biometric data. U.S. Patent No 5,991,431. 23 November 1999
10. Tait B, Von Solms B (2008) Secure biometrically based authentication protocol for a public network environment. In: International conference on global e-Security. Springer, Berlin, Heidelberg
11. Tait BL (2014) The biometric landscape, towards a sustainable biometric terminology framework. Int J Electron Secur Digit Forensics 9 6(2):147–156
12. Liu S, Silverman M (2001) A practical guide to biometric security technology. IT Professional 3(1):27–32
13. Jain AK, Arun R, Salil P (2004) An introduction to biometric recognition. IEEE Trans Circuits Syst Video Technol 14(1):4–20
14. Garcia P (2018) Biometrics on the blockchain. Biom Technol Today 2018(5):5–7
15. Xia F et al (2012) Internet of Things. Int J Commun Syst 25(9):1101
16. Apple Watch Series 5 homepage. https://www.apple.com/apple-watch-series-5/health. Last accessed 5 Feb 2020
17. Samsung Facial recognition. https://www.samsung.com/global/galaxy/what-is/iris-scanning/. Last accessed 5 Feb 2020
18. Samsung Fingerprint recognition. https://www.samsung.com/global/galaxy/what-is/biometric-authentication/. Last accessed 5 Feb 2020
19. Samsung Facial recognition. https://www.samsung.com/ca/support/mobile-devices/galaxy-note10-facial-recognition/. Last accessed 5 Feb 2020
20. Samsung Handwriting homepage. https://www.samsung.com/sg/support/mobile-devices/how-do-you-convert-handwriting-to-text-and-other-formats-using-s-pen-and-samsung-notes/. Last accessed 5 Feb 2020

21. Apple iPhone XS facial recognition. https://support.apple.com/en-za/HT208108. Last accessed 5 Feb 2020
22. Guo Z et al (2016) Hardware security meets biometrics for the age of IoT. In: 2016 IEEE international symposium on circuits and systems (ISCAS). IEEE
23. Clarke J, Christopher C (2014) The potential subversion of biometric measurements. SSRN 2381614
24. De Freitas P et al (2013) Can face anti-spoofing countermeasures work in a real world scenario? In: 2013 international conference on biometrics (ICB). IEEE
25. Adra B (2018) Facing the facts on biometric phone locks: your face and thumb not secure. U Ill JL Tech Pol'y 407
26. Samsung phones can be unlocked with any fingerprint. https://www.independent.co.uk/life-style/gadgets-and-tech/news/samsung-galaxy-s10-fingerprint-sensor-security-privacy-a9160486.html. Last accessed 5 Feb 2020
27. Researchers bypass Apple Faceid. https://threatpost.com/researchers-bypass-apple-faceid-using-biometrics-achilles-heel/147109/. Last accessed 5 Feb 2020
28. Toth B (2005) Biometric liveness detection. Inf Secur Bull 10(8):291–297
29. Samsung's Galaxy S10 fingerprint sensor fooled by 3D printed fingerprint. https://www.theverge.com/2019/4/7/18299366/samsung-galaxy-s10-fingerprint-sensor-fooled-3d-printed-fingerprint. Last accessed 5 Feb 2020
30. The reason behind the Samsung fingerprint spoof. http://www.digitaljournal.com/tech-and-science/technology/the-reason-behind-the-samsung-fingerprint-spoof/article/560407. Last accessed 5 Feb 2020
31. Jahromi MNS et al (2018) Automatic access control based on face and hand biometrics in a non-cooperative context. In: 2018 IEEE winter applications of computer vision workshops (WACVW). IEEE
32. Hong Lai LQ, You J (2020) A novel method for touchless palmprint ROI extraction via skin color analysis. In: Recent trends in intelligent computing, communication and devices. Springer, Singapore, pp 271–276
33. World Health Organization. https://www.who.int/emergencies/diseases/novel-coronavirus-2019/advice-for-public. Last accessed 5 Feb 2020
34. Bowyer KW (2004) Face recognition technology: security versus privacy. Technol Soc Magaz IEEE 23(1):9-19. ISSN: 0278-0097. IEEE Society on Social Implications of Technology
35. Woodward Jr JD, Orleans MN (2003) Identity assurance in the information age—biometrics. ISBN 0-07-222227-1
36. UK Border control. https://workpermit.com/news/uk%E2%80%99s-home-office-announces-cutting-edge-iris-recognition-technology-modernize-uk-border-control. Last accessed 5 Feb 2020
37. Retail Bank Opts for Biometric ATMs. https://findbiometrics.com/archive/south-africas-fastest-growing-retail-bank-opts-for-biometric-atms/. Last accessed 5 Feb 2020
38. Liu S, Silverman MA (2001) Practical guide to biometric security technology, pp 27–32. ISBN 1520-9202. IEEE Computer Society
39. Tait BL, Von Solms SH (2009) BioVault: biometrically based encryption. In: Conference on e-Business, e-Services and e-Society. Springer, Berlin, Heidelberg
40. Rehman YAU, Lai MP, Mengyang L (2018) LiveNet: improving features generalization for face liveness detection using convolution neural networks. Expert Syst Appl 108:159–169

Evaluating Multi-layer Security Resistance to Adversarial Hacking Attacks on Industrial Internet of Things Devices

Hussain Al-Aqrabi and Richard Hill

Abstract A primary concern of Industrial Internet of Things (IIoT) users is the threat of loss of valuable Intellectual Property (IP) through insecure operational device security. Whilst robust levels of security are technically possible, the approaches taken to ensure resistance to adversarial attacks can lack practicality in terms of implementation. IIoT devices use constrained hardware which can limit the extent to which data can be stored, processed or communicated and this can potentially increase the vulnerability of a system as additional IIoT devices are introduced. We explore the use of a multi-layer approach to security that produces an exhaust-trail of digital evidence at different levels, depending on the characteristics of the system attack. This approach is then evaluated with respect to common categories of system breach, and a set of characteristics and considerations for system designers is presented.

Keywords Multi-layer · Security · Digital forensics · Industrial Internet of Things · Cloud computing

1 Introduction

The adoption of new business models to take advantage of Industrial Internet of Things (IIoT) technology is subject to legitimate security and privacy concerns [1]. In particular, industrial users understand that a large proportion of the business value generated by such technology is directly related to the ongoing creation and ownership of Intellectual Property (IP) from industrial operations data. An infringement of security that could impair the exclusivity of IP poses a serious risk to an enterprise's underlying business model [2].

H. Al-Aqrabi (✉) · R. Hill
Department of Computer Science, University of Huddersfield, Huddersfield, UK
e-mail: h.al-aqrabi@hud.ac.uk

R. Hill
e-mail: r.hill@hud.ac.uk

R. Montasari et al. (eds.), *Digital Forensic Investigation of Internet of Things (IoT) Devices*, Advanced Sciences and Technologies for Security Applications,
https://doi.org/10.1007/978-3-030-60425-7_8

As IoT [3, 4] is emerging rapidly [5], IoT applications may greatly improve industrial efficiency and human behaviour [6]. The development process can be handled more precisely and dynamically by embedding and incorporating IoT devices into industrial systems. IoT is a way to connect physical objects and devices to the Internet to form omnipresent networks which allow the sensing of changing environments in response to dynamic impulses [7].

Although the IoT is developing rapidly, IoT security is becoming extremely prevalent, primarily for IIoT applications [8]. The perception layer is the root of all of IoT data [9] and the basis of all of the IoT architecture. The physical protection of sensor layer devices would be more at risk than the IoT application [10, 11] and transportation layers. As the sensors are common in industrial and agricultural settings, confidential information can be collected directly by an adversary if nobody monitors the sensors for long periods of time [12]. Thus, many circulating IoT devices are vulnerable to cyber attacks.

Analysis of incidents and forms of evidence in criminal investigations and digital forensics are essential procedures. The purpose of reviewing IoT incidents is because of the large number of cyber attacks that have been identified [13]. Other than that, due to the vast IoT world introduced by various multi-level third-party vendors, products, protocols, operating systems and IoT facilities, the issue of the necessary procedures for forensic investigations is challenging [14].

Although IoT and cloud computing are an instance of how innovations and business models may incorporate new business capabilities [15], enterprises are still threatened by new risks which are directly due to the use of flexible, frequently multi-tenant IoT cloud services.

However, the research literature discloses multiple IoT threat models focused on IoT properties, none of which implemented a robust IoT attack model and undermined security objectives for a highly complex model [16].

For this reason, IoT security is a major concern for individuals, users, organisations, industrial sector and government entities who want to prevent their objects from being compromised or hacked [17].

1.1 Cloud Computing in IIoT

Cloud computing virtualises storage and enables the sharing of server computing resources with other devices such as personal computers and smartphones, with data centres [2]. The use of the same technology in the IoT environment enables sensor nodes to send and receive data from or to other networks in a central location that many different sensors can access [18].

Both Cloud computing and IoT work to improve the efficiency and interdependence of daily activities. In various industries, IoT is used as an approach in business settings [19]. In this respect, Internet infrastructure is constantly under strain due to the volume of big data generated by IoT. Cloud technology is laying the ground work for transporting and managing this data [15]. In that way, businesses and organisations pursue an alternative to ease the burden.

Improved collaboration allows cloud computing to improve innovation significantly. IoT businesses can now collect immense amounts of data by storing it in the cloud. It also allows businesses to use computer resources as utilities, such as a virtual machines (VM), rather than installing (on premises) a computing infrastructure [20].

1.2 Advantages of IoT in an Industry Context

The IoT brings the advantages of IoT at a greater level as well as in industries in which massive threats can result from human error. The degree of precision obtained by IIoT is a huge benefit and one of the most promising developments of IoT in this field. In order to be safer and more efficient, IIoT provides a new way for industries to enhance their processes. IIoT systems enhance smart communication between devices or machines, increasing performance and profitability in several sectors, from manufacturing to healthcare. However, as with everything linked to the Internet, IIoT devices are vulnerable to cyber attacks.

In particular, recent attacks on industrial system control, for instance, supervisory control and data acquisition (SCADA) systems, as well as distributed control systems, and programmable logic controllers, have been used as gateways to orchestrate cyber attacks.

Several researchers have contributed to strengthening security, privacy [21], scalability and data processing by integrating algorithms from various technologies [22, 23]. The remainder of this article is organised as follows. In Sect. 2, we highlight the IIoT communication protocols and the most common attacks in an IIoT environment. Section 3 describes a multi-layer security framework. In Sect. 4, we discuss and evaluate IIoT Security threats. Finally, we conclude in Sect. 5.

2 Cybersecurity and IIoT

The industry-standard guidelines for adopting IIoT technologies in industrial environments are proposed to smooth integration of various services, which form the pillars of the fourth industrial revolution [24].

As a result of various government-supported initiatives, the IIoT is expected to gain attention as it serves as a new manufacturing paradigm that guarantees adaptability and flexibility. Governments of different countries are committed to implementing IIoT technologies to optimize their economic development.

Such advancements in technology considerably boost IIoT's ability to sense and recognise things or the environment.

2.1 IIoT Communication Protocols

IIoT communication protocols are communication mechanisms which protect data exchanged between connected IoT devices and achieve maximum protection. The requirements for communication differ enormously between various kinds of IIoT networks and are extremely different in terms of their functional- ity and resource limitations. Such networks have a variety of features including management and security.

IIoT systems can support wireless or wired communication protocols. There are some protocols and standards for IIoT communication. Several wireless protocol technologies, such as Bluetooth Low Energy (BLE), can easily communicate with modern mobile platforms, using short-range radio protocols.

ZigBee involves the sharing of limited amounts of data in a confined region. It is often used for communication between the sensor nodes and controllers when the sensors are remotely positioned outside of the controller. The Mobile Radio Frequency Identification (RFID) systems is used extensively in mobile identification.

In longer-range radio protocols such as LoRaWAN protocol, which are designed to have low power consumption and broad network availability, this communication protocol is used in energy management and smart cities. The SigFox protocol provides a reliable transfer of remarkably small amounts of data over long distances. The Sigfox communication protocol is standard in intelligent metering and environment sensors.

Wired communication protocols, for instance, Ethernet, USB, etc., as well give access to devices.

Various protocols are classified by a layer of communication. The session layer set protocols that permit messaging between the IoT communication subsystem components. For instance, the Constrained Application Protocol (CoAP), and the MQ Telemetry Transport (MQTT) protocol [25].

The network layer comprises of two tiers. The first layer manages the exchange of packets through a routing layer and the encapsulation layer focuses on creating packets.

The data link layer manages the communication protocol between different IoT devices via a physical connection, e.g. between IoT sensors and the gateway, connecting a collection of IoT sensors to the Internet, either wireless or wired.

2.2 Cyber Threats to IIoT

Increased reliance on intelligent, interconnected devices can put billions of intelligent communicating "things" that could significantly endanger personal privacy and threaten the security of the public in every aspect of our lives.

The numerous interconnections and the heterogeneity of a large number of devices, and technologies and systems in IIoT [26], create potential cyber physical security (CPS) [27] weaknesses, which can be exploited later by an adversary [28]. Here are the most common IoT attacks:

- **Denial-of-service (DoS) attack**—DoS cyber attacks occur when the attackers try to block legitimate users from accessing the IIoT service. The adversary begins the attack by first leveraging device vulnerabilities and then installing malware in their hardware and/or software. Multiple compromised IIoT devices are called botnets. The goal is to attack through huge amounts of requests from numerous IIoT devices in various locations.
- **Jamming attack**—This type of attack is carried out by jamming communication between wireless IIoT devices to jeopardise it. An intruder unexpectedly transmits a radio signal, as sensor nodes send communication signals. This causes the network is jammed such that no message can be sent or received.
- **Man-in-the-middle (MitM) attack**—MitM is an attack where an intruder intercepts a communication between two users (the sender and the receiver) to eavesdrop or alter traffic between them. For instance, these attacks induced by an attacker attempting to force the sensor node to transmit misleading data.
- **Side channel attack**—This attack is carried out when the attacker might to jeopardise the IIoT cloud and position a malicious virtual machine near the IIoT cloud server and then initiate a side channel attack. Also, the intruder gathers information on the encryption keys by monitoring the signal leakage, such as 'side channels'. Therefore, it is important to design a secure framework.
- **Social engineering attack**—This attack is an attack vector that involves human interaction. Attackers use various ways to collect basic knowledge about the target. Phishing is the most common type of attack by the social engineering. Phishing attacks target human errors to seize credentials or distribute malware. usually if a user is clicking on the malicious attachment or link that then installs malware or additional dangerous software.
- **Malware attack**—The malware specifically designed to attack IIoT devices is called IIoT malware. The first significant attack attempt is to inject potentially malicious services or virtual machine into the IIoT cloud network. This adversary attack establishes a virtual machine instance or malicious service and attaches it to the IIoT network. For instance, Ransomware is a form of malware that uses encryption to prevent users from accessing their device and locking personal user files.
- **Sybil attack** is an attack to create a deep illusion of traffic congestion; for instance, the attacker uses several fake identities to communicate between the network and adjacent nodes.
- **A replay attack** takes place when a hacker exploits and intercepts a secure network communication, and then slows or maliciously redirects the receiver to do as the hacker intends.
- **Exploit kits** where the hacker uses a malicious script for leveraging badly patched IIoT system vulnerabilities.

- **Forged malicious** device occurs when attackers disable legitimate IIoT devices when they have physical access to the network.

Al of the above attack vectors are common to network enabled systems and numerous strategies and frameworks have been developed to detect and limit the detrimental effects of such attacks [29]. However, the emerging societal environment is becoming increasingly reliant on "connected-ness" and from an industrial perspective this creates additional challenges for the secure operation of its operations [30].

Many industrial processes involve the conversion of materials from one state to another, or the transport of physical goods.

While developments in technology have improved commercial efficiencies through the use of industrial automation, it is only recently that such automation has been connected via high speed networks. These new developments are referred to as Cyber Physical Systems (CPS), where the automation of physical actuation is controlled either autonomously, or at least remotely via a communication network [11]. CPS is central to the Industry 4.0 movement, otherwise referred to as Digital Manufacturing or Industrial Digital Technologies (IDT) [24] (Table 1).

These systems are of considerable interest to adversarial attacks for the purposes of industrial espionage, and therefore the attack vectors characterised in this section are potential weak points (particularly in relation to the scheduling of utility resources [31]) for the safe operation of industrial processes.

3 Multi-layer Security Framework

The flexibility of cloud computing model driven us to recognise each layer as a separable cloud containing countless hardware resources, including servers arrays, IoT devices, storage arrays and so on [32].

The primary approach to managing connectivity in a highly equipped setting, for instance, a smart industry, is to identify and manipulate different security vulnerabilities. As a result, the attack surface continues to change. So the only feasible solution is multi-layer security.

While multi-layered security reflects the concept that vast array security measures will prevent threats from occurring on your system. In order to genuinely achieve the multi-layer security approach, a framework is needed to break down and align business IT security problems.

The authors have developed a security framework that can be deployed in a smart factory. The framework divides a range of control positions into multiple layers, inspired by a fundamental principle of cloud computing resource abstraction.

Figure 1 illustrates a multi-layered security cloud framework, with one of the privacy/security services offered by each layer, to support the development of IIoT cloud applications.

Table 1 Characterising IIoT security threats

Security attack	Impact	Probability	Threat level	Countermeasures	Affected
Denial-of-service attack	High	Very likely	High	Yes	Application and services platform and backend IoT devices
Jamming attack	High	Very likely Very likely	High	Yes	Communication information IoT devices
Man-in-the-middle attack	High	Very likely	High	Yes	Communication information IoT devices
Side channel attack	Medium	Likely	Medium	Yes	Communication information IoT devices
Social engineering attack	Medium	Likely	Medium	Yes	Communication information IoT devices
Malware attack	Medium	Likely	Medium	Yes	Platform and backend IoT devices
Sybil attack	Medium	Likely	Medium	Yes	Communication information IoT devices
A replay attack	Medium	Likely	Medium	Yes	Information decision making IoT devices
Exploits kits attack	Medium	Likely	Medium	Yes	Infrastructure IoT devices
Forged malicious devices	Medium	Likely	Medium	Yes	Platform and backend IoT devices

The IIoT Cloud consists of 7 layers. The layered service is provided by servers that are deployed in it to guarantee that performance bottlenecks are not present. Each layer can be seen by itself as a cloud.

The sessions are reviewed by each layer and require further communications with the next level above them. The sessions eventually enter the IIoT cloud layer running IIoT applications and more connectivity to other devices after a range of further examinations.

The tenant sessions access the IIoT Cloud via firewalls which proceed through a sequential series of firewalls serving as gateways.

The firewalls operate as firewalls in the network, transportation and applications. It can also enable itself through access control lists (ACLs) for exemplary services and applications, for instance, based on protocols, TCP/UDP ports, IP addresses etc.

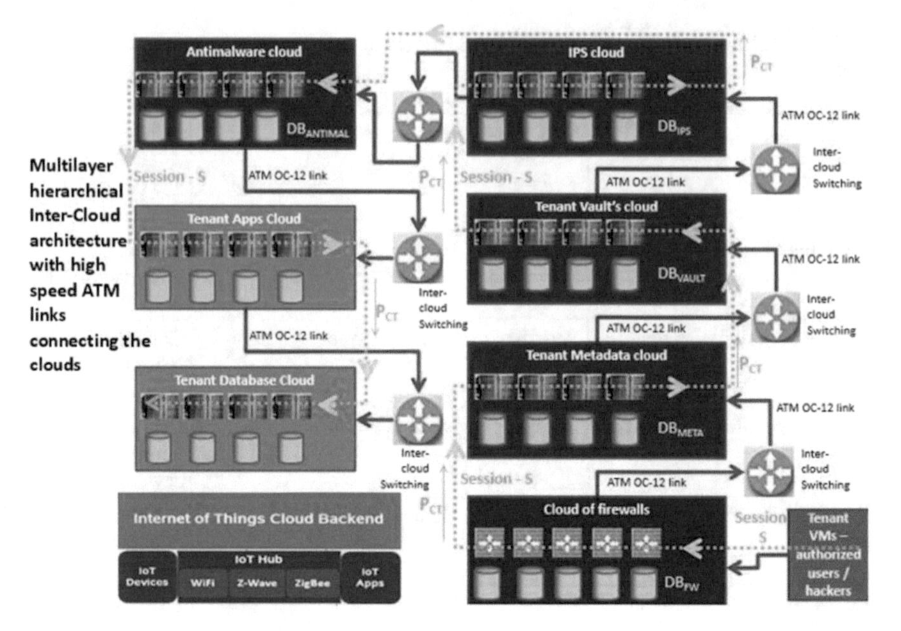

Fig. 1 Multi-layer security framework in IIoT cloud [9]

3.1 Adversarial Attacks

In this section, we shall describe an adversary attack on the model and illustrate how the system offers protection from such a scenario from happening. The scenario is that an attacker wishes to gain access to an OLAP system, which is a common business scenario. Many schemes have been developed to protect such systems, including methods to preserve privacy in OLAP system design [33, 34].

Now if we take into account an adversary attack on the model, referring to Fig. 2 which demonstrates the scenario where an adversary attacker tries to penetrate the system and gain entry to the virtual machine (VM) of the approved tenant.

Although the IIoT cloud provider needs authentication to be able to register and control remote services, what seems to be a valid user (tenant) may in fact be an attacker pretending as a legitimate user with the intention of accessing the cloud and then targeting another user VMs from inside the cloud itself.

Exploitation tools like Metasploit (https://www.metasploit.com), the attacker may use to automate the distribution of exploits easily. This could make it possible for a false tenant to establish a subversive mechanism of trying to expose confidential data, unidentified to any other group or party.

If the adversarial agent is a valid tenant, they would then target a VM that is approved and stored in the IoT cloud. This will not require traditional cloud protection mechanisms to identify such an operation.

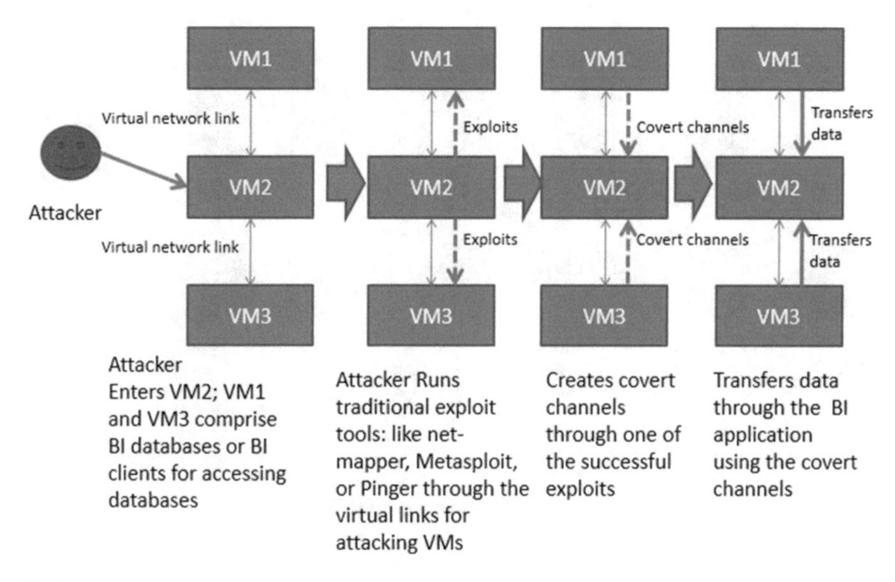

Fig. 2 Attack scenario for a multi-layer IIoT security architecture [9]

In addition, the number of VMs in a cloud environment obscures the operation. This poses a major challenge for cloud service providers, especially as Microservice Architectures orientation becomes increasingly prevalent [35]. As we apply this to the IoT environment, we want to bundle features in applications that can be deployed on distributed hardware. Consequently, a core aspect of this research is the opportunity to tackle this issue.

When an adversarial attacks the cloud with a subscription to (second) VM2, an adversarial attacker has the ability to use cross-channel attacks against different virtual machines like (first) VM1 and (third) VM3, thus leveraging the existence of simulated links between specific VMs. Cloud security checks are usually applied to deter external threats instead of internal threats.

A hierarchical framework described in Fig. 2, as a solution to the attack scenario. This framework reduces the ability to perform further hacks, as virtual links prohibit the attacker from going to the next control. The only alternative is to use a real network link to connect with the (fourth) VM4, (fifth) VM5 and sixth VM6, through demanding a session in Control A (for instance, Tenant Metadata inspection (TMI). Although Control A is still in operation, the intruder should effectively fulfil the TMI to begin the attack.

A coordinated and persistent attack would, of course, cause one to believe that an adversary would have credible credentials, whether by pretending or else as a legal occupant.

In Control B, it would be necessary for the adversarial attacker to conquer the whole cloud layer prior to the IoT analytics interface could be reached.

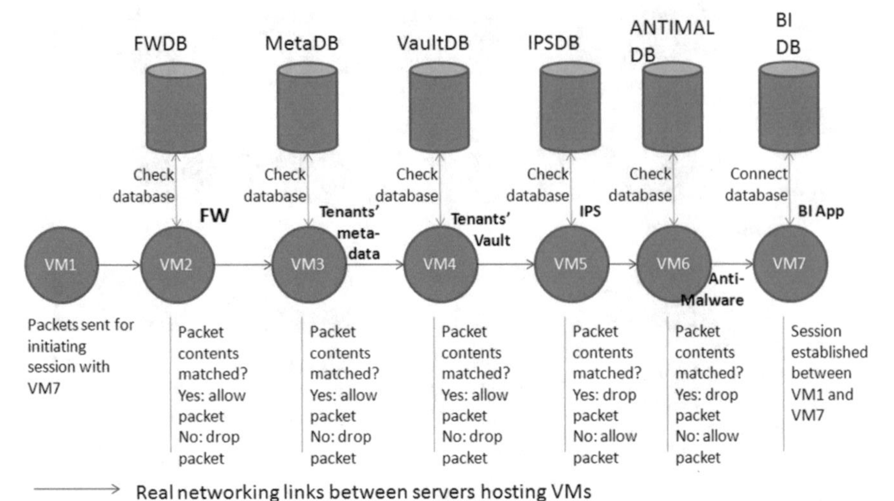

Fig. 3 Hierarchy of controls [9]

When the intruder has fulfilled the cloud authentication and the (TMI) layers, surely the best step ahead now is to hack the plant with the expectation that they will be completely unnoticed. Nevertheless, both the tenant anti-malware (TAM) layer and the tenant Intrusion Protection (TIP) layer provide adequate security and protection in defence from the inside for malicious and covert attacks.

Even when adversarial attacks are launched by what appears to be legitimate subscribers of service, our proposed model prevents data violations. A sequence with various security controls can be found in Fig. 3.

In order to achieve the objectives sought by the infrastructure provider, The system or host system security policies decision must notify the order in which controls are carried out. How a tenant is channelled via the various VMs to access the related IoT analytic services is also apparent. The result is that the model prevents data breaches, even when adversarial attacks are launched from what appears to be genuine service subscribers.

We can see in Fig. 3 a sequence in which various security controls might be instantiated. The security policies of the host system (or systems) will inform the order in which controls are implemented, to suit the goals desired by the infrastructure provider.

It is also evident how a tenant's session is routed through the various VMs in order to access the relevant analytics services.

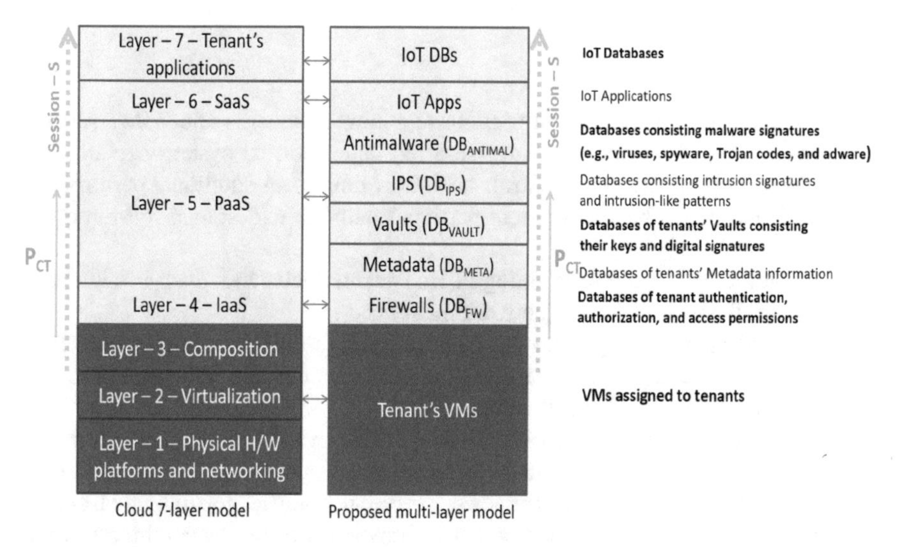

Fig. 4 Mapping of the proposed framework with the NIST seven-layer [32]

3.2 Session Flow

The stated *S* session workflow is illustrated in Fig. 4, the validations are concurrent and are clarified in relation to the positioning of controls also illustrated in Fig. 3.

The suggested framework lies on layer four as the infrastructure Service Model (IaaS) and five as the Platform of Service (PaaS) for the NIST seven-layer framework. Each VM in a set of scales obtains an instance ID that identifies it uniquely.

According to the feature of inspecting VM instance IDs, all firewalls are classified as IaaS and often refer these inspections to the authorization data provided by a tenant to prevent IoT cloud access. Whereas VM IDs in layers two and three are allocated, access controls on layer four are allocated.

For a given session *S*, concurrent controls are delegated outside the VM layer and are highly concerned with the inspection of the session packets.

Data is requested explicitly from the tenant to fulfil the checkpoint controls of tenant's metadata database (DB_{META}) and tenant's vault database (DB_{VAULT}) and is augmented by intrusion prevention database (DB_{IPS}), and anti-malware database ($DB_{ANTIMAL}$) controls, which is known as Platform as a Service PaaS controls, that serve the session packet inspection against malware signatures and other attacks.

Controls on the application layer are also likely to take place. For example, a Software as a Service (SaaS) instance would therefore necessitate tenant authentication just as a specific business application would require. The tenants may also have an extra level of authentication system set up to obtain information based on their organisational position.

4 Discussion

Both systems administrators and users have genuine concerns about IIoT security mechanisms and their potential adverse affect upon normal system operation. In particular, additional security controls inevitably creates an additional performance overhead, and this is becoming more pertinent with the widespread roll-out of 5G infrastructure.

5G is attractive for the retro-fitting of IIoT technologies to industrial settings as it eliminates the need for wired networking access.

Faced with such developments, we must consider the attack vectors characterised in Sect. 3.1, especially in light of the physical actuation that industrial processes involve and the potential harm to human life.

Multi-layer security for IIoT offers considerable advantages for the identification and management of emerging attacks as adversaries become more sophisticated.

First, each of the layers of the framework serves not only to stop individual attacks at various levels, but it also protects against compound attacks that combine a variety of different techniques to exploit system vulnerabilities.

Second, the process of identifying different stages of attacks also serves to characterise the separate elements of an attempt to broach a system. If the exploit is successful, the multi-layer architecture thus provides an improved and enriched evidence base by identifying the constituent parts of an attack, which facilitates the forensic re-construction of an event in order to further harden the system in the future.

As it stands, the multilayer architecture benefits from the elastic scalability of utility computing in that as new threats are identified, they can either be added to existing threat detection layers, or additional layers can be rapidly provisioned on demand. This elasticity is embraced by emerging software development approaches such as the use of Microservices to compartmentalise functionality into robust, reusable units [35], together with Software Defined Virtualization Functionality to facilitate rapid scalability [36–38].

Related work [39] describes a situation where authentication protocols are enhanced to enable secure access to a system that is marshalled by multiple parties. In particular, simulation studies have identified that while a performance overhead is added to initial checks of the participating agents, the actual effect of such overheads is acceptable in terms of system performance.

Similarly, the provisioning of new layers to protect against new threats also increases the communication overhead between layers and this is one aspect that requires further study to assess how scalable it is. The cloud-inspired architecture does mean that layers can be instantiated on a case-by-case basis, and the automation of this functionality is another potential avenue to explore.

Another aspect that is particular to IIoT, is that most of the 'smart' hardware is severely constrained in terms of hits storage and computation capabilities [40]. This limitation has driven adversaries to adopt distributed attacks that either compromise a large number of edge nodes (a DOS attack), or a large number of nodes are recruited to donate many small contributions of resource to support more sophisticated and comprehensive attacks across the network.

We see the multi-layer approach as providing a scalable approach to threat management, with different layers not only residing upon multiple clouds, but also across multiple IIoT nodes. Some nodes have limited functionality, such as reporting temperature, humidity or vibration for instance, for other processes to utilise as part of their operations. In such cases not all layers of the multilayer architecture are required and therefore a sub-set can be deployed without detriment to the system. This will also reduce the additional of redundant functionality that will adversely affect system performance.

A crucial part of the effective deployment of multi-party architecture is having a rigorous approach to the modelling and specification of the systems to be connected. Traditionally, large-scale systems have been designed at least as sub-systems, with controls and measures put in place to apply testing methods for the eventual development of program code [1, 41, 42].

However, IIoT systems are inherently flexible and elastic in nature, leading to the use of cloud technologies for potential solutions. This does impact upon the modelling of such situations, that are inherently multi-agent based, with complex communication requirements, the specification of which may only emerge after deployment.

We have thus considered the use of social network communication models and also related work in the secure communication of sensitive healthcare data in distributed care provision environments, where numerous agents are required to collaborate, cooperate and exchanged data on a need-to-know basis [43–46]. This has strong parallels with the IIoT environment, where commercially-sensitive process data is captured and shared with designated, authenticated stakeholders to improve the efficiencies of manufacturing value-chains.

5 Conclusions and Future Directions

A key concern for IIoT users is the risk of loss of precious IP by insecure security of operational devices. IIoT devices use restricted hardware to limit the extent to which information is stored, analyzed or exchanged, potentially raising a network vulnerability as IIoT devices are implemented.

Although robust security standards are theoretically feasible, there is a lack of practicality in terms of implementation in order to guarantee resistance to adverse attacks. The proposed work demonstrates the ability for attack prevention from both external and internal attacks, which are mainly of relevance in the context of IIoT.

A broad variety of surreptitious operations can be separated to withstand a number of attack vectors by taking protection at any layer of cloud abstraction into account.

This can further be enhanced by monitoring the client IIoT side channel parameters for added security. This convergence of software and hardware technologies guarantees that Cloud infrastructure is secure and stable during the IIoT revolution.

References

1. Alrawais A, Alhothaily A, Hu C, Cheng X (2017) Fog computing for the Internet of Things: security and privacy issues. IEEE Internet Comput 21(2):34–42
2. Al-Aqrabi H, Hill R (2018) Dynamic multiparty authentication of data analytics services within cloud environments. In: Proceedings of the 20th international conference on high performance computing and communications, 16th international conference on smart city and 4th international conference on data science and systems, HPCC/SmartCity/DSS2018. IEEE Computer Society, pp 742–749
3. Ashton K (2009) That "Internet of Things" thing. RFiD J
4. Gubbi J, Buyya R, Marusic S, Palaniswami M (2013) Internet of Things (IoT): a vision, architectural elements, and future directions. Futur Gener Comput Syst 29(7):1645–1660
5. Gartner (2015) Gartner says 6.4 billion connected things will be in use in 2016, up 30 percent from 2015? Gartner website, https://www.gartner.com/newsroom/id/3165317. Accessed 10 Nov 2015
6. Chen D et al (2019) A multi-layer hardware Trojan protection framework for IoT chips. IEEE Access J 7:23628–23639
7. Atzori L, Lera A, Morabito G (2010) The Internet of Things: a survey. Comput Netw 54(15):27872805. ACM
8. Al-Aqrabi H et al (2020) Hardware-intrinsic multi-layer security: a new frontier for 5G enabled IIoT. Sensors 20(7):1963
9. Al-Aqrabi H, Hill R (2018) A secure connectivity model for Internet of Things analytics service delivery. In: 2018 IEEE smartworld, ubiquitous intelligence and computing, advanced and trusted computing, scalable computing and communications, cloud and big data computing, internet of people and smart city innovation (SmartWorld/SCALCOM/UIC/ATC/CBDCom/IOP/SCI). IEEE, pp 9–16
10. Nastase L (2017) Security in the Internet of Things: a survey on application layer protocols. In: 2017 21st international conference on control systems and computer science (CSCS), pp 659–666
11. Tedeschi S, Mehnen J, Roy R (2017) IoT security hardware framework for remote maintenance of legacy machine tools. In: Proceedings of the second international conference on Internet of Things and cloud computing, ICC 2017, Cambridge, United Kingdom, March 22–23, 2017, pp 43:1–43:4
12. Al-Aqrabi H, Liu L, Hill H, Cui L, Li J (2013) Faceted search in business intelligence on the cloud. In: Proceedings of GREENCOM-ITHINGS-CPSCOM. IEEE, China, Beijing, pp 842–849
13. Hossain MD et al (2015) Towards an analysis of security issues, challenges, and open problems in the Internet of Things. In: IEEE world congress on services, pp 21–28
14. Perumal S, Norwawi N, Raman V (2015) Internet of Things (IoT) digital forensic investigation model: top-down forensic approach methodology. In: IEEE fifth international conference on digital information processing and communications (ICDIPC), pp 19–23
15. Hwang K, Chen M (2017) Big-Data analytics for cloud, IoT and cognitive computing. Wiley
16. Mohsen A, Jha NK (2016) A comprehensive study of security of Internet-of- Things. IEEE Trans Emerg Top Comput 99
17. European Research Cluster on The Internet of Things (IERC) (2015) Internet of Things: IoT governance, privacy and security issues. In: European research cluster on the Internet of Things, pp 128
18. Shahid R et al (2017) SecureSense: end-to-end secure communication architecture for the cloud-connected Internet of Things. Future Gener Comput Syst 77:40–51
19. Farhan MS, Marie ME, El-Fangary LM, Helmy YK (2012) Transforming conceptual model into logical model for temporal data warehouse security: a case study. Int J Adv Comput Sci Appl 3(3):115–122

20. Al-Aqrabi H, Lane P, Hill R (2019) Performance evaluation of multiparty authentication in 5G IIoT environments. In: Cyberspace data and intelligence, and cyber-living, syndrome, and health, pp 169–184
21. Sun G et al (2017) Efficient location privacy algorithm for internet of things (iot) services and applications. J Netw Comput Appl 89:3–13. Emerging Services for Internet of Things (IoT)
22. Vazquez JI et al (2009) An architecture for integrating wireless sensor networks into the Internet of Things. In: c3rd symposium of ubiquitous computing and ambient intelligence 2008. Springer, Berlin, Heidelberg, pp 219–228
23. Uckelmann D, Harrison M, Michahelles F (2011) An architectural approach towards the future Internet of Things. Springer, Berlin, Heidelberg, pp 1–24
24. Hill R, Devitt J, Anjum A, Ali M (2017) Towards in-transit analytics for industry 4.0. In: IEEE international conference on Internet of Things (iThings) and IEEE green computing and communications (Green-Com) and IEEE cyber, physicaland social computing (CPSCom) and IEEE smart data (SmartData). IEEE Computer Society
25. Roy DG, Mahato B, De D, Buyya R (2018) Application-aware end-to-end delay and message loss estimation in Internet of Things (IoT)—MQTT-SN protocols. Future Gener Comput Syst 89:300–316
26. Li X, Li D, Wan J, Vasilakos A, Lai C, Wang S (2015) A review of industrial wireless networks in the context of industry 4.0. In: Wireless networks, pp 1–19
27. Shu Z, Wan J, Zhang D, Li D (2015) Cloud-integrated cyber-physical systems for complex industrial applications. In: Mobile network applications, pp 1–14
28. Singh A et al (2018) A cybersecurity framework to identify malicious edge device in fog computing and cloud-of-things environments. Comput Secur 74:340– 354
29. Priebe T, Pernul G (2000) Towards OLAP security design—survey and research issues. In: Proceedings of third ACM international workshop on data warehousing and OLAP (DOLAP 2000), November 10, 2000, McLean, VA, USA. ACM, pp 33–40
30. Mahmoud M, Saputro N, Akula P, Akkaya K (2016) Privacy-preserving power injection over a hybrid AMI/LTE smart grid network. IEEE Internet Things J 1
31. Sotiriadis S, Bessis N, Antonopoulos N, Hill R (2013) Meta-scheduling algorithms for managing inter-cloud interoperability. Int J High Perform Comput Netw 7(2):156–172
32. Al-Aqrabi H et al (2019) A multi-layer security model for 5G-enabled industrial Internet of Things. In: 7th international conference on smart city and informatization (iSCI 2019), Guangzhou, China, November 12–15 2019, Lecture notes in computer science, Switzerland, 8. Springer International Publishing AG
33. Cuzzocrea A, Bertino E, Sacca D (2012) Towards a theory for privacy preserving distributed OLAP. In: PAIS'12, March 30, 2012, Berlin, Germany. ACM, pp 1–6
34. Agrawal R, Srikant R, Thomas D (2005) Privacy preserving OLAP. In: SIGMOD 2005, June 14–16, 2005, Baltimore, Maryland, USA. ACM, pp 1–12
35. Shadija D, Rezai M, Hill R (2017) Towards an understanding of microservices. In: Proceedings of the 23rd international conference on automation & computing, University of Huddersfield, 7–8 September. IEEE
36. Baker C, Anjum A, Hill R, Bessis N, Liaquat Kiani S (2012) Improving cloud datacentre scalability, agility and performance using OpenFlow. In: Proceedings of the 4th international conference on intelligent networking and collaborative systems (INCoS).IEEE, pp 1–15
37. Ndiaye M, Hancke GP, Abu-Mahfouz AM (2017) Software defined networking for improved wireless sensor network management: a survey. Sensors 17(5):1–32
38. Modieginyane KM, Letswamotse BB, Malekian R, Abu-Mahfouz AM (2017) Software defined wireless sensor networks: application opportunities for efficient network management: a survey. Comput Electr Eng: 1–14
39. Al-Aqrabi H, LaneP, Hill R (2019) Performance evaluation of multiparty authentication in 5G IIoT environments, cyberspace data and intelligence, and cyber-living, syndrome, and health. Springer, Singapore, pp 169–184
40. Bessis N, Xhafa F, Varvarigou D, HillR, Li M (eds) Internet of Things and inter-cooperative computational technologies for collective intelligence. In: Studies in computational intelligence. Springer

41. Barni M et al (2010) Privacy-preserving fingercode authentication. In: Proceedings The 12th ACM workshop on multimedia and security—MM Sec '10. ACM Press, New York, New York, USA, p 231
42. Al-Aqrabi H, Hill R, Lane P, Aagela H (2019) Securing manufacturing intelligence for the industrial internet of things. In: Proccedings of the Fourth International Multi-party trust for IIoT applications Congress on Information and Communication Technology, London, UK, February 27, 28, 2019, pp. 267–282
43. Hill R, Polovina S, Beer MD (2005) From concepts to agents: towards a framework for multi-agent system modelling. In: Dignum F, Dignum V, Koenig S, Kraus S, Singh MP, Wooldridge M (eds) Proceedings of the fourth international joint conference on autonomous agents and multi-agent systems (AAMAS 05), Utrecht, The Netherlands, July 25–29. ACM Press, pp 1155–1156. https://doi.org/10.1145/1082473.1082670, ISBN: 1-59593-093-0
44. Hill R, Polovina S, Shadija D (2006) Transaction agent modelling: from experts to concepts to multi-agent systems. In: Proceedings of the fourteenth international conference on conceptual structures (ICCS '06): conceptual structures: inspiration and application, Aalborg, Denmark, July 16–21. Lecture notes in artificial intelligence (LNAI), vol 4068. Springer, pp 247–259
45. Beer MD, Huang W, Hill R (2003) Designing community care systems with AUML. In: IEEE international conference on computer, communication and control technologies (CCCT2003)
46. Beer MD, Hill R, Huang W, Sixsmith A (2003) An agent-based architecture for managing the provision of community care–the INCA (Intelligent Community Alarm) experience. AI Commun 16:179–192. IOS Press
47. Kalra S, Sood SK (2015) Secure authentication scheme for IoT and cloud servers. Pervasive and Mobile Computing, Elsevier 24:210–223
48. Al-Aqrabi H, Liu L, Hill R, Antonopoulos N (2015) Cloud BI: future of business intelligencein the Cloud. J Comput Syst Sci. Elsevier
49. Ferrag MA, Maglaras L, Argyriou A, Kosmanos D, Janicke H (2018) Security for 4G and 5G cellular networks: a survey of existing authentication and privacy-preserving schemes. J Netw Comput Appl 101:55–82
50. Xu LD, He W, Li S (2014) Internet of Things in industries: a survey. IEEE Trans Ind Inf 10(4):2233–2243
51. Hosek IJ (2016) Enabling technologies and user perception with integrated 5g-iot ecosystem
52. Ishaq I et al (2013) IETF Standardization in the field of the internet of things (iot): a survey. J Sensor Actuator Netw 2(2):235–287
53. Akpakwu GA et al (2017) A survey on 5g networks for the Internet of Things: communication technologies and challenges. In: IEEE Access
54. Elkhodr M, Shahrestani S, Cheung H (2016) The Internet of Things: new interoperability, management and security challenges. Arxiv:1604.04824
55. Al-Aqrabi H, Liu L, Hill R, Antonopoulos N (2014) A multi-layer hierarchical inter-cloud connectivity model for sequential packet inspection of tenant sessions accessing BI as a service. In: Proceedings of 6th international symposium oncyberspace safety and security and IEEE 11th international conference on embedded software and systems, France, Paris, March 20–22. IEEE, pp 137–144
56. He D (2012) An efficient remote user authentication and key agreement proto- col formobile client and server environment from pairings. Ad Hoc Netw 10(6):1009–1016
57. Deng Y, Fu H, Xie X, Zhou J, Zhang Y, Shi J (2009) A novel 3GPP SAE authentication and key agreement protocol. In: Proceedings of international conference networks infrastructure and digital content. IEEE, pp 557–561
58. Karopoulos G, Kambourakis G, Gritzalis S (2011) PrivaSIP: ad-hoc identity privacy in SIP. Comput Stand Interfaces 33(3):301–314

59. Alrawais A, Alhothaily A, Hu C, Cheng X (2017) Fog computing for the Internet of Things: security and privacy issues. IEEE Internet Comput 21(2):34–42
60. Ma C-G, Wang D, Zhao S-D (2014) Security flaws in two improved remote user authentication schemes using smart cards. Int J Commun Syst 27 (10)

Establishing Trustworthy Relationships in Multiparty Industrial Internet of Things Applications

Oghenefejiro Bello, Hussain Al-Aqrabi, and Richard Hill

Abstract The uptake of smart devices in the manufacturing industry is accelerating as technological advancements enable hardware to become cheaper and more accessible. A primary concern for manufacturing companies, as well as those in the associated logistics supply chains, is how to establish trust between smart devices, such that the delegation of transactional responsibility and accountability, which is required for Industry 4.0, can be facilitated in a secure and sustainable manner. Trustworthy systems enable enhanced manufacturing operations to occur securely, while also providing a robust audit trail of digital evidence to support any future investigations into allegations of system breaches. This chapter examines a specific type of trust relationship that regularly occurs in supply chains—multiparty authentication—and proposes a framework that encompasses both the human and technical factors that must be considered to engender trustworthy relationships between IIoT devices and organisational operations technology.

Keywords Industry 4.0 · Trust · Digital forensics · Security · Industrial internet of things (IIoT)

1 Introduction

The Industry 4.0 movement [1] is a major driving force for the uptake of Internet of Things (IoT) technologies [2–5], commonly cited as the Industrial Internet of Things (IIoT), or Industrial Digital Technologies (IDT).

O. Bello (✉) · H. Al-Aqrabi · R. Hill
Department of Computer Science, University of Huddersfield, Huddersfield, UK
e-mail: o.bello@hud.ac.uk

H. Al-Aqrabi
e-mail: h.al-aqrabi@hud.ac.uk

R. Hill
e-mail: r.hill@hud.ac.uk

R. Montasari et al. (eds.), *Digital Forensic Investigation of Internet of Things (IoT) Devices*, Advanced Sciences and Technologies for Security Applications,
https://doi.org/10.1007/978-3-030-60425-7_9

Ensuring the repeatability of industrial processes is a key part of Quality Assurance mechanisms, in which the manufacturing industry is well versed. As manufacturers continue to discover new value in their products and services, there is a constant demand to improve the efficiency of operations and eliminate all forms of wasteful activity.

The realisation of Industry 4.0 goals is only possible through a significant and concerted effort to automate activities that are either manual at present, partly automated, or they are fully automated but not yet coordinated at a much larger scale. For instance, a computer controlled machine tool may already demonstrate an optimal set of efficiencies for a particular manufacturing plant, based on the constraints of knowledge and resources that are available to it [6, 7].

The use of IDT can raise efficiencies to new levels if data is shared between different, geographically-dispersed sites. This might mean the exchange of process set-up data, or information about the optimum tool selection for a particular type of raw material.

IDT facilitates this new era of information exchange by providing connectivity via computer communication networks, together with standards for file exchange, as well as the ability to collect, condition, aggregate and filter data at the point of origin. Alongside the continued reduction in cost of equipment, developments in the exploitation of IDT hardware are gaining momentum [8].

1.1 Manufacturing Value

The manufacturing industry needs to create value in order for the business to remain sustainable. As operations have become leaner, the concept of value-chains has become prevalent, as manufacturers and associated suppliers work together to coordinate and find new opportunities to create value for their respective enterprises.

Internet communication services have fundamentally accelerated the rate at which operations can be coordinated, and many businesses have taken advantage of this. As a business enabler, Cloud Computing has taken coordination and collaboration much further, by making computation, data storage and networking available as utilities that can be consumed as and when required [9]. The elasticity of cloud computing means that such services scale with the needs of a particular enterprise, make the services more cost-effective to consume as part of a leaner set of operations.

As such, it is the combination of communication and coordination that are at the core of facilitating business interactions, and the increased availability of IDT is supporting and extending this to new business models.

1.2 The Internet of Things (IoT)

For the purposes of this research, we consider IoT (and IIoT/IDT) as comprising discrete physical objects that contain sufficient computation, storage and networking to enable them to sense and react to the environment in which they are situated [10, 11].

As these objects proliferate, either through new products with enhanced, IDT capability, or via retro-fitted IDT technology to existing manufacturing plant, there are new challenges that emerge, including the extent to which devices must respond and reconfigure for new situations [12], as well as the sheer increase in volume of data that is generated and transported [3, 4].

Cyber Physical Systems (CPS) are pertinent examples of systems that possess 'smart' capabilities to enable them to be flexible for both forecasted and emerging requirements. A CPS may include one or more IDT devices that work together to provide the capabilities necessary for a given set of tasks.

Business processes are often collections of different steps required to complete a number of tasks that results in a goal being achieved. Such processes potentially require a multitude of IDT devices to help realise an automated version of that business transaction [13].

1.3 Security and Trust Concerns

It is normal for concerns around security to be expressed when considering communication networks. The security of communications is especially important for manufacturers who wish to protect the methods by which they create value, to maintain a sustainable enterprise [9].

A considerable quantity of Intellectual Property (IP) is concentrated within the design and operation of manufacturing processes, and traditionally this has been held mostly as tacit knowledge by human plant operators.

Industrial organisations are understandably nervous of any attempt to encode and transport that knowledge so that it can be stored remotely in a repository. Whilst the knowledge is captured and is stored for re-use, the repository is a source of vulnerability that does not exist with a disconnected set of services and processes.

The concept of trust is mature within computing networks and there has been much work using authentication mechanisms to enable new parties to establish a state of trust before they directly interact with a new system.

Traditionally, this has predominantly catered for situations where human users have been required access to a computer system, and a variety of procedures have been developed to facilitate this, typically employing the services of a certified authority to verify the credentials of a candidate.

However, the explosion in the use of devices such as smart phones to interact with computer systems has created considerable strain for certificate authority based validation, and when one considers the potential volume of IDT objects, authentication using certificates is not sufficiently scalable [14, 15].

1.4 Sharing Business Data

One example business scenario is that of the emerging need of enterprises to utilise predictive modelling, based upon their own operational data, to facilitate scenario planning and optimal resource allocation. Commonly referred to as 'data analytics' [16], this capability requires flexible processes that can harvest data from a range of sources to feed into a model so that a prediction for the future can be created.

If the enterprise wishes to consume data from a supplier, in order to collaborate and produce greater efficiencies, there is a need to share data that is relevant to a particular query. the source data is thus held and governed by at least two parties that maintain their own information systems and repositories.

Each enterprise maintains security policies and authentication mechanisms that are designed to suit the purposes of securing each enterprises' own repositories. Such systems are not normally designed around the need to collaboratively share data on a need-to-know basis, for the purposes of increased collaboration [17]. Therefore, to integrate such systems requires the assimilation of different security realms, across heterogeneous systems [18].

We consider a security realm to be the set of agents or objects that are considered to be trusted by the system, that is they are registered users of the system. The agent who oversees this trust is referred to as the *trusted principal.*

As shown by this explanation, authentication is vital for each security realm and before a principal can have a right to use the resources controlled by a security realm, verification of its identity must be confirmed by the authentication procedure of the security realm in order to ascertain the principal who it purports to be.

The analysis of a user's behaviour relating to trust is relevant for a successful network system [16]. Every contribution made by all users within the network needs to be reliable to achieve the set goal of the network. According to [19] "Trust in a person is a commitment to an action based on a belief that the future actions of that person will lead to a good outcome".

The two key words in this definition are: commitment and belief. Belief alone does not necessarily mean there will be trust. However, trust is created when the belief is the starting place for making a commitment to an action. This means a user or users can achieve trust over time as the relationships progresses.

Trust has also been defined as "a subjective degree of belief about agents or objectives on user's previous experiences and knowledge" [20]. This kind of trust is the belief in the users previous experiences in building trustworthy future relationships within the network.

2 Multiparty Authentication in Dynamic Settings

Due to the exponential growth of emerging technology and computing paradigms like cloud computing and the Internet of Things, significant security and privacy problems lie ahead [21, 22].

As the Internet of Things applications rise and develop, the transition from conventional communication services to the Internet becomes extremely important for group communication.

There are several new internet services and applications, including the Internet of things and cloud computing, which enables users to broaden their software, applications and hardware platforms [23]. In contrast the multi-tenancy setting of cloud computing is providing dynamic services that involves dynamic authentication interactions between many different agents.

These cloud architectures maximise resource sharing by separating approaches into different stages. Additionally, there are also many security and privacy risks associated with the increasing proliferation of services provided by IoT technologies [13].

The authentication systems can therefore not be static. That being said, Cloud Computing expertise allows us to appreciate how IoT applications can be exposed to several security risks, such as a variety of malicious attacks and other documented cloud obstacles.

The IoT user becomes dynamic within the domain of IoT cloud, and the system might have to update the product in order to stay up-to-date. Nonetheless, the IoT network is linked to greater security and privacy concerns because unauthorised users may be able to access confidential business information [24].

The authentication enables different IoT devices to be integrated into different situations. Given that services and enterprises can take a highly complicated and versatile collaborative mechanism, direct relationships between separate realms are not merely a way of linking the two collaborative realms. However, the biggest obstacle for any multiparty application is to authenticate users to ensure managed access to cloud-based data and information services.

A complicated and challenging application requires that one or more parties securely delegate access control systems, which in the effect can regulate the mechanisms that many other parties can authenticate over the services they intend to provide [6, 15].

Such challenges are becoming more complicated considering the possible proliferation of IoT apps. These systems may typically be mapped one by one among system in order to access IoT devices and sensors and cloud services.

2.1 A Multiparty Authentication Model

As described in [6, 14, 15], previous research presents a framework that solves the problems of achieving the requisite permission flexibility in a dynamic multiparty setting.

When participants from different security domains want to access distributed information via a trustworthy authority, a multi-party authentication model for dynamic authentication interactions is necessary. The authors [15] addressed issues about efficient data transmission processes in a secure manner needed for the mutual data processing networks of enterprises.

Moreover, the multi-party model would be used successfully to support distributed networks, e.g. where cloud participants require to authenticate their session members, thereby requiring more straightforward authentication procedures in multi-party sessions in the cloud domain.

This situation can be transferred directly to the scenarios in which significant numbers of sensor nodes generate the overwhelming volume of high-speed, which demands real-time analysis for signal conditioning, data acquisition, data cleaning, integration, local analysis processing, and so on [6]. To allow a service oriented co-operation of the sensor nodes that also includes computational services, requires a system in which trustworthy access to data can be made both in transit and when stored in a repository [25]. This scenario is directly transferrable to the IIoT situation where large numbers of sensor nodes are producing streamed data, that require real-time processing for the purposes of signal conditioning, data cleansing, localised analytics processing, etc. For the sensor and computational nodes to work together in a service oriented (such as microservices architectures [26]) way, there needs to be a mechanism where trusted access to data that is both in-transit and stored in a repository is feasible. We have therefore built this work to facilitate the advancement of, for example, unique use cases, for example, specific use cases where the establishing trustworthy relationships in multiparty Industrial Internet of Things applications can allow new business opportunities through enhanced performance. In this context, Figure below demonstrates the architecture for a *Session Authority Cloud* (SAC) implemented in this scenario as a certificate authority, while this may also be a centralised cloud.

As a component part of the proposed framework, the function of *SAC* is is to authorise the individual sessions demanded by any multiple parties (IoT clouds). Also, it does not rely on clouds and regulates any group that wants to enter the IoT cloud network. The *SAC* provides authentication details for all tenants, including cloud root keys.

2.2 Proposed Protocol

Throughout this section, Fig. 1 describes the proposed session approval protocol that addresses the potential IoT applications and data analytics obtained via IoT clouds whereby IoT individuals or users from multiple security realms may join distributed analytics services via such a trustworthy principle. The principle A (U_A) sends a request to enter the IoT Cloud *A* and Cloud *B* resources for a new session. The request for user keys is submitted at the request of the Multiparty Handler (F). The principle A (U_A) shares his certificate with (F) which contains a user's root key and subdomain key and is encoded with a private key for U_A. The (F) creates a new session ID, including the User's request and forwards it to the Session Authority Cloud (*SAC*) (The trustworthy principle). SAC then checks identity and approves a new session, U_A's public key uses SAC_DB if the user's identity is valid. *SAC* therefore creates a session key and sends to an IoT cloud. *SAC* sends a session approval response to (F), with a session key and available resource list, after receiving a response from resources. Then, F forwards a session permission reply to U_A in order to enter either Cloud *A* or *B*.

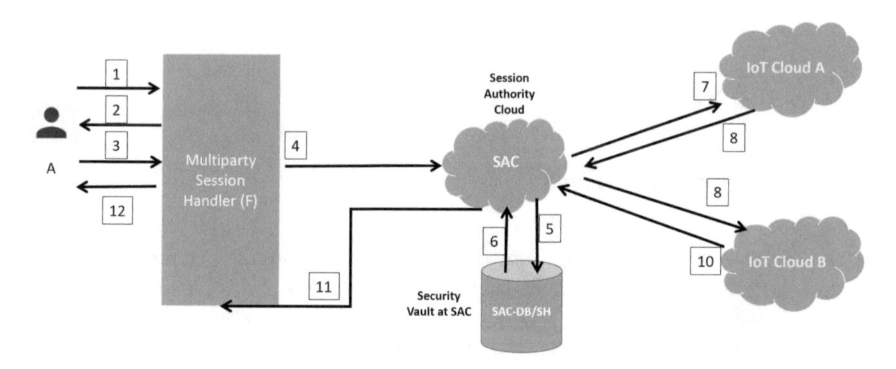

Fig. 1 Multiparty security framework [18]

Output A value in variable *Flag* to show that a session is granted ($Flag = 1$) or denied ($Flag = 0$).

Algorithm 1: Protocol for session approval [17].
Steps

1. U_A to F :request Access to IoT cloud
2. F to U_A : request for the identity ID
3. U_A to F : U_A sends the certificate CA to F
4. F to SAC : session request sent to SAC
5. SAC to $SAC\text{-}DB\text{-}SH$: verifies U_A identity
6. SAC to IoT cloud: Flag indicating U_A is authenticated or not. Sends the SessionID and UserID to the IoT Cloud CA if authenticated.
7. IoT Cloud to SAC: stores the session ID and key in its registry and then sends a reply
8. SAC to F : sends a reply for session approval to F for authenticated user U_A.
9. F to U_A : approves the decision to grand session for authenticated user U_A. Flag = 1 and exit.

3 Credibility

3.1 Introduction to Trust

Trust is a factor related to every topic or issue arising within the premise of relationships. User interactions come across trust daily as an aspect of dependable communication, which contributes to successful interactive phases within their communities. Amongst trust characteristics we identify these; Direct Trust, Transitivity Trust, Reciprocal Trust, and Recommender trust. To understand these trust characteristics, we will use a company warehouse scenario.

The warehouse operates using a multiparty network system for its suppliers. All suppliers are given access to the warehouse network partition that allows communication between the administrative staff in the warehouse and its suppliers. Network access is granted by the warehouse manager after due diligence is carried out on each prospective suppliers (A, B or C) for easy communication between both parties (Fig. 2).

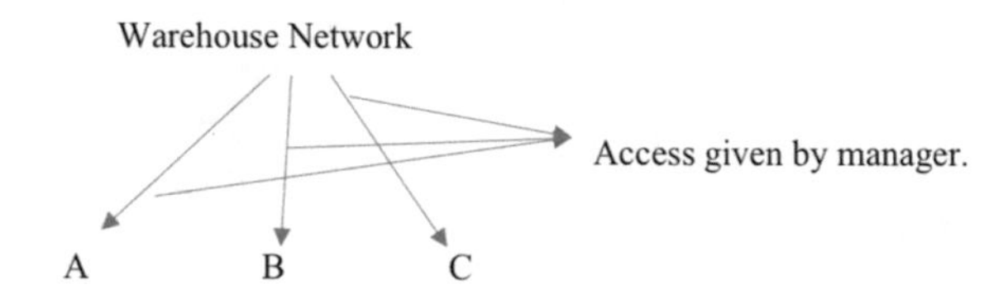

Fig. 2 Example of multiparty connection in warehouse scenario

- **Direct Trust**—This trust is system in a network that occurs between two users. Using our scenario, trust is given to a supplier based on direct association system between the manager and the supplier. No third party required because there is already an active relationship.
- **Transitivity Trust**—This characteristic of trust requires a third party. Simply put in order for access to be given to a user (supplier), the current trust in an existing user who has relations with the new user can be taken into account. Although this is not a common method within networks, trust could be passed from user to user in some circumstances and we can argue that a multiparty network system is no different [19].
- **Reciprocal trust**—In this relationship, trust is generated between two parties in exceptional cases. A user is deemed trustworthy based on the need for the product it offers the company, the user also deems the company trustworthy and a relationship is developed over time [16].
- **Recommender trust**—This trust incorporates all the aforementioned trust characteristics. Recommendation within a network works solely on a user's reputation, it signifies the objective opinions of the user's expertise from its relationships with other users [20].

Reputation mechanisms are used to decide the credibility status of users in many network systems. It functions in two ways, according to [27], one is automatically identify reputation based on user experience in the market and the second: as a filtering tool to guide an administration's evaluation of the user's affiliations in other networks.

In e-commerce for example, online trading usually relies on trust [28] and according to [29], repeated transactions between sellers and buyers is encouraged by reputations harnessing truthful behaviour reviews and customer ratings, as is the case in many online marketplaces e.g. eBay.

3.2 Credibility in Networks

Credibility is defined as believability (perceived trust) and those who are credible are believable people [30], which means when an entity or network is trusted they have characteristics of credibility that makes them believable. Credibility in computing was first discussed early in 1999 by [30]. Here they asked questions relating to the use of computers:

> What is credibility? What makes computers credible? And what can we, as computer professionals, do to enhance the credibility of the products we design, build and promote?

Today we can ask the same questions as it relates to networks and the use of cloud technology; how do we not only enhance the credibility of the products we design, build and promote but how can we make them credibility enough to protect the systems from third party misappropriation.

3.3 Types of Credibility

We now consider how trust can be categorised. Based on the hierarchy of Fig. 3 there are four distinct layers as follows:

- **Presumed Credibility**—This portrays how people take in information created by an individual, out of assumption such as, for example, when an article is published on a BBC web page, it comes with the assumption that the editor is a renowned journalist in a reputable organisation and, as such, is trusted to report only credible information [30].
- **Reputed Credibility**—This is trust based on reputable third-party endorsement. Branded products receive these kinds of trust.
- **Surface Credibility**—This is assumption-based trust. The professional look of a user or Web-page assumes it credible status. Relying on what shows on the surface rather than what the content could hold.
- **Experienced Credibility**—This kind of credibility believes in something based on first-hand past experience, so trust based on user experience. Experience gained from years of existence comes into effect in this kind of credibility.

When a new or strange agent is encountered, the initial categorisation of the relationship is based on no transaction history. At this point the agent is *presumed* to have some *credibility*. In the future, the evidence created by transactions will then provide an indication as to whether the relationship can advance to one of greater trust.

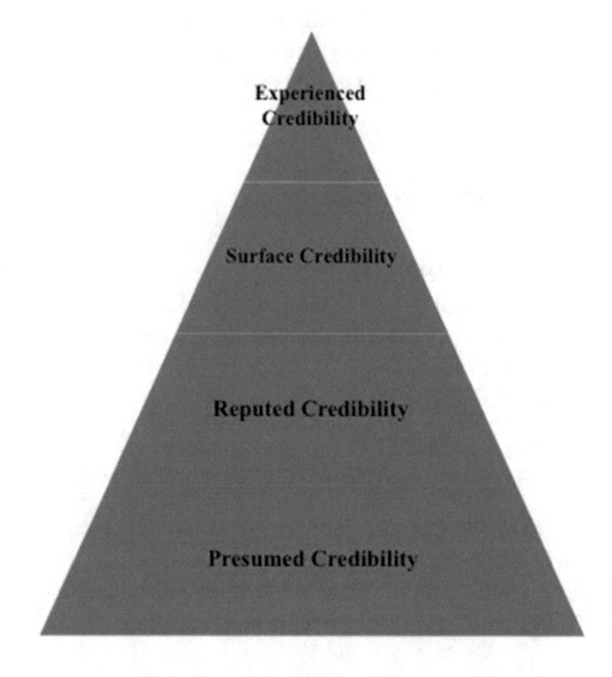

Fig. 3 Hierarchy (pyramid) showing the positions of credibility based on characteristics

Clearly, the objective for any incoming agent to a system is to establish trust as quickly as possible, to a level that facilitates the business goals that are mutually desired by both parties.

This categorisation of trust enables automation to take place by invoking system rules that can assess and be invoked as a transaction history is established. The automation of this assessment is a necessary prerequisite for IIoT enabled operations as the volume of transactions is likely to be massive and in the future, inconceivable.

One example of the challenging process of establishing trust in a sensitive process-oriented environment, is that of community healthcare delivery. Beer et al. [17, 31–33] describe the complexities of modelling and building secure multiagent communication networks that are required to exchange confidential patient data for the purposes of delivering personal healthcare services into the home environment.

Such relationships are initially established using personal identification mechanisms, and then advance as individual agents develop relationships through familiarity. This work mimicked the increased trust that could be built as a by-product of operating with a fully coordinated system, albeit at a very finegrained level due to the nature and sensitivity of the data.

Such sensitivity is unlikely to be present with the use of industrial data under the guise of current business models. However, as technology advances it is feasible that more control of the data utilised may be required in the pursuit of expedited trust transactions. In fact, the ownership of the data use to form the relationship may be distributed among a community of agents.

3.4 Rewiring

Rewiring is a tool for connecting with new network agents (users). In social networks rewiring is used to create relationships with new agents or disconnect with dormant agents in the network. It is a social instrument collaborative networks depend on to fulfil their tasks, it "*facilitates the emergence of norms from repeated interactions between members of a society*" [34].

A collaborative network is a kind of network that is formed specifically by members or an organisation to collectively achieve a task and could be termed 'private'. Unlike in public networks, a private network requires agent authentication to maintain the security if the Network and make sure data is not compromised in any way.

Since agents sometimes rely on the reputation knowledge of other agents for authentication, many malicious agents can pretend, for the sole purpose of gaining trust, to be given access.

Once trust is achieved these users begin their activities penetrating into the network gaining some level of credibility and, in order for information to be shared and utilized, interaction must take place, thus allowing for mutual access of resources [35].

Rewiring can be achieved by one of these procedures:

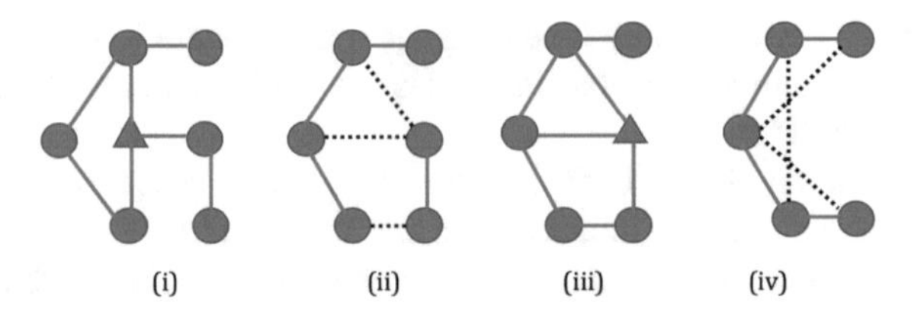

Fig. 4 Network adaptation with random rewiring: (i) shows the initial network with the targeted node outlined, (ii) shows the adapted network following the node removal with rewired links shown as dashed blue lines, (iii) shows the next targeted node and (iv) shows the subsequent adaptation [36]

- **Random Rewiring**—Selecting network agents randomly from a vast population of agents in a public network.
- **Neighbour Advice**—This is based on reputation and referral.
- **Global Advice**—Choosing an agent according to certain criteria or strategies that have been used in the past (Fig. 4).

The concept of rewiring is important in network management at all times because it forms the basis for continuity in network interactions and growth. However, in the process of connecting with new users, administrators can be tricked into authenticating pretentious users into their network.

Therefore, when choosing to authenticate in a multiparty system, care must be taken this is because data can either be altered maliciously by such pretentious users or shared outside the network without permission.

There are a vast number of cases of companies that have fallen into the hands of such agents in the real world resembling a case of corporate espionage.

In situations where the network fails to notice the presence of such activities, it is likely that it will continue undetected.

3.5 Forensic Investigation

The *Credibility Hierarchy* categorises agents based on their trust history in the network, keeping records of transactions authorised for users who have attained each level of *Credibility*. Where a suspicious operation has been attempted or carried out, the incident is recorded with an alert pointing investigators in the right direction.

Malicious agents (users) who have been deemed *reputed* or have achieved *surface credibility* are those in the network likely to have malicious intent. This is because, at this level in the hierarchy, their trust category gives them access to a wide range of sections in the network without requiring authorisation.

In the event of an investigation, forensics experts will scan the networks for suspicious activities. Activities such as illegal or unauthorised transactions that include but are not limited to; frequent attempts to access sections of the system only administrators are allowed to access, file uploads masked with malware aimed at harvesting sensitive credentials.

The use of Phishing or spear-phishing emails should also be considered as transactions that would have taken place. Collaborative operations are inspected because the possibility of multiple transactions completed by agents combining their resources is also highly likely.

Since the list of suspicious activities are unlimited, it is therefore essential that the system is screened for unauthorised activities as often as possible to identify security breaches and reduce the risk of a hijacked network.

The trust hierarchy facilitates timely categorisation of illegal activity, which is a necessary precursor for IIoT devices.

4 Conclusions and Future Directions

The assessment of trust is a complex scenario to manage. As technologies such as IIoT become an intrinsic part of industrial environments, the need to scale trust authentication will become unmanageable unless it can be automated.

As RFID and 'smart' devices proliferate, businesses shall be looking for ways to integrate supply-chain relationships to remove wasteful processes and transactions. Automated trust formation allows such supply-chains to be rationalised in order to create the most business value.

To some extent, traditional methods of security have relied upon certificate authorities which works up to a point. When the authentication is limited to human agents, there is still a finite number of agents to be verified and this has been managed with cloud-based scalable utilities that allocate the necessary computational resource on demand.

However, such is the potential scale of IIoT, that there shall be far more devices than there is in the global human population, and therefore the current models lack scalability. Automating this is a considerable problem, yet automation is logically the only way forward.

This work brings together the emergence of technical solutions towards the mechanisms for managing multi-party relationships, which are an inherent part of the need for agile, lean business processes for Digital Manufacturing, together with established work in the social networks domain, to offer a route forward for the classification of stages along the route to trusted relationships.

Representing nodes in an IIoT network as agents permits behaviours to be modelled and thus the interactions between each entities can be scrutinised. This is important if the nature of each interaction is to be managed so that it can be verified in the context of trustworthiness.

Additionally, the real-world scenario of IIoT nodes entering and leaving networks (due to their mobility) presents a significant challenge in terms of the opportunities for nefarious activities to take place, as a result of the increased traffic and resultant strain on authentication mechanisms.

There is a necessity to remove the requirement for centralised authentication for every transaction, in order that trust can be delegated in a controlled way to trusted parties. We see the trust hierarchy as a means of facilitating such delegation within industrial multi-party authentication scenarios, particularly as the computational capabilities of small IIoT devices increases.

We see this work developing in a number of ways, and the most important next steps are as follows:

- **Quality of Service evaluation**—simulation studies so far have established that low-latency, high speed networks such as 5G can facilitate much higher data rates and the multi-party architecture can scale much further with IIoT device volume. However, such authentications have been relatively simple and the next stage is to test for deadlocks in much more dynamic networks.
- **Trust scale granularity**—The current hierarchy offers a straightforward scale for use within the existing multi-party scheme. We see the level of granularity increasing to cater for a wider range of trust relationships. For example, the work so far has assumed that each party maps to at least one human agent. This now needs extending to a many to one scenario where a human (or enterprise) agent is responsible for a multitude of IIoT nodes, who all interact with another network in different ways. A finer grained control of trust will help prevent nefarious parties from masquerading as a trusted party, and this is a monitoring function that can only be considered feasible if it is automated. Thus, a more detailed set of trust characteristics must be established.
- **Trust/capability mapping**—This is related to the previous point, in that an assessment of devices' capabilities is an important consideration during the authentication process, along with the credibility and reputation of the parties involved in the transaction. Such capabilities—and therefore trust levels—may be transient, which again requires a dynamic solution that maintains an overall QoS that is acceptable.
- **Learning mechanisms for trust assessment**—such is the potential scale of IIoT devices, there is a clear need to automate and accelerate the assessment of reputation based on a node's behaviour. It is impractical to assume that trust formation has to commence with no data, and in the social context, human agents use data acquired from other sources when forming trust assessments. Machine learning approaches are an obvious first step, while being conscious of the ability to poison ML models to mislead third parties. This is one area where distributed ledger technologies such as IOTA may have a role to play, to support the safe formation and consumption of reputation data. This would also directly support forensic techniques and approaches.

To a certain extent, the work exploring trust and delegated trust in devices such as the IIoT, since equipment is being rapidly deployed without an appreciation for valid vulnerabilities that exist. Businesses are motivated to adopt technology that will create new business value, but at present the awareness about adequate security is lacking.

This challenge is compounded by the fact that there is a trend towards distributed, socially inspired communications between equipment, which accelerates the scaling of systems, but again this development introduces weaknesses into systems that may already be secure.

References

1. Hill R, Devitt J, Anjum A, Ali M (2017) Towards in-transit analytics for Industry 4.0. In: IEEE international conference on internet of things (iThings) and IEEE green computing and communications (GreenCom) and IEEE cyber, physical and social computing (CPSCom) and IEEE smart data (SmartData). IEEE Computer Society
2. Bessis N, Xhafa F, Varvarigou D, Hill R, Li M (eds) (2013) Internet of things and inter-cooperative computational technologies for collective intelligence. In: Studies in computational intelligence. Springer
3. Ashton K (2009) That "internet of things" thing. RFiD J
4. Gubbi J, Buyya R, Marusic S, Palaniswami M (2013) Internet of things (IoT): a vision, architectural elements, and future directions. Future Gener Comput Syst 29(7):1645–1660
5. Gartner (2015) Gartner says 6.4 Billion connected things will be in use in 2016, up 30 percent from 2015? Gartner website: https://www.gartner.com/newsroom/id/3165317, 10 Nov 2015
6. Al-Aqrabi H et al (2020) Hardware-intrinsic multi-layer security: a new frontier for 5G enabled IIoT. Sensors 20(7):1963
7. Tedeschi S, Mehnen J, Roy R (2017) IoT security hardware framework for remote maintenance of legacy machine tools. In: Proceedings of the second international conference on internet of things and cloud computing, ICC 2017, Cambridge, United Kingdom, pp 43:1–43:4, 22–23 Mar 2017
8. Mahmoud M, Saputro N, Akula P, Akkaya K (2016) Privacy-preserving power injection over a hybrid AMI/LTE smart grid network. IEEE Internet Things J 1
9. Al-Aqrabi H, Hill R, Lane P, Aagela H (2019) Securing manufacturing intelligence for the industrial internet of things. In: Proceedings of the fourth international congress on information and communication technology, London, UK, pp 267—282, 27–28 Feb 2019
10. Jara AJ, Zamora-Izquierdo MA, Skarmeta AF (2013) Interconnection framework for mHealth and remote monitoring based on the internet of things. IEEE J Sel Areas Commun 31(9):47–65
11. Ikram A, Anjum A, Hill R, Antonopoulos N (2015) Approaching things (IoT): a modelling, analysis and abstraction framework. Concurr Comput: Pract Exp 1966–1984
12. Chui M, Loffler M, Roberts R (2010) The internet of things
13. Al-Aqrabi H, Liu L, Hill H, Cui L, Li J (2013) Faceted search in business intelligence on the cloud. In: Proceedings of GREENCOM-ITHINGS-CPSCOM. IEEE, China, Beijing, pp 842–849
14. Al-Aqrabi H et al (2019) A multi-layer security model for 5G-enabled industrial internet of things. In: 7th International conference on smart city and informatization (iSCI 2019), Guangzhou, China. Lecture notes in computer science, Switzerland, vol 8. Springer International Publishing AG, 12–15 Nov 2019
15. Al-Aqrabi H, Hill R (2018) Dynamic multiparty authentication of data analytics services within cloud environments. In: 20th IEEE international conference on high performance computing and communications (HPCC-2018), Exeter 28-30

16. Wang H, Wang F, Liu J (2012) Accelerating peer-to-peer file sharing with social relations: potentials and challenges. In: INFOCOM, 2012 proceedings IEEE. IEEE, pp 2891–2895, Mar 2012
17. Beer MD, Hill R, Huang W, Sixsmith A (2003) An agent-based architecture for managing the provision of community care—the INCA (Intelligent Community Alarm) experience. AI Commun 16:179–192 (IOS Press)
18. Al-Aqrabi H, Lane P, Hill R (2019) Performance evaluation of multiparty authentication in 5G IIoT environments. In: Cyberspace data and intelligence, and cyber-living, syndrome, and health. Springer, pp 169–184
19. Golbeck J, Parsia B (2006) Trust network-based filtering of aggregated claims. Int J Metadata Semant Ontol 1(1):58–65
20. Kim YA, Le MT, Lauw HW, Lim EP, Liu H, Srivastava J (2008) Building a web of trust without explicit trust ratings. In: IEEE 24th international conference on data engineering workshop, 2008. ICDEW 2008, pp 531–536, Apr 2008
21. Sotiriadis S, Bessis N, Antonopoulos N, Hill R (2013) Meta-scheduling algorithms for managing inter-cloud interoperability. Int J High Perform Comput Netw 7(2):156–172
22. Elkhodr M, Shahrestani S, Cheung H (2016) The internet of things: new interoperability, management and security challenges. arxiv:1604.04824
23. Baker C, Anjum A, Hill R, Bessis N, Liaquat Kiani S (2012) Improving cloud datacentre scalability, agility and performance using OpenFlow. In: Proceedings of the 4th international conference on intelligent networking and collaborative systems (INCoS). IEEE, pp 1–15
24. Al-Aqrabi H, Liu L, Hill R, Antonopoulos N (2015) Cloud BI: future of business intelligence in the cloud. J Comput Syst Sci (Elsevier)
25. Al-Aqrabi H, Hill R (2018) A secure connectivity model for internet of things analytics service delivery. In: 2018 IEEE SmartWorld, ubiquitous intelligence and computing, advanced and trusted computing, scalable computing and communications, cloud and big data computing, internet of people and smart city innovation. IEEE, pp 9–16
26. Shadija D, Rezai M, Hill R (2017) Towards an understanding of microservices. In: Proceedings of the 23rd international conference on automation & computing, University of Huddersfield. IEEE, 7–8 Sept 2017
27. Wang JC, Chiu CC (2008) Recommending trusted online auction sellers using social network analysis. Expert Syst Appl 34(3):1666–1679
28. Hogg T (2009) Security challenges for reputation mechanisms using online social networks. In: Proceedings of the 2nd ACM workshop on security and artificial intelligence. ACM, pp 31–34, Nov 2009
29. Klein DB (1997) Reputation: studies in the voluntary elicitation of good conduct. University of Michigan Press
30. Tseng S, Fogg BJ (1999) Credibility and computing technology. Commun ACM 42(5):39–44
31. Beer MD, Huang W, Hill R (2003) Designing community care systems with AUML. In: IEEE international conference on computer, communication and control technologies (CCCT2003)
32. Hill R, Polovina S, Beer MD (2005) From concepts to agents: towards a framework for multi-agent system modelling. In: Dignum F, Dignum V, Koenig S, Kraus S, Singh MP, Wooldridge M (eds) Proceedings of the fourth international joint conference on autonomous agents and multi-agent systems (AAMAS 05), Utrecht, The Netherlands. ACM Press, pp 1155–1156, 25–29 July 2005
33. Hill R, Polovina S, Shadija D (2006) Transaction agent modelling: from experts to concepts to multi-agent systems. In: Proceedings of the fourteenth international conference on conceptual structures (ICCS '06): conceptual structures: inspiration and application, Aalborg, Denmark. Lecture notes in artificial intelligence (LNAI), vol 4068. Springer, pp 247–259, 16–21 July 2006

34. Villatoro D, Sabater-Mir J, Sen S (2011) Social instruments for convention emergence. In: The 10th international conference on autonomous agents and multiagent systems-volume 3. International Foundation for Autonomous Agents and Multiagent Systems, pp 1161–1162, May 2011
35. Bentahar J, Meyer JJC, Moulin B (2007) Securing agent-oriented systems: an argumentation and reputation-based approach. In: Fourth international conference on information technology, 2007. ITNG'07, pp 507–515
36. Tran HT, Domerçant JC, Mavris DN (2016) A Network-based cost comparison of resilient and robust system-of-systems. Procedia Comput Sci 95:126–133

IoT Forensics: An Overview of the Current Issues and Challenges

T. Janarthanan, M. Bagheri, and S. Zargari

Abstract The pursuit of cybercrime in an IoT environment often requires complex investigations where the traditional digital forensics methodology may struggle to support the forensics investigators. This is due to the nature of the technologies such as RFID, sensors and cloud computing, used in IoT environments together with the huge volume and heterogeneous information and borderless cyber infrastructure, rising new challenges in modern digital forensics. In the last few years, many researches have been conducted discussing the challenges facing digital forensic investigators and the impact of these challenges bring upon the field. Some of these challenges include the ambiguity of data location, data acquisition, diversity of devices, various data types, volatility of data and the lack of adequate forensics tools. Moreover, while there are many technical challenges in IoT forensics, there are also non-technical challenges such as determining what are IoT devices, how to forensically acquire data and secure the chain of custody among other unexplored areas, including resources required for training or the type of applied forensics tools. A profound understanding of the challenges found in the literature will help the researchers in identifying future research directions and provide some guidelines to support forensics investigators. This study presents a succinct overview of IoT forensics challenges focusing on a typical smart home investigation and a comparison of the existing frameworks to conduct forensics investigations in the IoT environment.

Keywords Digital forensics · IoT forensics · Internet of things · Smart homes · Cyber security

T. Janarthanan · M. Bagheri · S. Zargari (✉)
Faculty of Science, Technology and Arts, Sheffield Hallam University, Sheffield, England
e-mail: S.Zargari@shu.ac.uk

T. Janarthanan
e-mail: Tharmini_1@hotmail.com

M. Bagheri
e-mail: Maryam.Bagheri@shu.ac.uk

R. Montasari et al. (eds.), *Digital Forensic Investigation of Internet of Things (IoT) Devices*, Advanced Sciences and Technologies for Security Applications,
https://doi.org/10.1007/978-3-030-60425-7_10

1 Introduction

The Internet of Things (IoT) refers to connecting any device to the internet and it is one of the most explored topics by researchers at present. This is due to the incredible capabilities this technology has provided. The Internet of Things (IoT) is defined as the interconnection of uniquely identifiable embedded computing devices within the existing Internet infrastructure [1]. In simple, it involves things or objects such as sensors, actuators, RFID tags and readers to interact and coordinate with each other thereby reducing human intervention in basic everyday tasks [2]. Conversely, the number of human interactions with these IoT systems creates a new paradigm for evidence-based data. With the current advancement in networks and communication systems, IoT enables billions of growth and connectivity. Tech analyst company IDC predicts that in total there will be 41.6 billion connected IoT devices or "things" by 2025. In addition, Gartner predicts that the enterprise and automotive sectors will account for 5.8 billion devices this year, up almost a quarter in 2019 [3].

While IoT has increased productivity for businesses, it has also introduced new risks and threats such as security and privacy issues. IoT devices contain sensitive and valuable data and it has become one of the main sources of attacks and cybercrimes. The complexity of IoT in terms of the integration of different communication technologies, devices, protocols and standards makes it difficult to ensure public or private security. Moreover, protecting data of IoT devices has been challenging because of the heterogeneous and dynamic features of the IoT. Even if precautions are carefully taken to secure data, the level intelligence exhibited by cyber-attackers is undoubtedly great. Attacks can be crafted not just from public networks but from private sources, such as cars, smartphones, and even smart homes [4]. As a result, cyber-attacks can have a significant socio-economic impact on both global businesses and individuals.

Besides that, digital forensics investigation is one of the important areas that require additional work. Despite the numerous benefits provided by IoT in various applications, the modern infrastructures are becoming complex and virtualized whereby digital forensics investigators are required to acquire and analyse evidence coming in many forms and different scenarios. Unlike computer-based investigation where there exists the ACPO (Association of Chief Police Officers) [5] guidelines in order to make sure the correct procedure has been employed, for the IoT environment such as smart homes there is not a formal integrated guide to obtain legally and analyse the evidence.

Recently, there has been research conducted discussing the challenges facing digital forensics investigators and the impact of these challenges bring upon the field. Some of these challenges include the ambiguity of data location, data acquisition, diversity of devices, various data types, volatility of data and the lack of adequate forensics tools [6–8]. In the IoT environment, data is mostly stored and processed on the cloud environment. The acquisition of access to data for investigation purposes becomes difficult for IoT forensic investigators due to the constraints of service level agreements and volatility of this data. While there are many technical challenges in IoT forensics, there are also non-technical challenges such as determining what are IoT devices, how to forensically acquire data and secure the chain of custody among other unexplored areas, including resources required for training or the type of applied forensics tools [9].

1.1 Aims and Objectives

This research aims to overview the current IoT forensic issues from the literature. It also discusses and compares the existing developed frameworks to conduct forensics investigations in the IoT environment. It will help the researchers in identifying future research directions and provide some guidelines to support forensics investigators. The rest of this chapter is organised as follows: Sect. 2 provides a background to Internet of Things and the challenges that brings to forensics investigators. It also reviews the current research and studies on the traditional and IoT forensics investigation, current forensic tools and legal considerations carried out by the other researchers. Section 3 describes the current proposed forensics investigation frameworks and identifies the research gaps. In order to explore a feasible solution for conducting forensics investigations in the IoT environment complying with the legal requirements, the proposed frameworks will be compared and analysed critically. Finally, this study draws some conclusions and recommendations for future research.

2 Literature Review

2.1 Internet of Things

The Internet of Things (IoT) has been leveraged in many industries. For instance, "A smart city uses digital technology to connect, protect, and enhance the lives of citizens. IoT sensors, video cameras, social media, and other inputs act as a nervous system, providing the city operator and citizens with constant feedback so they can make informed decisions" [10]. Cities use sensors to control many of their infrastructure systems such as water distributions, traffic management, energy management, parking and street Lighting [11].

According to the report carried out by Philips Lighting and Smart Cites World [12], Barcelona, Singapore and London are three remarkable examples of the smart cities which use sensors to control many of their infrastructure systems such as water distributions, traffic management, energy management, parking and street Lighting [13]. It also shows how IoT has brought a variety of benefits to the cities. For example, Barcelona's smart city project has created 47,000 jobs, saves $58 m on water, and generates an extra $50 m a year through smart parking.

Leveraging IoT into the cities has a huge impact on the economy. For example, finding a parking space is a critical issue for some major cities. Smart parking generates $41 billion revenue and provides drivers with real-time information on the availability of the parking space across the city [14]. Smart building reduces the energy consumption by automating and controlling lighting, heating, ventilation, conditioning and security in the buildings and generates $100 billion revenue.

The Internet of Things has also redefined the health care systems and had a profound impact on the patient experience and treatment. It has reduced in-person visits and allowed patients to manage their care from home. For instance, IoT-enabled devices such as wearables can collect and analyse critical data from patients and diagnose various health issues such as blood pressure, heart rate, brain waves, temperature, physical condition, number of steps and breathing pattern. Specialists can remotely monitor the patient's data and provide the possible treatments.

Since the number of objects equipped with network connectivity and intelligence, are growing fast and it has been predicted that, this number will be 50 billion by the end of 2020 which will result in $19 trillion in profits and cost savings [14], more and more industries such as Transport, smart home, automotive, manufactures are deploying IoT to redefine their operations (Fig. 1).

Tech company IDC suggests industrial and automotive equipment represent the largest opportunity of connected "things,", but it also sees strong adoption of smart home and wearable devices in the near term. In contrast, Garner suggests utilities will be the highest user of IoT due to continuing rollout of smart meters. Security devices, in the form of intruder detection and web cameras will be the second biggest use of IoT devices. Building automation such as connected lighting will be the fastest growing sector, followed by automotive (connected cars) and healthcare (monitoring of chronic conditions) [3].

Over the years IoT has changed the way businesses interact with people and brought a variety of benefits to both people and industries. It allows industries to

Fig. 1 IoT application in Industries

understand consumer needs in real time, to become more responsive, to improve machine and system quality, to streamline operations and to discover innovative ways to operate as part of the digital transformation efforts [11].

Fortune Business Insights report says the global \$ 190 billion IoT market is expected to reach \$ 1.11 trillion (\$ 1111.3 billion) in annual growth in 2018 by 2026 of 24.7%. The banking and financial services sector is expected to be the largest market share segment [15].

2.2 Smart Home

One of the most widely used applications of IoT is smart homes. In a smart home, all devices—lights, locks, refrigerators, coffee makers, heating/cooling systems and cameras are connected and controlled by a central device through Wi-Fi, Bluetooth, X10, UPB, INSTEON, Z-Wave and Zigbee [16]. It enables people to control and monitor objects remotely from their smartphone and to accomplish personal tasks more easily and faster. It also offers many benefits to the homeowner including energy saving, money saving and increasing security.

For example, Smart lighting system is an integral part of a smart home and is a great way of controlling the ambiance of the home. They can be easily controlled through simple voice command or mobile apps. They can be programmed to turn on and off when users enter or leave the room so users do not need to be worried about wasting energy.

Nest thermostat is a Wi-Fi-based thermostat that allows users to control the heating and air conditioning system with an app or voice command. It learns automatically from the user's behaviour and adjusts itself accordingly. Nest Thermostat saves home-owners about 10–12% on heating and 15% on cooling. This translates to a savings of about \$140 per year [17].

Maximizing home security is another amazing benefit of smart home device. By installing smart cameras, users can monitor their home anywhere anytime and receive security alerts on their mobile phone. Smart door locks also reduce the risk of being locked out from home. The users can secure and lock the door from anywhere with the internet access.

Smart Home devices are divided into the smart appliances, security, control and connectivity, home entertainment, energy management, and comfort and lighting [18]. Many companies and vendors are invested in smart home devices and the smart home market is expected to reach \$ 141 billion by 2023 [19].

Smart devices usually connect to either each other or a central control hub via home's Wi-Fi network (Fig. 2). Many companies develop smart hubs and smart-phone apps to control their own devices. Different hubs support different connectivity protocols such as Wi-Fi, Bluetooth, X10, UPB, INSTEON, Z-Wave and Zigbee.

X10 is an automation protocol which was developed in 1975 for home automation. It uses home's existing electrical wiring to send the signals. Although X10 devices are outdated but X10 protocol provided the foundation for wired technology such

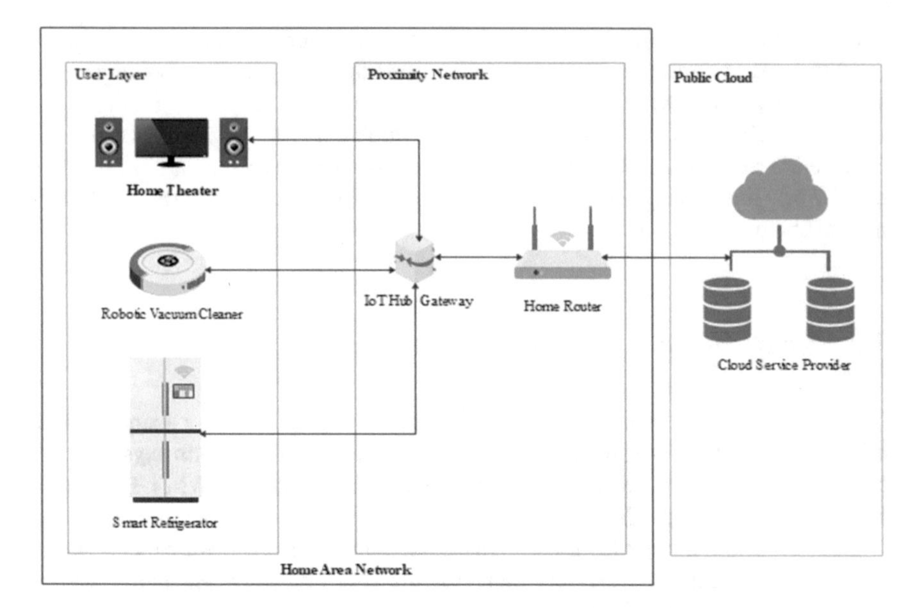

Fig. 2 A typical Smart Home layout

as Universal Powerline Bus (UPB). INSTEON uses both wired power lines and wireless technologies to communicate with other devices. When a problem occurs, it switches from one communication channel to another one thus enhancing both speed and reliability over older technology. INSTEON devices wirelessly connect to every other device, creating a mesh network. In a mesh network, each device communicates with other devices directly without using a central hub so the device can independently transmit the data.

Z-Wave and Zigbee are newer wireless technologies that create a mesh network between each connected device. Zigbee can be built in smart devices such as door locks, lights, thermostats, and more.

After connecting smart devices to the network, a controller can be used to control the devices. The simplest type of controller is a smartphone app such as Apple's Home app. Apple Home Kit lets control smart home devices all in one place. It allows people to adjust smart thermostat, turn lights on and off, control locks and more in multiple rooms. Devices can also be controlled remotely through this app.

Although smart home has brought many benefits to people's lives, they lack technical standards and heterogeneous platforms. A few companies accepted industry standards which lead to having multiple incompatible platforms and technologies. Most smart home devices by the manufacturers and vendors are generally not built with strong security controls in mind. Smart devices and sensors collect a lot of information about people to learn and predict their behaviour. To automate a task, they need to know what, where and when people do a task. Smart devices know in which room and when to turn the lights on or off. Therefore, connecting these devices to wireless networks and to the Internet makes users vulnerable to malicious attacks further resulting in security and privacy threats such as identity theft and data leakage [20].

2.3 IoT Security Challenges

The evaluation of IoT from limited access networks to a distributed public network increased the needs for security alarms to protect interconnected IoT devices from intrusions such as data modifications, malicious code injection, sniffing, and Denial of Service (DoS) and many other threats [21]. SonicWall reported that IoT malware attacks increased 215.7% to 32.7 million in 2018 compared to 10.3 million in 2017. The first two quarters of 2019 exceeded 55% in the first two quarters of 2018. If this rate continues, it will be another record year for IoT malware attacks [15]. Tabane E. et al. highlighted that though there are existing technologies and protocols dealing with issues of threats to security, the limitations on the IoT devices and network prevents a straightforward adaptation and implementation of IoT solutions in the new arising sets of security scenarios [22].

At present, the adopted security protocol and cryptographic setting requires a lot of resources and IoT devices such as smartphones, tablets, PCs, routers, active sensors or passive RFID tags, have very limited resources and capabilities to support the implementation and adaptation of traditional security protocols solutions. Hence, the implementation and adaptation of traditional security protocols solutions still remain as a challenge making it difficult to provide confidentiality of data transmission. Since unattended IoT devices are not supervised because they operate in a self-support manner with limited maintenance (e.g. monitoring) this further leads to concern in terms of data integrity (trust). As a result, the data obtained from IoT devices is likely to be of low quality or corrupted (e.g. data tempering) [23, 24].

There are various security challenges and limitations related to IoT, which are affecting large scale adoption. In this section, these challenges and limitations have been discussed in detail:

A. **Privacy**
   User privacy and data protection is an important issue in IoT security taking into consideration the ubiquitous characteristics of the IoT environment. The ability of the IoT sensors and devices to sense, collect and transmit data over the internet pose a threat to individuals' privacy. IoT nodes are known to collect people's private data without them even noticing [25]. Koien et al. [26] mentioned although an abundance of research has already been proposed with respect to privacy, many topics still need further investigation.
   According to a report by Aaron in 2015, Nest thermostat which is one of the most secure IoT devices, can be hacked and controlled while the device boots up. Hackers can load their custom software onto it which would stop thermostat data from being sent back to Nest's servers [27]. The compromised Nest Thermostat will then act as a jumping off point to take control of other devices in a home

which allows hackers to access sensitive information about people such as their presence in the house or their sleeping schedule.
Smart device apps can also be as vulnerable as the device itself. A study by the security research team at Checkmarx showed how attackers bypass user permissions and take control of Google and Samsung camera apps. Attackers are able to remotely take photos, record video, spy on conversations, identify people's location, and more [28].

B. **Authentication**
The identification and authentication of objects could be challenging because of the nature of the IoT environment. It is essential to consider managing identity authentication in the IoT, as multiple users and devices need to authenticate each other through trustable services [26]. In addition, efficient key deployment and key management is a challenge in IoT devices as it could cause overhead on IoT nodes [29]. Moreover, in the absence of a guaranteed Certificate Authority (CA), other mechanisms are required for validating cryptographic keys and ensuring integrity of key transfer [4].

C. **Heterogeneity**
IoT devices connected to different types of entities with varying capabilities complexity and vendors. These devices come with different configurations, dates, release versions and the use of technical interfaces which are designed for altogether different functions. Thus, the requirement to develop protocol to work with all the different devices is required [30–32]. Mahmoud et al. [33] mentioned that one more challenge that must be considered in IoT is the dynamic environment, at one time a device might be connected to a completely different set of devices than in another time; thus to ensure security optimal cryptography system is needed with adequate key management and protocols.

D. **Policies**
Current policies that are implemented in computer and network security may not be applicable for IoT due to its heterogeneous and dynamic nature. Hence, there must be policies and standards developed to ensure that the data will be managed, protected and transmitted in an efficient way. This includes a mechanism to enforce such policies is needed to ensure that every entity is applying the standards. Similarly, for every IoT service involved a Service Level Agreement (SLAs) must be clearly identified to introduce trust by human users in the IoT environment which will further results in its growth and scalability [33].

Most of the technical security concerns are related to manufacturing standards, update management, physical hardening, user's knowledge and awareness [34]. Weak and guessable default passwords, hardware issues, unpatched embedded operating systems and software, insecure data transfer and storage and Lack of encrypted firmware updates by companies could allow the device to be compromised. Many IoT devices have operational limitations such as low processing power and small memory which is just enough to perform the allocated tasks and they can't handle proper software updates.

Due to lack of awareness and user's ignorance, factory default passwords are usually forgotten to be changed. Some devices are set with a poor password which is easy to be breached for malicious purposes. Many well-known companies recently provide two-factor authentication (2FA) to eliminate the risk of security challenges but still millions of IoT devices do not support this feature.

Changing factory default passwords, installing necessary updates, disable remote access to IoT devices when not needed, disable features that are not being used can also reduce the risk of being compromised. Wi-Fi networks are also one of the first points of security attacks which make the entire network vulnerable. Setting strong passwords and encryption methods for Wi-Fi networks, can mitigate the risk of security attacks.

2.4 *Digital Forensics Investigation*

Digital forensics is the process of identifying digital evidence in its most original form, collecting, examining, analysing and presenting the evidence to a court of law. In recent years with the rapid increase in the use of IoT technology, the forensics investigators are facing new challenges where the traditional digital forensics is inapplicable for conducting forensics investigations and more research has to be carried out in order to develop frameworks and guidelines for practitioners in such a volatile environment. The traditional digital forensics mainly deals with evidence sources such as computers, mobile devices, servers and gateways whereas the evidence sources for IoT forensics include home appliances, actuators, sensor nodes, medical devices and a multitude of other smart devices. From a legal perspective, jurisdictional and ownership issues are essentially similar but then from a technical perspective, there are many areas that require further research and development. The obvious example is the lack of forensics tools capable of supporting various IoT devices in the market due to a wide range of proprietary designs, unclarity of the network boundaries or uncertainty of the location of stored data [35].

2.4.1 Traditional Forensics Investigation

Traditional forensics investigation is a relatively mature field having formal standardisation of key processes to carry out investigation. The data acquisition in traditional forensics deals with sources such as hard drives, RAM, system logs or any peripheral storage [36] and for deeper investigation, the examiner can use techniques such as file carving in unallocated space. The traditional forensics also includes the detection of malicious network activities where the network traffics are collected and examined. In addition, currently, most crimes include mobile phone investigation which has its own challenges such as preserving the evidence in a volatile environment or bypassing the passcode and encryption. After the data acquisition, the collected artefacts are analysed from a technical and legal perspective, and presented as evidence supporting a crime during the court proceedings [36]. In simple words, it can be said that the traditional forensics is a subarea of forensics investigation in IoT because the latter consists of examining more variety of digital devices which are intercommunicating and data syncing among each other as well as the cloud servers. One of the major complexities in this situation is maintaining the chain and custody and legal requirements.

2.4.2 Forensics Investigation in IOT/Smart Home

The proliferation of IoT devices and the increase in the number of cybersecurity crimes have given rise to enhance forensics investigation techniques in IoT. Smart homes can be counted as a simple form of IoT environment which can be a good research starting point to explore the challenges of conducting forensics investigations in an IoT environment. Some of the main challenges that the forensics examiners have to overcome in any forensics investigation exist in the data acquisition stage and the data analysis stage where the proper and suitable forensics tools play an important role in supporting the forensics examiners in the investigations. In terms of identifying the sources of evidence in smart homes, the IoT devices, the home and hub gateways, the mobile devices on which the IoT applications are installed and the cloud servers are to be the main sources of evidence in any typical smart home investigations. However, it is important to consider situations where some of these IoT devices may not be present in the crime scene at the time of seizure, such as wearables or mobile phones. The data from these sources can be extracted from the local storage of IoT device(s), the user applications' data stored on the mobile device(s), the incoming and outgoing network traffic via the home and hub gateways, and the cloud servers that are holding the users' data on their personal accounts. This might look an easy task but actually one of the main challenges in conducting such investigations is maintaining the chain of custody because at the time of seizure, these devices are actively intercommunicating among themselves including the cloud servers.

In the acquisition phase, the data extraction of IoT devices depends on a few factors such as the manufacturers' hardware design of IoT devices, the capabilities of the forensics tools and the familiarity and expertise of the forensics investigator with such devices.

The acquisition of network traffic in smart homes can be done via the home and hub gateways. In general, the IoT devices in a smart home are often connected to a smart hub gateway whose sole purpose is to act as a base station for their particular radio standard and then, the hub gateway is to be connected directly to the home router. However, more advanced home routers are now integrating these radio standards to be more appropriate with standards such as ZigBee, Thread or Bluetooth which is an easy solution to reduce the use of smart hubs. This will be more environment friendly and less confusing for the customers because the current smart hub gateways are proprietary vendor designed. This integration also could reduce the possibility of different IoT hubs using the same radio frequencies and networking protocol, which would create the potential for unreliable connectivity due to overlapping networks [37].

Therefore, the acquisition of network traffic would be less complicated if these advanced home routers are used in smart homes which shows that the level of complexity of the forensics investigation process depends heavily on the design of the IoT devices and architecture. This demonstrates that a collaboration among government, academia and industry is vital in order to regulate and standardise the IoT industry from a security perspective (i.e. secure by design) by which the forensics investigations would subsequently be leveraged (i.e. forensics readiness) [38].

The forensics investigation in the IoT environment can be divided into three forensics zones; traditional forensics, network forensics and cloud forensics. The traditional forensics investigation zone includes the forensics analysis of the local storage of the IoT devices and any other digital devices connected to the smart home network such as computers and mobile phones whereas the network forensics investigation zone covers the forensics analysis of the network traffic of the IoT devices, the smart hub gateway and the home router. These first two zones may not require much cooperation from any third parties such as the Cloud Service Providers but the forensics investigation of the cloud servers will definitely necessitate the collaboration with the Cloud Service Providers while overcoming the jurisdiction challenges from legal perspective [35].

Some of the challenges in the acquisition stage are related to the fact that there are many types of IoT devices in the market, using specific vendor designs and proprietary interfaces which might lead to difficulty accessing stored values, causing the investigator to perform a non-negligible reverse-engineering attempt [39]. In addition, there is no forensics readiness when it comes to monitoring the network traffic in a smart home which can be developed and integrated in the home routers. This preparation would assist the forensics investigator in preserving and collecting data for further examination in the event of an incident as a part of forensics readiness [40].

On the other hand, the installed applications on the user's mobile phone/computer that are used to operate the IoT devices in a smart home generate user-specific data where some of the data are stored on the local storage of the mobile phone device (assuming the suspect mobile phone device was present at the crime scene to be seized) and the rest of the data could be stored on the cloud servers. The data stored on the cloud will not be accessible to law enforcement agencies unless the Cloud Service Providers would be under some legal obligations to do so, such as issued court warrants for specific users account holders which can be a lengthy process, presuming bypassing the encryption challenge [21]. It is understandable that the Cloud Service Providers would be reluctant to dedicate their resources for conducting forensics investigations unless some incentives are provided. Therefore, this study proposes ***IoT Forensics as a service*** to be offered by the Cloud Service Providers in order to support law enforcement agencies in their forensics investigations when needed. However, there are some technical and legal challenges for offering such services which require more research and investment. For example, some of the legal challenges related to privacy and data protection might be resolved by exploring the options and updating the customers' service legal agreements (SLA).

(a) **Current Digital Forensics Tools**

Digital Forensics relies on scientifically derived and proven digital evidence collection methods and validated tools used by professional forensic experts [41]. Digital forensics tools are used to identify, preserve, examine and present the digital evidence in investigations.

One of the problems facing IoT forensics is the shortage of digital forensics tools available to perform investigations due to its limitations and inability to cope with the current development in the IoT environment [35]. When compared to traditional digital forensics techniques, IoT forensics faces several challenges due to the versatility and complexity of the IoT devices. The following are some of the challenges that may be faced in an investigation [42]:

- Variance of the IoT devices
- Proprietary Hardware and Software
- Data present across multiple devices and platforms
- Data can be updated, modified, or lost
- Proprietary jurisdictions for data are stored on the cloud.

Therefore, IoT forensics is multidisciplinary in approach and often a combination of tools is required to collect and analyse data from various sources such as the smart IoT devices, network traffic and the cloud servers.

The sensors and actuators in smart devices tend to generate data autonomously and in response to human behaviour such as motion detection. This makes them an excellent source of digital evidence. Although some commercial tools such as Encase and FTK may be used to collect evidence effectively, it is evident that there is no one tool capable of doing everything or is capable of doing it very well [43, 44]. In addition, customised or specialised tools are required to acquire data from the proprietary hardware or software applications of the smart IoT devices [42].

For example [45], developed a plugin in two parts for Autopsy as well as standalone python script to parse information related to the iSmartAlarm device [46]. In their research used an open source tool, Nmap to discover ports that were open on the Amazon Echo device. Putty was used as a serial terminal to read the boot logs of the Echo. The authors had proposed the use of reverse engineering techniques such as eMMC Root, JTAG and debug ports to gain access to the filesystem of the Echo. Further, it is important to note that with every new generation of devices, the structure and hardware design are changed as well [44]. Therefore, new tools and techniques are required to be developed to facilitate investigation within these devices.

In the IoT network layer, network forensics tools and methods can be applied to analyse traffics between the IoT devices and the servers. For instance, [46] used Wireshark to analyse traffic between the Echo device and the Amazon server. Conversely, [47] proposed an automated forensic management system (FEMS) that was developed to collect data from perception, network, and application architecture layers of

IoT. Nonetheless, in dynamic IoT networks, it is difficult for FEMS to examine all IoT devices.

In addition, most of the data on IoT devices is stored in the cloud, forensic investigators face challenges in physically accessing sources of evidence [35]. A survey conducted by Wu et al. [9] determined research should specifically focus on developing tools in IoT forensics to identify and acquire data from the cloud. At present, the developed forensics tools include cloud data collection forensic tools that are able to extract some of the data requiring the user's login details. However, these tools and techniques have only been developed and tested on specific IoT devices such as the Amazon Alexa and Google OnHub. Chung et al. [48] proposed using unofficial APIs technique to acquire cloud artefacts from the server. However, a challenge experienced by the authors within the past is that unofficial APIs are subject to change without warning which could require revising of code if the functionality is still available. This makes the extraction methods unlikely to be forensically sound.

Based on previous literature and current challenges faced by digital forensic investigators, it is crucial that future research needs to concentrate on the development of IoT forensic tools that would work effectively across a wide range of devices [49]. Many businesses in industry that rely on sensitive data for real-time decision making are prone to cyber-attacks therefore in the next few years, the demand for IoT security and forensics experts and resources will rise sharply [50].

Further the development of the anti-forensic techniques such as encryption and activities to overwrite data and metadata or hiding information as defensive measures are increasingly successful. These include encryption, obfuscation, and camouflage techniques, and hiding information [39]. Yildirim et al. [51], had conducted an analysis on Amazon Alexa Echo and Google Home Mini by creating anti-forensics fake activities (e.g. modifying device name, creating routine and developing custom skills) to deceive the forensic investigators. The authors determined that illogical requests with custom skills or acts allow users to perform various operations and generate fake activity history records. Other techniques include using the "TimeStomp" tool to overwrite the timestamps in NTFS system [52].

(b) **Legal Considerations/Jurisdiction**

The use of IoT devices poses a wide range of issues and concerns from a regulatory and legal point of view. The rise in IoT devices brings about new legal and regulatory issues and privacy concerns in addition to the existing issues that are already present in the traditional devices. As it is known, the use of IoT devices has potential benefits to law enforcement and the data produced by these devices can be used as evidence to investigate crimes. However, the digital forensic investigator will have to take into consideration the legal and privacy implications when conducting IoT forensics investigation.

The digital forensics methodology provides a framework consisting of procedures and processes that should be in line with standards and guidelines such as ACPO guidelines [5] to ensure maintaining the chain of custody. The forensic investigator guarantees that the legal requirements have been met at every stage of the investigation including identification, seizure, data collection, analysis, interpretation and presentation of the evidence. However, in IoT forensics the complexity involved and lack of unified standards hinder the digital investigation process and the law enforcement from acquiring evidence in a forensic manner [21]. Besides that, the issues pertaining to cross border data flows prove to be a challenge when acquiring data which is an existing issue in cloud forensics. When IoT devices gather data of individuals within one jurisdiction and then the data are stored in another jurisdiction (by the cloud storage service providers) with different data protection laws for processing, it will be a challenge for digital forensic investigators to get access to such data (chain of custody).

Even access to such data is obtained, the capability of IoT devices to autonomously make decisions makes it a challenge to determine accountability, responsibility and liability for actions taken. As the devices exchange data between themselves and storing data could be in multiple locations, there are many stakeholders and partners involved whereby several data processors may have access to the data. Basically, the service provider being the data controller would essentially determine the scope, extent, manner and purpose of the use of personal data. The service provider may also have different third-party data processors processing the data on behalf of the control of the data controller. Therefore, clarity in the ownership of data needs to be established and looked at very carefully. Legal frameworks must be updated alongside the development of digital forensics techniques to ensure that the data gathered by the IoT is not misused [53, 54].

Another major challenge from a legal perspective is developing and enforcing a privacy standard that relates to the current laws as it is different in each country. Moreover, in some circumstances the law may differ in various states and provinces within those countries. There is currently no universal privacy standard model, although many attempts have been made [55].

On a security perspective, there are proven incidents whereby the IoT devices developed have security flaws. A follow-up research on the security of IoT devices revealed that vulnerabilities in IoT devices have doubled since 2013 [56]. In 2018, hackers had abused Alexa and Google Home smart assistance to eavesdrop on users without their knowledge. This includes tricking users into revealing personal information [57]. Though both manufactures respectively have made great effort to deploy updates every time, it seems that newer ways to hack apps have started to emerge [58]. Nevertheless, attempts are being made to introduce legislation to combat weak security on IoT devices. For example, the state of California has passed a law (Senate Bill 327 [SB-327]) that came into effect on 1st January 2020 to ban pre-installed and hard-coded default passwords such as “admin” and “passwords” [59]. However, the law drew criticism from the security community which appreciated the first move but said that the law did not go far enough to control IoT security.

Similarly, the UK Government introduced “Secure by Design Code of Practice” for consumer IoT Security for manufacturers in 2018 which provides guidance for consumers on smart devices at home. A document entitled “Code of Practice for Consumer IoT Protection” was published by the Department of Digital, Culture,

Media and Sport (DCMS) in collaboration with the National Cyber Security Center (NCSC). The Code was first released as part of the Safe by Design study in the draft in March 2018 [60, 61]. However, this guidance does not include penalties for those manufacturers who do not comply as the UK government prefers to take the approach of collaborating with industry on a voluntary basis. The UK government aims to enforce "IoT Security -by-Design" law and is holding ongoing discussions with all parties involved to continue improving the legislation, no deadline has been set [62].

Overall, it is evident that efforts have been made to develop and improvise legislations on IoT Security. However, there is no effort to update cyber security legislations directly related to IoT forensics. In a survey conducted by Wu T. et al. [9], majority of the cyber forensics' respondents believe strongly that the current cyber security legislations regarding IoT forensics are not up to date which is one the significant challenges in digital forensics.

3 Digital Forensics Frameworks

In the last decade, researchers have developed new process models and solutions to improve digital forensics investigation. This has helped significantly progress not only in the field of technology but also in methodology improvement. Digital forensics has become prevalent as the modern infrastructures are becoming complex and virtualised whereby digital forensics investigators are required to acquire and analyse evidence coming in many formats on various platforms not just computer systems. While computer forensics is defined to focus on specific methods of extracting evidence from a particular platform, digital forensics must be designed in a manner such that it can encompass all types of digital devices as well as future technology. Different investigators use different methods of conducting investigation depending on the area of investigation and type of cases, thus there is no standard framework for an investigation process. This is said to be problematic because evidence must be obtained using methods that are proven to reliably extract and analyse evidence without bias or modification [63].

Recently, there have been various frameworks proposed in the field of digital forensics which attempt to refine a particular methodology for a specific case (see Table 1). Some of the digital forensics' methodologies only focus on specific stages of the digital forensics' framework such as identification, collection, preservation and examination stages [64–66] and the triage framework [67, 68] that attempts to address time sensitive applications, accelerating digital forensics investigation process.

According to Alkhanafseh et al. [69], if the employed framework contains a few stages, then this framework will not provide much guidance for the investigation process. A framework that contains many stages in which each stage has substages, with its usage scenario being more limited, may prove more useful. Therefore, it is essential to analyse various known forensics frameworks and compare their advances properly. Various frameworks have been proposed for each forensics area such as

Table 1 Overview of IoT Forensics frameworks with the main stages involved

Authors	Framework/Model names	Identification stage	Initialisation stage	Planning/Preparation stage	Preservation stage	Collection stage	Authentication stage	Evidence reduction stage
Oriwoh et al. (2013) [76]	1-2-3 Zones of Digital Forensics	✓		✓	✓	✓		
Oriwoh et al. (2013) [76]	Next Best Thing (NBT) Triage	✓		✓	✓	✓		
Perumal S. et al. (2015) [1]	Top-down approach methodology	✓						
Zawoad S. et al. (2015) [6]	FAIoT			✓	✓	✓	✓	✓
Kebande V. R. and Ray I (2016) [21]	DFIF-IoT	✓	✓		✓	✓		
Meffert C. et al. (2017) [79]	Forensic State Acquisition from Internet of Things (FSAIoT)					✓		✓
Nieto A. et al. (2018) [8]	PRoFiT			✓	✓	✓		
Kebande V. R. et al. (2018) [77]	IDFIF-IoT	✓	✓	✓	✓	✓		

(continued)

Table 1 (continued)

Authors	Framework/Model names	Identification stage	Initialisation stage	Planning/Preparation stage	Preservation stage	Collection stage	Authentication stage	Evidence reduction stage
Al-Masr E. et al. (2018) [71]	FoBI	✓			✓	✓		✓
Hossain M. et al. (2018) [74]	FIF-IoT	✓				✓		
Goudbeek A. et al. (2018) [80]	Home Automated System (HAS) Framework			✓	✓	✓		
Sathwara S. et al. (2018) [81]	Digital investigation framework for IoT systems	✓			✓			
Hossain M. et al. (2018) [82]	Probe-IoT				✓	✓		
Cebe M. et al. (2018) [73]	Block4Forensic: An Integrated Lightweight Blockchain Framework for Forensics Applications of Connected Vehicles					✓	✓	✓

(continued)

Table 1 (continued)

Authors	Framework/Model names	Identification stage	Initialisation stage	Planning/Preparation stage	Preservation stage	Collection stage	Authentication stage	Evidence reduction stage
Le D. et al. (2018) [72]	BIFF: A Blockchain-based IoT Forensics Framework with Identity Privacy					√	√	√
Ryu J. H. et al. (2019) [75]	Blockchain based framework			√	√	√	√	√

Authors	Documentation stage	Examination stage	Transportation stage	Analysis stage	Storage and archive stage	Presentation stage	Reporting stage	Review stage	Process
Oriwoh et al. (2013) [76]		√		√	√	√	√		N/A
Oriwoh et al. (2013) [76]		√		√	√	√	√		To be used in conjunction with the 1-2-3 zones of Digital Forensics
Perumal S. et al. (2015) [1]		√		√	√				Based on Triage model and 1-2-3 zone model
Zawoad S. et al. (2015) [6]									N/A

(continued)

Table 1 (continued)

Authors	Documentation stage	Examination stage	Transportation stage	Analysis stage	Storage and archive stage	Presentation stage	Reporting stage	Review stage	Process
Kebande V. R. and Ray I (2016) [21]	✓	✓	✓	✓		✓	✓		Based on ISO/IEC 27,043:2015 international standard
Meffert C. et al. (2017) [79]		✓		✓	✓				N/A
Nieto A. et al. (2018) [8]				✓		✓		✓	Based on ISO/IEC 29,100:2011
Kebande V. R. et al. (2018) [77]	✓	✓	✓	✓		✓	✓		Based on ISO/IEC 27,043:2015 international standard
Al-Masr E. et al. (2018) [71]		✓		✓	✓	✓			Based on the principle of 1st Digital Forensics Research Workshop in 2001
Hossain M. et al. (2018) [74]		✓		✓		✓			N/A

(continued)

Table 1 (continued)

Authors	Documentation stage	Examination stage	Transportation stage	Analysis stage	Storage and archive stage	Presentation stage	Reporting stage	Review stage	Process
Goudbeek A. et al. (2018) [80]				✓					N/A
Sathwara S. et al. (2018) [81]				✓					N/A
Hossain M. et al. (2018) [82]				✓					N/A
Cebe M. et al. (2018) [73]	✓	✓		✓		✓			N/A
Le D. et al. (2018) [72]				✓					N/A
Ryu J. H. et al. (2019) [75]			✓	✓			✓		N/A

computer forensics, mobile forensics, network forensics, cloud forensics and IoT forensics. These frameworks can be distinguished from one another in terms of number of stages, methods used to collect evidence and digital forensics approach such as being active or passive.

Palmer [66], defined Digital Forensics Framework as a structure to support a successful forensics investigation. This implies that the conclusion reached by one digital forensics expert should be the same as that of any other person who conducted the same investigation.

A standardised digital forensics framework consists of 9 stages which are outline as below [70]:

1. **Identification**: This stage includes recognising an incident from indicators and determining its type.
2. **Preparation**: This stage includes preparing tools, techniques, search warrants and monitoring authorisation and management support.
3. **Approach strategy**: This stage includes dynamically formulating an approach based on potential impact on bystanders and the specific technology in question.
4. **Preservation**: This stage includes isolating, securing and preserving the state of physical and digital evidence.
5. **Collection**: This stage includes recording the physical scene and duplicate digital evidence using standardise and accepted procedure.
6. **Examination**: This stage includes in-depth systematic search of evidence relating to the suspected crime.
7. **Analysis**: This stage includes determining significance, reconstructing fragments of data and drawing conclusions based on evidence found.
8. **Presentation**: This stage includes summarising and providing explanations of conclusions.
9. **Returning evidence**: This stage includes ensuring physical and digital property is returned to the proper owner as well as determining how and what criminal evidence must be removed.

The section below provides an overview of IoT Forensics Framework and outlines the limitation of some of these frameworks to identify the research gap.

3.1 Overview of IoT Forensics Framework

Advances in the digital system, together with the rapid growth in the IoT era, have caused a crucial period in digital forensics. Mauro al. [4] identified that there is no documented method or reliable forensic tool to collect forensics sound artefacts from a device. The diversity of the IoT environment has made it difficult for forensics investigators to acquire and analyse data using traditional methods. The IoT devices are known to have customised operating systems or file structures and number of wireless protocols. The lack of appropriate tools and methods makes it difficult to identify and acquire data from the IoT devices.

In the recent years, there have been attempts by various researchers to develop IoT frameworks to facilitate digital forensics investigation in the IoT environment as well as ensure that the evidence is acquired in a forensic manner. An overview of some of the known IoT frameworks that were proposed in the last few years are demonstrated in Table 1. This table outlines the main stages of each of these frameworks, the names of the original frameworks on which the proposed frameworks are based.

A new integration between digital forensics and new technology such as mining of algorithms, security algorithms and data integrity that have been used by researchers to propose new frameworks to address some of the challenges in IoT forensics. This includes integration of fog computing proposed in [71] and blockchain technology proposed in [72–75 to preserve privacy, authenticity and collection of evidence. Oriwoh et al. [76] proposed a systematic approach to identify sources of evidence within the IoT environment using three zones. Zone 1 emphasises on the internal network such as hardware, software and network connections. Zone 2 focuses on the peripheral devices such as IDS/IPS, Firewalls or Gateway. Zone 3 focuses on the hardware and software outside the network such as cloud and internet service providers. Further, they also presented a Forensics Edge Management system (FEMS) to provide an autonomous forensics service within a smart home. A layering approach has been proposed to collect data from the sensor via a network layer, which is then managed by the perception layer, and the application used to interface with the end users [47]. However, this proposed process coverage within the framework is limited to partial artefacts identification.

In 2015, Perumal et al. [1], proposed an integrated model, designed based on the triage model and 1-2-3 zone model for volatile based data preservation [76]. The proposed IoT digital forensic model includes the following processes authorization, planning, chain of custody, analysis and storage. However, it did not address the digital forensic readiness process and the research work was presented in a shallow manner. Conversely, Zawoad et al. [6] proposed a centralized trusted evidence repository in the Forensics Aware IoT (FAIoT) conceptual model which is aimed at giving support in executing digital forensics investigation in the IoT environment by providing an analysis of the existing challenges. Their proposed approach is to constantly monitor registered IoT devices and provide access to evidence through the use of API services to law enforcement authorities. This paper served as an introduction to the IoT forensic domain and a high-level investigation model was presented with partial artefacts acquisition.

Kebande and Ray [21] proposed a framework that complies with the ISO/IEC 27043: 2015 which is an international standard for information technology, security techniques, incident investigation principles, and processes. However, the proposed framework is generic and the effectiveness of the framework was not tested. In 2018, the authors proposed an IDFIF-IoT Framework [77]. This framework was an extension of an initially proposed generic Digital Forensic Investigation Framework for the IoT environment which was to address the lack of IoT digital forensics investigation standardisation. This enables the analysis of Potential Digital Evidence (PDE) generated by the IoT ecosystem. However, the framework lacks ground details that would facilitate similar adaptation to different scenarios without changing any main components or processes [78].

Conversely, Meffert et al. [79] proposed the FSAIoT framework which comprises a centralised Forensics State Acquisition Controller (FSAC) employed in three collection modes known as IoT device controller, cloud controller and controller to controller. Nevertheless, the authors did not explore the forensics soundness of the implemented IoT acquisition controller and did not take into consideration the accessing of historical data and deleted data when developing the framework. Nieto et al. [8] proposed a privacy-based model called PRoFIT to address issues related to extracting evidence data without violating users' right of privacy. This framework was based on the international standard ISO/IEC 29100:2011 requirement. It is important to mention that this model limits cases with the information voluntarily provided by the users.

In Table 2 the contribution and limitation of the above proposed frameworks are outlined. As it can be seen in this table, these proposed frameworks are only focusing on one or more stages of a digital forensics investigation not addressing the process challenges as a whole. For example, Oriwoh et al. in his work is considering only the artefact identification whereas Zawoad et al. is only considering the artefact acquisition. Some of these proposed frameworks are based on theories and they were not tested in the real environment so the effectiveness of these proposed frameworks is in question. One of the proposed frameworks requires users to give explicit consent to the collection and processing of their data in order to prevent the privacy issue of the participants. It might not be practical in a real digital forensics' investigation [8].

In summary, most of the current proposed IoT forensics frameworks implemented in pilot IoT environments have both strengths and limitations. Although these frameworks may be viable theoretically but they may not be practical solutions in a realistic IoT environment where an industrial collaboration is required to overcome the potential challenges. In addition, the focus in developing the IoT forensics frameworks should be on the entire forensics' stages rather than a part of the digital forensics' investigation.

4 Conclusion and Recommendation

The variance of IoT devices, proprietary hardware and software along with different storage devices and platforms alongside intercommunication among IoT devices have presented new challenges in the IoT forensics investigation. Some of these challenges are exacerbated by the lack of appropriate frameworks and IoT forensics tools as well as the legal and privacy issues.

In this research, the current IoT forensic solutions and frameworks proposed in the previous studies were reviewed. The strengths and limitations related to these frameworks were critically analysed in order to provide a clear direction for future studies.

Table 2 The contribution of each framework and their limitations

Authors	Framework/Model names	Contribution and comments	Limitations
Oriwoh et al. (2013) [76]	1-2-3 Zones of Digital Forensics	Provides a structured approach to systematically reduce complexity of investigations in IoT environments	The proposed process coverage is limited to partial artefact identification
Oriwoh et al. (2013) [76]	Next Best Thing (NBT) Triage	Assists with the identification of additional potential evidence sources when primary source is unavailable	The proposed process coverage is limited to partial artefact identification
Perumal et al. (2015) [1]	Top-down approach methodology	Provides guidance in investigation of IoT devices and addresses issues relating to volatile data preservation	The process did not address the digital forensic readiness process and the research work was presented in a shallow manner
Zawoad et al. (2015) [6]	FAIoT	Addresses lack of standardization in the IoT ecosystem using a centralized and secure evidence logging preservation and provenance service	The proposed process coverages are limited to partial (artefacts acquisition)
Kebande and Ray (2016) [21]	DFIF-IoT	Proposed a generic and holistic framework for a specific domain: Digital Forensics Investigation in IoT settings	The proposed framework lacks ground details that would facilitate similar adaptation to different scenarios without changing any main components or processes
Meffert et al. (2017) [79]	Forensic State Acquisition from Internet of Things (FSAIoT)	Proposed a general framework that focuses on IoT devices acquisition	The proposed model did not consider accessing historical data and deleted data and did not explore the forensic soundness of the implemented IoT acquisition controller

(continued)

Table 2 (continued)

Authors	Framework/Model names	Contribution and comments	Limitations
Nieto et al. (2018) [8]	PRoFiT	Proposed privacy-based model to address issues related to extracting evidence data without violating users´ right of privacy	This model limits the case with the information voluntarily provided by the users
Kebande et al. (2018) [77]	IDFIF-IoT	The IDFIF-IoT framework is an extension of an initially proposed generic Digital Forensic Investigation Framework for IoT environment (DFIF-IoT) and as proposed to address the shortcomings of lack of IoT digital forensics investigation standardisation	The proposed framework lacks ground details that would facilitate similar adaptation to different scenarios without changing any main components or processes
Al-Masr t al. (2018) [71]	FoBI	Proposed a Fog based IoT framework that is suitable for IoT systems that are data intensive and have a large number of deployed IoT devices	Requires further research
Hossain et al. (2018) [74]	FIF-IoT	Proposed a public digital ledger (block-chain) based framework that addresses issues on collecting evidence and a tamper-evident scheme to store evidence in a trustworthy manner	Requires further research
Goudbeek et al. (2018) [80]	Home Automated System (HAS) Framework	Proposed a seven phase forensics investigation framework to guide investigation of Home Automated System (HAS)	Requires further research

(continued)

Table 2 (continued)

Authors	Framework/Model names	Contribution and comments	Limitations
Sathwara et al. (2018) [81]	Digital investigation framework for IoT systems	Proposed an IoT Framework that focuses on helping investigators on information gathering	The proposed framework lacked ground details that would facilitate similar adaptation to different scenarios without changing any main components or processes and the research work was presented in a shallow manner
Hossain et al. (2018) [82]	Probe-IoT	Proposed Probe-IoT to addresses faced in evidence acquisition and integrity of the evidence during investigation	Requires further research
Cebe et al. (2018) [73]	BlockForensic: An Integrated Lightweight Blockchain Framework for Forensics Applications of Connected Vehicles	Proposed a framework to facilitate accident investigations and preserve the privacy of users	Requires further research
Le et al. (2018) [72]	BIFF: A Blockchain-based IoT Forensics Framework with Identity Privacy	Proposed a framework to enhance the integrity, authenticity and non-repudiation properties for the collected evidence	Requires further research
Ryu et al. (2019) [75]	Blockchain based framework	Proposed a blockchain based investigation framework focusing on data integrity preservation method	Requires further research

Some of these frameworks concentrated on time sensitive applications and accelerating digital forensics investigation processes whereas the others only focused on specific stages of digital forensics frameworks such as identification, collection, preservation and examination stages. The presence of limitations in some of these frameworks makes it unsuitable to be implemented in a real IoT environment.

A comparison among the proposed frameworks revealed that the 1-2-3 Zones of Digital Forensics [76], the Next Best Thing (NBT) Triage [76], the DFIF-IoT [21] and the IDFIF-IoT [77] frameworks are considered to be the most completed frameworks as they cover most of the stages of a digital forensics investigation. The 1-2-3 Zones of DF and the NBT Triage frameworks are limited to partial artefacts identification whereas the DIFI-IoT and the IDFIF-IoT frameworks lack ground details that would facilitate similar adaptation to different scenarios without changing any main components or processes. Most of these frameworks are based on theories so it is not certain they can be implemented in a real IoT environment. Therefore, this study focused on the simplest form of the IoT environment, smart home, to create a better picture of the challenges in IoT forensics. The challenges were discussed in Sect. 2.4 and it was recommended that there is a need for a collaboration among the government, industry and academia in order to develop a robust IoT forensics framework.

Moreover, it was discussed that the Cloud Service Providers can play an important role in assisting the forensics practitioners in IoT investigations however, due to the limitation of resources, the Cloud Service Providers might be reluctant to cooperate fully in the investigations. Therefore, in order to provide some incentives, this study suggests the ***IoT Forensics as a service*** to be offered by the Cloud Service Providers, empowering the ability for forensics readiness.

References

1. Perumal S, Norwawi NM, Raman V (2015) Internet of Things (IoT) digital forensic investigation model: Top-down forensic approach methodology. In: 2015 fifth international conference on digital information processing and communications (ICDIPC), Sierre, pp 19–23. https://ieeexplore.ieee.org/stamp/stamp.jsp?tp=&arnumber=7323000&isnumber=7322996
2. Vashi S, Ram J, Modi J, Verma S, Prakash C (2017) Internet of Things (IoT): a vision, architectural elements, and security issues. In: 2017 international conference on I-SMAC (IoT in social, mobile, analytics and cloud) (I-SMAC), Palladam, pp 492–496. https://ieeexplore.ieee.org/stamp/stamp.jsp?tp=&arnumber=8058399&isnumber=8058234
3. Ranger S (2020) The Internet of Things explained. What the IoT is, and where it's going next. Zedge. [Online]. https://www.zdnet.com/article/what-is-the-internet-of-things-everything-you-need-to-know-about-the-iot-right-now/. Accessed 15 Dec 2019
4. Mauro C, Dehghantanha A, Franke K, Watson S (2018) Internet of Things security and forensics: challenges and opportunities. Futur Gener Comput Syst 78, Part 2. https://www.sciencedirect.com/science/article/pii/S0167739X17316667
5. ACPO Good Practice Guide for Digital Evidence. Association of Chief Police Officers of England, Wales & Northern Ireland. [Online]. http://library.college.police.uk/docs/acpo/digital-evidence-2012.pdf. Accessed 15 Dec 2019
6. Zawoad S, Hasan R (2015) FAIoT: towards building a forensics aware eco system for the Internet of Things. In: 2015 IEEE international conference on services computing, New York, NY, pp 279–284. https://ieeexplore.ieee.org/stamp/stamp.jsp?tp=&arnumber=7207364&isnumber=7207317
7. Hegarty RC, Lamb DJ, Attwood A (2014) Digital evidence challenges in the Internet of Things. In: Proceedings of the 10th international network conference, INC 2014, pp 163–172. https://www.researchgate.net/publication/288660566_Digital_evidence_challenges_in_the_internet_of_things
8. Nieto A, Rios R, Lopez J (2018) IoT-forensics meets privacy: towards cooperative digital investigations. Sensors (Basel) 7;18(2):492. https://www.ncbi.nlm.nih.gov/pmc/articles/PMC5856102/

9. Wu T, Breitinger F, Baggili I (2019) IoT ignorance is digital forensics research bliss: a survey to understand IoT forensics definitions, challenges and future research directions. In: Proceedings of the 14th international conference on availability, reliability and security (ARES '19). Association for Computing Machinery, New York, NY, USA, Article 46, pp 1–15. https://dl.acm.org/citation.cfm?id=3340504
10. What Is a Smart City? Cisco, 2020. [Online]. https://www.cisco.com/c/en/us/solutions/industries/smart-connected-communities/what-is-a-smart-city.html. Accessed 08 May 2020
11. Introduction to IoT. Cisco, 2019. [Online]. https://www.netacad.com/courses/iot/introduction-iot
12. Simpson P (2020) Smartcitiesworld.net. [Online]. https://smartcitiesworld.net/AcuCustom/Sitename/DAM/012/Understanding_the_Challenges_and_Opportunities_of_Smart_Citi.pdf. Accessed 08 May 2020
13. The Internet of Things (IoT)—What it is and why it matters. SAS, 2020. [Online]. https://www.sas.com/en_us/insights/big-data/internet-of-things.html. Accessed 15 Dec 2019
14. Hanes D, Salgueiro C, Grossetete P, Barton R, Henry J (2017) IoT fundamentals: networking technologies, protocols, and use cases for the Internet of Things
15. Crane C (2019) 20 Suprising IoT statistics you don't already know. Security Boulevard. [Online]. https://securityboulevard.com/2019/09/20-surprising-iot-statistics-you-dont-already-know/. Accessed 15 Dec 2019
16. Gomez C, Paradells J (2010) Wireless home automation networks: a survey of architectures and technologies. IEEE Commun Mag 48(6):92–101. https://ieeexplore.ieee.org/document/5473869
17. 10 reasons to use the nest learning thermostat | Service champions. Service Champions NorCal, 2017. [Online]. https://www.servicechampions.net/blog/10-reasons-use-nest-learning-thermostat/. Accessed 08 May 2020
18. Smart Home—worldwide | Statista market forecast. Statista, 2020. [Online]. https://www.statista.com/outlook/279/100/smart-home/worldwide. 08 May 2020
19. Smart home report 2019. Statista, 2019. [Online]. https://www.statista.com/study/42112/smart-home-report/. Accessed 17 Feb 2020
20. Davis BD, Mason JC, Anwar M (2020) Vulnerability studies and security postures of IoT devices: a smart home case study. IEEE Internet Things J 7(10). https://ieeexplore.ieee.org/abstract/document/9050664
21. Kebande VR, Ray I (2016) A generic digital forensic investigation framework for Internet of Things (IoT). In: 2016 IEEE 4th international conference on future Internet of Things and Cloud (FiCloud), Vienna, pp 356–362. https://ieeexplore.ieee.org/stamp/stamp.jsp?tp=&arnumber=7575885&isnumber=7575827
22. Tabane E, Zuva T (2016) Is there a room for security and privacy in IoT? In: 2016 international conference on advances in computing and communication engineering (ICACCE), Durban, pp 260–264. https://ieeexplore.ieee.org/stamp/stamp.jsp?tp=&arnumber=8073758&isnumber=8073703
23. Liu X, Zhao M, Li S, Zhang F, Trappe W (2017) A security framework for the Internet of Things in the future internet architecture. Future Internet. 9. 27. www.mdpi.com/1999-5903/9/3/27/pdf
24. Mendez D, Papapanagiotou I, Yang B (2017) Internet of Things: survey on security and privacy. https://www.researchgate.net/publication/318259049_Internet_of_Things_Survey_on_Security_and_Privacy
25. Lopez J, Rios R, Bao F, Wang G (2017) Evolving privacy: from sensors to the Internet of Things. Futur Gener Comput Syst 75:46–57. https://www.sciencedirect.com/science/article/abs/pii/S0167739X16306719?via%3Dihub
26. Abomhara M, Koien G (2014) Security and privacy in the Internet of Things: current status and open issues. https://ieeexplore.ieee.org/document/6970594
27. Tilley A (2015) How hackers could use a nest thermostat as an entry point into your home. Forbes. [Online]. https://www.forbes.com/sites/aarontilley/2015/03/06/nest-thermostat-hack-home-network/#6266ed343986. Accessed 08 May 2020

28. Winder D (2019) Google confirms android camera security threat: 'Hundreds of Millions' of users affected. Forbes. [Online]. https://www.forbes.com/sites/daveywinder/2019/11/19/google-confirms-android-camera-security-threat-hundreds-of-millions-of-users-affected/#c9d5dc4f4e12. Accessed 08 May 2020
29. Yang Y, Cai H, Wei Z, Lu H, Choo KKR (2016) Towards lightweight anonymous entity authentication for iot applications, pp 265–280. Springer, Cham. https://link.springer.com/chapter/10.1007%2F978-3-319-40253-6_16_16
30. Zhao K, Ge L (2013) A survey on the Internet of Things security. In: International conference on computational intelligence and security (CIS), pp 663–667. https://ieeexplore.ieee.org/document/6746513
31. Leo M, Battisti F, Carli M, Neri A (2014) A federated architecture approach for Internet of Things security. In: Euro med telco conference (EMTC), pp 1–5. https://ieeexplore.ieee.org/document/6996632
32. Roman R, Zhou J, Lopez J (2013) On the features and challenges of security and privacy in distributed Internet of Things. Comput Netwo 57:2266–2279. https://www.sciencedirect.com/science/article/abs/pii/S1389128613000054
33. Mahmoud R, Yousuf T, Aloul F, Zualkernan I (2015) Internet of things (IoT) security: current status, challenges and prospective measures. In: 2015 10th international conference for internet technology and secured transactions (ICITST), London, pp 336–341. https://ieeexplore.ieee.org/document/7412116
34. Top 10 IoT security issues: ransom, botnet attacks, spying. Intellectsoft Blog, 2015. [Online]. https://www.intellectsoft.net/blog/biggest-iot-security-issues/. Accessed 08 May 2020
35. Alabdulsalam S, Schaefer K, Kechadi T, Le-Khac NA (2018) Internet of Things forensics: challenges and case study. https://www.researchgate.net/publication/322851720_Internet_of_things_forensics_Challenges_and_Case_Study
36. Bakhshi T (2019) Forensic of Things: revisiting digital forensic investigations in Internet of Things. In: 2019 4th international conference on emerging trends in engineering, sciences and technology (ICEEST), Karachi, Pakistan, pp 1–8. https://ieeexplore.ieee.org/abstract/document/8981675
37. Forrest S (2017) Smart architectures for smart home gateways. MIPS, [Online]. https://www.mips.com/blog/smart-architectures-for-smart-home-gateways/. Accessed 22 April 2020
38. Government response to the Regulatory proposals for consumerInternet of Things (IoT) security consultation. gov.UK, 2020. [Online]. https://assets.publishing.service.gov.uk/government/uploads/system/uploads/attachment_data/file/862953/Government_response_to_consultation__Regulatory_proposals_for_consumer_IoT_security.pdf. Accessed 9 May 2020
39. Caviglione L, Wendzel S, Mazurczyk W (2017) The future of digital forensics: challenges and the road ahead. IEEE Secur Priv Mag 15. https://doi.org/10.1109/MSP.2017.4251117. https://ieeexplore.ieee.org/document/8123473
40. Kent K, Chevalier S, Grance T, Dang H (2006) Guide to integrating forensic techniques into incident response. NIST. [Online]. https://nvlpubs.nist.gov/nistpubs/Legacy/SP/nistspecialpublication800-86.pdf. Accessed 4 May 2020
41. Chernyshev M, Zeadally S, Baig Z, Woodward A (2018) Internet of Things forensics: the need, process models, and open issues. IT Prof 20(3):40–49. https://ieeexplore.ieee.org/document/8378977
42. IoT forensics: security in connected world | Packt Hub. Packt Hub. [Online]. https://hub.packtpub.com/iot-forensics-security-connected-world/. Accessed 08 May 2020
43. Alenezi A, Atlam H, Alsagri R, Alassafi M, Wills G (2019) IoT forensics: a state-of-the-art review, challenges and future directions. https://www.researchgate.net/publication/333032591_IoT_Forensics_A_State-of-the-Art_Review_Challenges_and_Future_Directions
44. Pawlaszczyk D, Friese J, Hummert C (2019) "Alexa, tell me …"—a forensic examination of the Amazon Echo Dot 3 rd generation. Int J Comput Sci Eng 7(11):20–29. https://www.researchgate.net/publication/337681675_D_Pawlaszczyk_J_Friese_C_Hummert_Alexa_tell_me_-_A_forensic_examination_of_the_Amazon_Echo_Dot_3_rd_Generation_International_Journal_of_Computer_Sciences_and_Engineering_Vol7_Issue11_pp20-29_2019

45. Servida F, Casey E (2019) IoT forensic challenges and opportunities for digital traces. Digit Investig 28:S22–S29. https://www.researchgate.net/publication/332614704_IoT_forensic_challenges_and_opportunities_for_digital_traces
46. Clinton I, Cook L, Banik S (2016) Survey of various methods for analyzing the Amazon Echo. https://www.semanticscholar.org/paper/A-Survey-of-Various-Methods-for-Analyzing-the-Echo-Clinton/47647a865622106c024d42e680acbb726aeea69d
47. Oriwoh E, Sant P (2013) The forensics edge management system: a concept and design. In: 2013 IEEE 10th international conference on ubiquitous intelligence and computing and 2013 IEEE 10th international conference on autonomic and trusted computing, Vietri sul Mere, pp 544–550. https://ieeexplore.ieee.org/stamp/stamp.jsp?tp=&arnumber=6726257&isnumber=6726171
48. Chung H, Park J, Lee S (2017) Digital forensic approaches for Amazon Alexa ecosystem. Digit Investig 22:S15–S25. https://www.sciencedirect.com/science/article/pii/S1742287617301974
49. Li S, Choo KR, Sun Q, Buchanan WJ, Cao J (2019) IoT forensics: Amazon echo as a use case. IEEE Internet Things J 6(4):6487–6497. https://ieeexplore.ieee.org/document/8672776
50. Chi H, Aderibigbe T, Granville BC (2018) A framework for IoT data acquisition and forensics analysis. In: Proceedings—2018 IEEE international conference Big Data, pp 5142–5146. https://ieeexplore.ieee.org/document/8622019
51. Yildirim I, Bostanci E, Guzel M (2019) Forensic analysis with anti-forensic case studies on Amazon Alexa and Google assistant build-in smart home speakers, pp 1–3. https://ieeexplore.ieee.org/abstract/document/8907007
52. TimeStomp—Metasploit Unleashed. Offensive Security. [Online]. https://www.offensive-security.com/metasploit-unleashed/timestomp/. Accessed 18 April 2020
53. Industrial IoT—legal and regulatory aspects. IIoT World. [Online]. https://iiot-world.com/connected-industry/industrial-iot-legal-and-regulatory-aspects/. Accessed 05 May 2020
54. India: legal issues pertaining to Internet of Things (IOT). Mondaq. [Online]. https://www.mondaq.com/india/privacy-protection/691560/legal-issues-pertaining-to-internet-of-things-iot. Accessed 05 May 2020
55. Fabiano N (2017) Internet of Things and the Legal Issues related to the Data Protection Law according to the new European General Data Protection Regulation. Athens J Law 3:201–214. https://www.athensjournals.gr/law/2017-3-3-2-Fabiano.pdf
56. Coble S (2020) Vulnerabilities in IoT devices have doubled since 2013. InfoSecurity. [Online]. https://www.infosecurity-magazine.com/news/vulnerabilities-in-iot-devices/. Accessed 06 May 2020
57. Cimpanu C (2019) Alexa and Google Home devices leveraged to phish and eavesdrop on users, again. Zedge. [Online]. https://www.zdnet.com/article/alexa-and-google-home-devices-leveraged-to-phish-and-eavesdrop-on-users-again/. Accessed 06 May 2020
58. Top 5 shocking IoT security breaches of 2019. PentaSecurity, 2019. [Online]. https://www.pentasecurity.com/blog/top-5-shocking-iot-security-breaches-2019/. Accessed 06 May 2020
59. SB-327 information privacy: connected devices. California Legislative Information, 2017–2018. [Online]. https://leginfo.legislature.ca.gov/faces/billNavClient.xhtml?bill_id=201720180SB327. Accessed 06 May 2020
60. Secure by Design. gov.UK 2019. [Online]. https://www.gov.uk/government/collections/secure-by-design. Accessed 08 May 2020
61. Code of practice for consumer IoT security. gov.UK, 2019. [Online]. https://www.gov.uk/government/publications/code-of-practice-for-consumer-iot-security. Accessed 08 May 2020
62. Truta F (2020) UK to mandate IoT security-by-design in upcoming legislation. Bitdefender Box, 2020. [Online]. https://www.bitdefender.com/box/blog/iot-news/uk-mandate-iot-security-design-upcoming-legislation/. Accessed 08 May 2020
63. Cisar P, Cisar SM (2011) Methodological frameworks of digital forensics. In: 2011 IEEE 9th international symposium on intelligent systems and informatics, Subotica, pp 343–347. https://ieeexplore.ieee.org/stamp/stamp.jsp?tp=&arnumber=6034350&isnumber=6034292
64. Kruse W, Heiser JG (2002) Computer forensics: incident response essentials. Addison-Wesley. [Online]. =https://books.google.com.my/books/about/Computer_Forensics.html?id=nNpQAAAAMAAJ&redir_esc=y. Accessed 08 May 2020

65. A guide for first responders. National Institute of Justice: Electronic Crime Scene Investigation, 2001. [Online]. https://www.ncjrs.org/pdffiles1/nij/187736.pdf. Accessed 08 May 2020
66. Palmer G (2001) A road map for digital forensics research-report from the first digital forensics. In: Research workshop (dfrws), || Utica, New York, 2001. [Online]. https://dfrws.org/presentation/a-road-map-for-digital-forensic-research/. Accessed 08 May 2020
67. Pilli ES, Joshi RC, Niyogi R (2010) Network forensic frameworks: survey and research challenges. Digital Investig 7(1–2):14–27. https://www.sciencedirect.com/science/article/abs/pii/S1742287610000113
68. Kohn M, Olivier MS, Eloff JH (2006) Framework for a digital forensic investigation. In: ISSA, pp 1–7. https://www.researchgate.net/publication/220803284_Framework_for_a_Digital_Forensic_Investigation
69. Alkhanafseh M, Qatawneh M, Almobaideen W (2019) A survey of various frameworks and solutions in all branches of digital forensics with a focus on cloud forensics. Int J Adv Comput Sci Appl. https://www.researchgate.net/publication/335694535_A_Survey_of_Various_Frameworks_and_Solutions_in_all_Branches_of_Digital_Forensics_with_a_Focus_on_Cloud_Forensics
70. Reith M, Carr C, Gunsch G (2002) An examination of digital forensic models international journal of digital evidence. https://www.just.edu.jo/~Tawalbeh/nyit/incs712/digital_forensic.pdf
71. Al-Masri E, Bai Y, Li J (2018) A fog-based digital forensics investigation framework for IoT systems. In: 2018 IEEE international conference on smart cloud (SmartCloud), New York, NY, pp 196–201. https://ieeexplore.ieee.org/stamp/stamp.jsp?tp=&arnumber=8513738&isnumber=8513698
72. Le D, Meng H, Su L, Yeo SL, Thing V (2018) BIFF: a blockchain-based IoT forensics framework with identity privacy. In: TENCON 2018—2018 IEEE region 10 conference, Jeju, Korea (South), pp 2372–2377. https://ieeexplore.ieee.org/stamp/stamp.jsp?tp=&arnumber=8650434&isnumber=8650051
73. Cebe M, Erdin E, Akkaya K, Aksu H, Uluagac S (2018) Block4Forensic: an integrated lightweight blockchain framework for forensics applications of connected vehicles. IEEE Commun Mag 56(10): 50–57. https://ieeexplore.ieee.org/stamp/stamp.jsp?tp=&arnumber=8493118&isnumber=8493098
74. Hossain M, Karim Y, Hasan R (2018) FIF-IoT: a forensic investigation framework for IoT using a public digital ledger. In: 2018 IEEE international congress on Internet of Things (ICIOT). https://ieeexplore.ieee.org/document/8473437
75. Ryu JH, Sharma PK, Jo JH, Park JH (2019) A blockchain-based decentralized efficient investigation framework for IoT digital forensics. J Supercomput. https://doi.org/10.1007/s11227-019-02779-9
76. Oriwoh E, Jazani D, Epiphaniou G, Sant P (2013) Internet of Things forensics: challenges and approaches. https://www.researchgate.net/publication/259332114_Internet_of_Things_Forensics_Challenges_and_Approaches
77. Kebande VR, Karie NM, Michael A, Malapane S, Kigwana I, Venter HS, Wario RD (2018) Towards an integrated digital forensic investigation framework for an IoT-based ecosystem. In: Proceedings—2018 IEEE International conference on smart Internet Things, SmartIoT 2018, pp 93–98. https://ieeexplore.ieee.org/document/8465532
78. Stoyanova M, Nikoloudakis Y, Panagiotakis S, Pallis E, Markakis EK A survey on the Internet of Things (IoT) forensics: challenges, approaches and open issues. IEEE Commun Surv Tutor. https://ieeexplore.ieee.org/stamp/stamp.jsp?tp=&arnumber=8950109&isnumber=5451756
79. Meffert C, Clark D, Baggili I, Breitinger F (2017) Forensic state acquisition from Internet of Things (FSAIoT): a general framework and practical approach for IoT forensics through IoT device state acquisition, pp 1–11. https://www.researchgate.net/publication/319045807_Forensic_State_Acquisition_from_Internet_of_Things_FSAIoT_A_general_framework_and_practical_approach_for_IoT_forensics_through_IoT_device_state_acquisition
80. Goudbeek A, Choo KR, Le-Khac N (2018) A forensic investigation framework for smart home environment. In: 2018 17th IEEE international conference on trust, security and privacy in

computing and communications/12th IEEE international conference on big data science and engineering (TrustCom/BigDataSE), pp 1446–1451. https://ieeexplore.ieee.org/document/8456070

81. Sathwara S, Dutta N, Pricop E (2018) IoT forensic a digital investigation framework for IoT systems. In: 2018 10th international conference on electronics, computers and artificial intelligence (ECAI), pp 1–4. https://ieeexplore.ieee.org/document/8679017
82. Hossain M, Hasan R, Zawoad S (2018) Probe-IoT: a public digital ledger based forensic investigation framework for IoT. In: IEEE INFOCOM 2018—IEEE conference on computer communications workshops (INFOCOM WKSHPS), Honolulu, HI, pp 1–2, https://ieeexplore.ieee.org/document/8406875

Making the Internet of Things Sustainable: An Evidence Based Practical Approach in Finding Solutions for yet to Be Discussed Challenges in the Internet of Things

Benjamin Newman and Ameer Al-Nemrat

Abstract The Internet of Things (IoT) is well on its way to forming a fully digitalised society. Whilst IoT provides opportunities which other technology cannot, the enormous amount of responsibility also means it can be the key to access critical infrastructure. IoT is insecure by nature, is a gateway to the network, can be deployed in safety-critical areas and can generate substantial amounts of detailed data. Traditional approaches to protecting IoT is inefficient as the very limitations means best practices and standards are ineffective when being applied to the IoT environment. This study argues that work needs to be shifted from security and privacy to consumer safety and software sustainability. The impact of IoT is largely uncertain and the technology is redefining new areas of research which have yet to be addressed. Standards and regulations are an essential part to the integrity of sustainability and safety, but it is clear that they are currently too fragmented and are not able to keep up with the emerging technology. This study aims to highlight the underlying issues which other studies have missed, and to provide solutions which can be applied in future work. In order to let IoT be beneficial we must force organisations and crowd-funded projects to employ secure-by-default into the design phase of their product and only then will IoT be able to thrive into what it should be.

1 Introduction

The Internet of Things (IoT) is a paradigm that is progressing society into becoming a fully digitalised environment, where everything is connected. At a conceptual level IoT refers to the interconnectivity among devices, along with the ability to abstract large amounts of data, also known as, Big Data. Gartner [40] predicts the total number of IoT devices will increase from 5 billion in 2015 to 25 billion in 2020, for this reason IoT is quickly becoming an important future technology that is being adopted by a wide range of industries. The recognised value of IoT is clear when connected

B. Newman · A. Al-Nemrat (✉)
University of East London, London, UK
e-mail: ameer@uel.ac.uk

R. Montasari et al. (eds.), *Digital Forensic Investigation of Internet of Things (IoT) Devices*, Advanced Sciences and Technologies for Security Applications,
https://doi.org/10.1007/978-3-030-60425-7_11

devices can communicate and be integrated with, smart cities, smart energy grids, smart homes, business intelligence and analytical applications [51]. Furthermore, as IoT technologies evolve, concerns and efforts to resolve security issues in IoT environments are increasing. That is, more active IoT-related researches and development are in various industries, the more crucial security in the IoT environment is needed.

As more safety-critical devices such as, pacemakers, insulin pumps, autonomous cars, and smart metres become connected to the internet, "security will be more about safety" [3]. However, most of the stories on IoT have been about privacy. The US government, more specifically James Clapper, US intelligence chief, has already acknowledged the use of IoT devices for use in spying on the masses [1]. Other events have seen children's toys and smart cameras banned in Germany, as they can also be used to spy on households [79]. IoT devices are typically seen as the "weakest link," as they are embedded into systems and are often relied upon, therefore they are attractive targets [90]. This is less of a hypothetical thought and instead a much more real concern. September 2016 saw the rise of the Mirai and Repear botnets, that had been conducted through poorly configured IoT devices [4]. A much recent event happened in 2018 when GitHub survived a 1.3TB Distributed Denial of Service (DDoS) attack. The size of these attacks is only going to grow each year, whilst more organisations release unsupported, poorly configured internet connected devices. It is only a matter of time until an attacker can target more critical infrastructures such as energy grids.

Standards and best practices are a large aspect of security and safety. The automotive industry is a fantastic representation into the benefits of proper procedural regulation handling, and how standards are directly correlated to safety. The US Department of Transportation (DOT) was born in 1966 which revolutionised the safety of driving. The Highway Safety Act established by NHTSA was solely responsible for dramatically reducing deaths and injuries resulting from motor vehicle crashes [74]. The implementation of standards caused car companies to apply for type approval, mandate recalls and coordinate the safety of the car with the design of the road. This had a dramatic change on the present-day ecosystem, where car manufacturers are now forced to spend millions on defects which are developed by no fault of the owner [70]. However, the introduction of low-power-wide-area-networks (LPWAN) has seen a dramatic increase in the ability to provide over-the-air (OTA) updates, this system will provide a dramatic reduction in costs as many original-equipment-manufactures will not have to recall their equipment and instead apply updates that fix the problems remotely. But OTA updates are difficult, they pose many risks and the complexity of the vehicles is only going to add to the problem. As the complexity grows the complications of successfully developing and deploying OTA updates also increases. Tesla have already shown that cars are capable of being autonomously driven, the problem is there are a range of critical systems, such as, ABS control, airbag deployment, collision detection system, which are also being autonomously controlled. Each system may depend on a different supplier that uses different methods and software to design their system, the coordination of OEM and supplier is only going to get harder and the more complex a system becomes the bigger the impact on safety. The goals of a cybersecurity regulator will be to mix security, safety, and privacy. Different intuitions will focus on different factors, the automotive industry focuses on security and safety whereas privacy will be more important for personal fitness wearables. However, there is a problem, NIST [77] provide a comprehensive list of all the current bodies which have developed some standards for IoT, the issue is, there are multiple bodies with similar objectives, the same applies for IoT platforms, there are over 400 platforms that provide a different solution to the same problem, eventually an entity will have to determine what needs to be regulated. Moreover, organisations do not want to be held accountable for liability for their device, a clear set of standards which provide structured processes of regulations would reduce the asymmetry in the plethora of bodies, but a large problem still stands where liability is still a tricky subject, as it's based on old IT practices and still has yet to be revisited [58]. It is clear there are dramatic differences between industries as to which approach should be taken but the concepts learned from other industries still apply. Despite the benefits of regulations, organisations can also be reluctant to adopt new processes, especially if they are more likely to be held liable or if the costs of production are increased.

Currently, technology relies on a system which supplies monthly distributed updates such as, Microsoft, with their "Patch Tuesday's." At present with this system, vendors tend to stop support for a product that is three years or older, for example, TomTom recently announced their promise of "lifetime" sat-nav updates will be stopped, even for models that had only recently, last year, stopped being sold, the reason as stated by TomTom came down to limitations in the hardware [82]. The problem becomes much more drastic in areas such as the automotive industry; where the average age of a car in America is 11.4 years and growing [50]. Tesla have already announced some of their older model S 75D vehicles are not capable of receiving a performance upgrade due to hardware differences [56]. In some situations, this system is beneficial, if a vulnerability is found in the software then a patch can be issued within 24 h [12], but the OTA process, which companies such as, Tesla, BMW, and Ford are utilising is difficult and largely unregulated.

To fix vulnerable systems, an ideal solution would be to employ the ability for rapid patching, however, OTA updates are difficult to implement as devices tend to outlive software updates [97]. Refrigerators, TV's and cars are all expected to last for decades but maintaining technical support for a product is costly and time consuming. IoT solutions which are designed for specific products that are applied as a means of continuing technical support are also difficult to maintain [67]. Some organisations will implement a range of products into their business infrastructure, and this adds complexity to the solution as more models will be introduced to maintain the running costs of the platform. Start-ups are exceptionally vulnerable, after selling devices they either drop their product or are bought by another company, which end up cancelling the original product [39]. Start-ups are also affecting the balance of security, there is a difference between a prototype implementation which can handle tens of devices, and one that has been designed to handle millions of devices [85]. It used to be very difficult to obtain PCB controllers but in the 21st century you can use services, such as, PCB shopper. Organisations are not helping the problem either, for example the EVM 430-F6779 model, from Texas Instruments is suggested to be used for the development of a smart metre but give no information on how to secure the device. The controller has a debugging feature, it is not farfetched to suggest that this feature could be left on during distribution, and thus has the potential to leak sensitive information. However, it is unreasonable to expect start-ups or even established companies to create software that is entirely bug-free [16], but it's also not unreasonable to suggest that organisations should take a proactive approach instead of a reactive approach.

Another challenge is getting the consumer to apply an update. Some IoT devices require users to manually install updates but getting the user to notice is a challenge. For example, thousands of baby monitors manufactured by Foscam, contained a remote vulnerability that allowed attackers to gain access, because Foscam had no central disclosure platform many of the devices were left vulnerable, as users were un-aware of the patch [45]. The answer would be to implement automatic updates, however as seen by the history of Microsoft, Patch Tuesday's have been causing more problems than they have been fixing them [57]. Software updates are usually deployed to either add new features or fix a bug, and as such these updates add modifications to the existing software [109]. The ability to apply updates is a factor of security and is currently the only method of which can be used once the software or device is deployed [36]. The importance of updates being distributed in a timely manner is extremely important for ensuring the protection of the device or software, when a vulnerability occurs, and it is publicly disclosed, the exploit rates increase by magnitudes [14]. Many organisations, unfortunately, take a reactive approach to vulnerability disclosures and only after a vulnerability is found does an organisation then apply a patch. Interestingly, systems that are regularly maintained are less likely to be compromised [54].

There is a negative view on software updates, Google Statistics [42] shows the problem with interoperability between versions. Oreo, Androids latest version distribution, as of this writing, has only 4.6% of the market share, with Nougat leading with 30.8% a version which released in 2016. The problem is, many Android users feel that updates cause more harm than good and therefore users tend to stay away from updating their device. Apple have also had a history of bad software updates, especially in 2017. First a vulnerability had been found in macOS High Sierra [112], in the same year another software update caused the file sharing feature to stop working [5]. This new update undid the original macOS High Sierra critical vulnerability [43]. The changes and problems that software updates bring are not only an annoyance for its user base, but it also shows even established organisations still fail to provide reliable software updates. With the IT ecosystem not being limited by battery life, memory size and duty-cycle it is sufficient in applying patches as many times throughout the devices or software's life. However, trying to apply this strategy to IoT is a task that is met with huge difficulty.

Attitudes towards adopting a sustainable business model is fraught with problems. As Atlas [9] has shown, the average cost of IoT sensors is failing, therefore applying any type of technical support for long periods of time is not within the scope of an organisation. It is therefore essential that organisations are forced into situations where they must apply support until the devices end-of-life. The problem worsens when you understand the severity of found vulnerabilities per year. Microsoft have shown in their annual security report, there are 5,000 to 6,000 vulnerabilities surfacing each year, working out to be about 15 per day [11]. Services such as Shodan, a search engine for vulnerable internet connected devices, can be used to take advantage of these vulnerabilities. Furthermore, the vulnerabilities raise another concern, poor security is allowing botnets like Mirai to scan the full IPv4 range in less than 6 min, showing that any new exploit can be rapidly deployed. The more alarming concern is the 3.7 s rate at which login requests occurred, added to the fact of poor security and poorly configured devices it would be practically impossible at these rates to deploy a patch in time to millions of devices [72], therefore we must tackle the problem at its heart, that is by designing frameworks which prevent these factors before they can cause harm. But we first must understand, in detail, why organisations fail to take a proactive approach and end up relying on a reactive approach. The state of the Internet of Things is still undecided but as can be seen from previous history it is largely insecure. A literature review has been carried out to find which factors have a direct impact on software and device sustainability and consumer safety. This information will then be used as a basis for choosing proper case studies where, hopefully, the underlying factors are found.

2 Definition of Safety-Critical

Before we begin the next sections of this work, it is critical that we define the term "safety-critical," as it is an important part of this study. The term is defined by TRAC [103], as "components that are critical to the safety of equipment." More importantly they perform actions which protect against harmful hazards that can occur when a function becomes faulty. TRAC specifically say the components "do not have to be mains connected to be safety critical," in the context of IoT this would apply.

3 Literature Review

The following literature survey will comprise of a detailed evaluation of essential literature, which is suited to perform as a source of knowledge which can be used to support key arguments and to underline the assumptions set out by this dissertation. The following survey structure will review three key topics, which, when combined form the basis of consumer safety in IoT.

i. Regulations and standards
ii. Liability and Transparency
iii. Software and Firmware updates

The Internet of Things (IoT) which can also be called the Internet of Everything or Industrial Internet, is an emerging paradigm which foresees global coverage of internet connected devices. Gartner [40] expects to see 20 billion IoT devices by 2020 and it can therefore be viewed as one of the most important areas of future technology and research. The design of advanced light-weight concepts that are infused with information technology and sensor systems, allows the widespread distribution of internet-connected devices. These devices have the capability for applications such as, e-medicine, implants, early warning and detections systems, smart metres, and population monitoring [37]. The innovation in sensor technology and the incorporation of Low-power-wide-area-networks (LPWAN) have a dramatic impact on the effect of consumer and business alike. For instance, the business sector can benefit most from IoT, where the collection and analysis of Big Data allows for better services, higher efficiency of production and a superior product [92].

There are many published surveys on IoT that focus on security and privacy. Sicari et al. [95] conducted a comprehensive survey on the most researched topics of IoT. They found seven categories: authentication and confidentiality, access control, privacy, trust, enforcement, secure middleware and mobile security, where each of these areas of study contain numerous ongoing projects which continue to focus on security and privacy as being the main issue. The problem is that most of these studies focus on the challenges and limitations that are a consequence of the restrictions of IoT, for example: battery life and computing power. Whilst these topics are the prevailing reason for security issues, the studies do not incorporate more modern developing concerns. Zhou et al. [120] published a paper that highlighted new threats, existing solutions, and future challenges of IoT. They analysed security issues from a new perspective expressing that many previous research papers lack the incorporation of diversity, interoperability, fragmentation and scalability into their solutions. They conclude their work by exaggerating the point that IoT is not limited to the factors they have provided and that new challenges will arise as technology advances.

A major issue of IoT is the dependency on a network connection. Each device is essentially a gateway into an infrastructure, whether that be a smart home or a business [44]. The heterogeneous network that IoT is, and the data it handles, provides severe risks to security and privacy of its users. For instance, CloudPets (2017) a smart teddy bear had been found to expose 750,000 children's voice messages, the teddy bear used the cloud to store intermit recordings between parent and child, and the information had been leaked by a publicly facing network. The integration of services in what was, originally, a mere toy has caused added complexity and shown new areas in which security must be addressed. Moreover, the embedded nature of IoT means many of the devices will be deep inside networks and therefore become appealing targets for hackers. Kovacs [55] gives a real example of this phenomenon, the researcher reported the use of handheld scanners that were used to gain entry to a logistic shipping firm, highlighting the potentially lack of security. According to Yuchen et al. [118] security and privacy is the largest issue for IoT devices. They express concerns that it is not just personal information that is being stored but many times the information collected, can be analysed to monitor user activities. A paper by Apthorpe et al. [6] discovered valuable information in the metadata that is generated from smart devices inside a smart home. They discovered that individuals could use passive network monitoring techniques to analyse "traffic flows" and the more devices inside a home the simpler the task. Their solution is to use traffic shaping as a method of protecting smart home privacy, however, this raises an interesting point, many investors of IoT devices rely on the one-time setup, and thus lack the knowledge on applying solutions such as: network VPN tunnels, whitelisting IP addresses or separating IoT devices from your main network. Interestingly Williams et al. [113] conducted a survey and addressed the issue that many IoT devices are less useable or familiar, and that we expect these devices to work without interaction, 42% of the people surveyed expressed functionality as a leading factor and only 9% raised concerns about a lack of privacy and security settings. It would therefore seem that it is up to the developer to apply security and privacy by default in their design, but security in constrained devices is one of the biggest challenges IoT is currently facing.

Current security in IoT is lack at best and holds a plethora of security issues that need to be addressed. Rose and Ramsey [86] presented their findings in Bluetooth low energy locks, 12 out of 16 smart locks were found to be hackable from a quarter of a mile away. They concluded "shoddy code" as the reason for the lack of security. Lv et al. [60], part of the Keen Security Lab was able to take remote control of a Tesla Model S from 12 miles away, it was possible by tricking the car to join a hostile Wi-Fi connection. The researchers were able to take control of the entire system, from the movement of the mirrors to the ABS system. Interestingly Tesla were able to provide an over-the-air update (OTA) 24 h later. Another example presented by Thomson [101] who reported on a story about a baby heart monitor by Owlet. The baby monitor, which contained a sensor that monitors a baby's heartbeat, had been found to send unencrypted data wirelessly to a nearby hub. The examiners also found that the device had no capability of receiving a software patch as it had not been implemented at the design stage. Researchers, Chen et al. [23], created a ghost traffic jam by attacking the traffic control system in the US, interestingly the Department of Transpiration had been using the system for trials since 2016. Called I-SIG (Intelligent Traffic Signal System), the system works by real-time tracking of vehicle trajectory which is fed into an algorithm that controls the traffic light system. The key issue is that many future Driver-less cars are expected to maintain a similar system, if after 2 years it is still vulnerable after numerous tests then what are the potential implications to consumer safety? Is there anything being done to mitigate these problems? It would seem once again a proactive approach is being taken, instead by tackling the underlying issues, identified as being the poor implementation of security, this situation is less likely to occur. National Audit Office [71] investigated the WannaCry cyber-attack on the NHS, it had been found that many of the systems were running unsupported operating systems (OS) and thus were incapable of receiving patches. The Owlet and Tesla cases raise an interesting point, there is a clear problem of liability and transparency between a large, reputable organisation like Tesla, and a small organisation such as Owlet. Tesla were able to deploy a patch within 24 h of disclosure. Whereas Owlet had been notified numerous times by the examiner without Owlet responding or disclosing the vulnerability to its customers. This raises an interesting question, how do smaller, lesser known organisations take a proactive approach and notify their customers of a critical vulnerability. Furthermore, without a central platform and no dedicated security team, Owlet baby monitors could still be in use today, where many owners could be unaware of the vulnerabilities their device has.

Some papers such as Zhou et al. [120] present modern topics and discuss the associated threats and challenges, the paper however, lacks depth by not understanding the underlying issues. For instance, there is no discussion of the TTN Fair Access policy that limits the duty-cycle of devices, there is mention of IoT devices being incapable of receiving updates, but there is little analysis as to why and how the leading industry platforms, SigFox and LoRaWAN are, in their current state, ineffective at being able to apply safety before availability. There is mention of interoperability, ubiquitous and diversity in the IoT environment but as stated above, the paper, once again, does not focus on the underlying issues. For example, Cisco have announced their intention to deploy infrastructure to get control over the Internet of Things [24, 25], but a detailed review of their history regarding security reveals a plethora of vulnerabilities, the CVE website, a central hub for IT cybersecurity vulnerabilities, lists over 40 vulnerabilities for the years 2017–2018 alone, with some scores being the max 10.00 [30]. There is a further lack of discussion on the effect of error filled software patches, something with which established companies such as, Microsoft, have yet to develop a reliable patching system.

Abandoned IoT devices is a particularly growing concern. This is not to be mistaken with abandonware, which will be reviewed at a later point in the literature review. Research has shown there are currently over 170 million exposed internet connected devices [46]. Many of these devices can and will be used for unintended purposes. Antonakakis et al. [4] investigated the Mirai botnet, where more than 600,000 IoT and embedded devices were used to conduct one of the largest Distributed Denial of Service attacks (DDoS). A particularly interesting discovery highlighted Mirai's "device composition was strongly influenced by… design decisions of a handful of consumer electronics manufacturers," many infected devices were taken over by simple dictionary attacks which the dictionary contained various simple passwords such as "Admin." This raises an interesting question of liability. These devices were essentially a perpetrator to the DDoS attack, and therefore who becomes liable for the devices actions and for the protection of the consumer? If the DDoS attack caused outages in hospitals or safety-critical infrastructure, then the impact could have been fatal. It is quite apparent that there is a serious amount of negligence to applying security into the design of an IoT device.

Many of the legal and contractual laws for product liability are not sufficient in the IoT environment [111]. For instance, the current EU Product Liability Directive 85/374/EEC fails to see the failure of a service or software as a means of being liable in the event where harm is caused to an individual. Leverett et al. [58] identified that firmware on a physical device is covered while, "the server software on which an IoT device resides could well count as a service." With that said by incorporating safety-critical services with the cloud, an OEM could potentially avoid liability altogether. Many times, organisations, particularly in the EU, are fined for security failings under the Data Protection Act 1998. For instance, TalkTalk (2016) had been given a record fine of £400,000 by not encrypting customers information. Despite multiple warnings TalkTalk failed to apply security best practices. Interestingly this case shows a further problem in product liability directives, TalkTalk were only criticised for their lack of best practices and instead the actual leakage of information is where they broke the law. The case law in UAW v Chao [105], found that courts do not see failure to follow best practices as a violation and thus this would mean there is no incentive for vendors to pursue secure-by-default. Internet-connected households also pose a potential problem, currently consumers can be confident that if they purchase an oven and it starts a fire a week later then they can bring forward a liability suit towards the company. This process however, does not pertain its stature when being applied to internet-connected homes [21]. Many IoT devices contain strict license agreements which protect the company from litigation if the device were to malfunction and cause fatalities [101]. If negligence, by not following security best practices, is not a criminal offence, then organisations do not have an incentive to apply security by design, if menial and simple license agreements allow organisations to defer them from any liability then it would suggest that the lack of liability directive could have an impact on maintaining the longevity of an IoT device.

To understand the problem of liability it is first beneficial to review literature about transparency. Schneier [89] mentions to keep IoT in check we must evaluate transparency and accountability, for when the security of the device is most important. Many of the organisations which develop IoT devices lack a background in security and have not been moulded by cybersecurity best practices unlike other organisations such as Microsoft, Linux or Apple. They therefore do not maintain platforms such as "Patch Tuesdays" and it can be argued that they are not designed by those who have the appropriate knowledge. For example, Miele Professional PG 8528 dishwasher was found to have a web server directory traversal vulnerability. Because Miele is an appliance vendor they do not have a structured process of reporting or disclosing security vulnerabilities. Despite numerous tries at contacting the OEM, Jens Regel of German Company Schneider-Wulf publicly disclosed the vulnerability in 2017. Another, more alarming example is famously dubbed as Devil's Ivy. Researchers from Senrio [91] discovered a vulnerability in the M3004-V network camera, which allowed attackers to hijack the feed. The flaw was deemed to not be Axis problem but a flaw in the widely distributed open-source library software gSOAP. The library managed by Genivia, with millions of downloads, has been patched, but this event raises serious concerns. Developers must now first patch their vulnerable code and issue new firmware updates to their devices, as seen previously in this literature, some devices are not capable of receiving updates, or developers have discontinued technical support, thus leaving millions of devices potentially vulnerable if this were to happen to devices which were not capable of receiving updates.

The CVE standard, although designed to be global is mainly used in the United States. It is a central hub where vulnerabilities are disclosed along with a CVSS v2 score; the higher the score the more critical the vulnerability and provides a set of mitigation techniques. IoT on the other hand, as of this writing, does not have a central hub for vulnerability disclosures, instead the vulnerability disclosure for devices such as the AuYou Wi-Fi Switch, are left as reviews on Amazon [20]. The clear lack of transparency is one of the prime reasons as to why there are so many vulnerable devices. An example is shown by Toll [102], a researcher from Northeastern University Global Resilience Institute, who used the Shodan search engine to discover vulnerable devices. The researcher was able to find a PLC unit that had been discontinued since 2014. Shodan revealed this product to be running an outdated firmware version, interestingly sixth months prior to its discontinue date the OEM issued six vulnerabilities, one allowed a remote attacker to put the device into "defect mode" which can cause the device to shut down, this will have affected anything that had been connected to the PLC unit. Not only was the device incapable of being updated as it was never developed with over-the-air updates, but the mitigation set out by the OEM was to purchase the new version. One of the key points is that this device had been deployed in an industrial factory, meaning it could be controlling critical working equipment, it is not uncommon for internet connected devices to be the cause of serious damage. In 2014, a German factory suffered massive physical damage, the reason, determined by German researchers was a badly configured internet connected device which had not been secured properly [18].

A potential solution to the problem of mass vulnerable devices is to tackle one of the root causes, outdated software. An interesting study by Wash et al. [110] investigated automatic updates by conducting: interviews, surveys, and logging computer data from numerous Windows users. It is mentioned early on that to improve security many designers remove users from the software update equation and tend to only give them information on what the update entails, specifically mentioning Microsoft as an example. But the researchers do mention the side effect is intrusiveness and that in some situations humans should still be involved in some way or another. Cranor [29] identifies the reasoning for keeping humans in the loop as: some updates will inevitably fail and therefore some updates cannot be installed without user interaction, updates can add or remove features and causing some users to avoid the update, some updates depend on a reboot to proper configure the update, although now less prevalent in the 21st Century it is still common in firmware updates. Wash et al. [110] also expressed that users are the "weak link" in security, adding a point made by Zurko [121] that "security only becomes apparent to end users when something has already gone wrong." The points made by the above researchers are interesting as they give a strong argument for automatic updates, however the point made by Cranor [29] "some updates will inevitably fail," is interesting, a review of academic literature regarding the effects of faulty software or firmware updates is missing. It is simple to see that even reputable and established organisations such as Windows, Apple and Tesla have yet to figure out a reliable way to deliver patches. For instance, in early 2017 Microsoft automatically issued 16 patches, within weeks they released statements exaggerating "known issues" [68, 69], some of these patches have caused organisations the inability to use applications such as Outlook. June 2017 saw another bad update for Internet Explorer which broke the use of printers [68, 69]. Although some of these seem menial, they do directly affect organisations and question as to whether smaller companies with less resources could provide software update functionality in an industry which has yet to be fully understood. The problem is software updates can also "brick" devices, causing a stop in functionality. Malwarebytes, AV protection software for desktop users, issued an update in January 2018. The update turned user machines into unusable machines [63], interestingly, although they have issued upwards of 20,000 web updates, they still caused huge destruction to many of its users [24, 25], showing that there will inevitably be a problem. Being able to update without limitations is a luxury which the IoT environment does not have and therefore measures need to be in place before they can happen, especially when numerous IoT devices are deployed in safety-critical areas.

Safety by providing consistent support to an IoT device is often a skipped research topic. There is a growing concern that IoT abandonware could cause numerous amounts of problems when organisations discontinue a product. As seen in 2014 when Nest, a subsidiary of Google, bought the company Revolv; a smart home start-up which concentrated on hub gateways; connecting all devices to one single device. In less than 2 years after acquisition, before the devices end-of-life, Nest disabled the ability for consumers to use the hardware, leaving customers with a useless product as of May 2016 [39]. Another case, the Rdio service, which used another discontinued device, Aether Cone, had been purchased in 2015 by Pandora, within the same year it was announced that the Rdio service will be stopped [81]. This raises an interesting question, when an organisation stops supporting the hardware does it mean the manufacture can disable the device without consequence, if organisations are not going to be held liable for the maintenance of a smart device then there is no incentive for an organisation to carry on maintaining the service or device until it's end-of-life. A problem therefore arises when an OEM discontinues support of a safety-critical device such as, a pacemaker or the ABS system in a car, not only is the safety of the consumer jeopardised but the vehicle in which the ABS system resides is now no longer supported and could potentially have an underlying harmful vulnerability that will never be fixed.

There is a lack of suggested solutions which address the issue of abandonware and there is still much to be discussed. One solution is to force the OEM to put aside a sum of money which can then be used to maintain the device until it's end of life, this ensures both security and safety is reassured, however, there are a few issues to this approach and still requires a detailed analysis as to whether it is viable. The most reliable path, suggested by Daley [31], is to force the OEM to make their proprietary software open source, after it has been announced the discontinuation of their service. The researcher cites Hadoop and Kafka as being prime examples of success. However, there is a looming problem, the preliminary issue stems from the fact that the software is, proprietary, enforcing an OEM to disclose their source code would be incredibly difficult. Moreover, you must hope that someone picks up the project, with over 400 IoT platforms, offering a variety of solutions, fragmentation will be a restrictive hurdle when trying to apply regulation [115]. The process of making software open source still does not solve the issue of applying updates, especially when embedded IoT devices rely on firmware-over-the-air (FOTA) to supply updates. Additionally, even though the source code is now open to everyone, it still does not mean the OEM kept the repository or the server that provided the network for FOTA updates. If the software is DRM dependable such as a connection to the Cloud, then the shutdown will not only cause a disease in technical support, but also any device that was connected is now no longer useable, it is therefore impossible for that type of service to exist with an open source model, unless those who took over the project had the funds to continue the servers. Potentially a more threatening situation is when the organisation still gives a service but stops technical support. Pogoplug, a cloud service, stopped support in 2016, the service is still accessible, but they no longer provide any support for bugs or vulnerabilities, however, within months the service has become uncooperative [78].

There is rapid development to combine the technologies; Cloud computing and IoT together. The cloud provides advantages such as: resource availability, cost efficiency, scalability and industrial technology, which allow the infrastructure of IoT to thrive [13]. The integration of the cloud can be seen to fill in the inefficiencies of IoT. The merge of both technologies develops new characteristics which are seen to be "storage over internet, service over internet, applications over internet, energy efficiency and computationally capable" [99]. As seen in the paper by Stergiou et al. [99], there is a plethora of papers which give a solution to the secure integration of the cloud and IoT, but many times the concern of reliability is often overlooked. A rising problem is the dominance in one cloud provider, according to Coles [28], an industry analyst for Gartner, Amazon Web services hold 47.1% of the cloud market, some safety-critical organisations such as Centrica (British Energy Supplier and the UK's Financial Conduct Authority depend on AWS for the continuing effectiveness of their services [19]. The question therefore arises as to whether safety-critical systems should be dependent on the cloud. A brief look at history would show that the cloud is far from secure or reliable. Tsidulko [104] reported on cloud outages which reputable names had suffered: IBM, GitLab, Facebook, AWS, Microsoft Azure, Microsoft Office 365 and even Apple iCloud. Interestingly many of these organisations also provide some type of cloud service for IoT, Amazon provide AWS for IoT and Microsoft have Azure for IoT. Other organisations in other industries are also developing platforms for the automotive industry. The Volkswagen Group introduced a plan to create a private cloud service that other car manufactures can use. Although this approach is possible the automotive industry still maintains gaps in automotive security and privacy [65]. The issue stems from allowing safety-critical systems access to a network, this is unfortunately a very real threat to this day. Researchers Palanca et al. [80] were able to disable services inside the car such as the airbag, parking sensors or any safety system. The interesting point to this case is that the attack is "indefensible," it is incredibly difficult to resolve and not even an OTA update will resolve the issue. Furthermore, these types of events show a lack of standards and regulations, specifically for the automotive industry [62].

Other solutions such as Edge and Fog computing still rely on a single gateway, just by reviewing the history we can infer that gateways are also prone to an absence of security [97]. It can therefore be suggested; some organisations will not follow security best practices and end up with another event like Mirai but on a larger scale. It would seem there is a trend to these issues, the current IT ecosystem contains respected and well-rehearsed security best practices, why are these not being implemented into the IoT environment?

The IT ecosystem is dependent on best practices which are provided and maintained by governing bodies such as ISO, HIPAA, NIST, SANS and even the EU. Many of their security suggestions rely on the use of static perimeter network defences, end-host defences and software patches. Whilst these are workable solutions for IT security, they are however, "fundamentally ill-equipped to handle IoT deployments [119]." The very means of IoT, specifically scalability, heterogeneity, geolocation, device and vendor constraints mean conventional approaches are not implementable [119]. For example, CYREN provides an Embedded Antivirus solution which requires 35 MB storage and <2 MB available space for updates, whereas most IoT devices use micro-controllers such as the 8051, MSP430 AMTEL series with <2 MB RAM and 20 KB of FLASH storage. For this reason, many IoT devices rely on other means as protection. A comprehensive study carried out by NIST [77] reviewed the status of Cybersecurity Standardisation for IoT applications, in their findings they found many of the core areas of cybersecurity; connected vehicles, consumer IoT, health IoT, smart buildings and smart manufacturing, all lack IT system security. Furthermore, commercial off the shelf software is becoming a larger concern, specifically for e-medicine devices, the rise of error prone software patching is starting to become apparent, for instance early 2018 saw a vulnerability in Cisco's Talos Natus XItek EEG medical products, namely the XItek EEG range that included the EEG32U Electroencephalography brain recorder. Although NIST state medical devices should require "timely security software patches" regulatory bodies make it very difficult to determine whether a patch would invalidate a medical device, also the multi-step process required to re-validate a device could affect the time that an IoT device is left vulnerable [114]. If for instance a pacemaker required an update to fix a software bug, but the update changed how features work on the device, does this mean new validation is required? If true, this will also be a factor into why many organisations are reluctant to maintain IoT devices. NIST [77] conclude by stating more work needs to be done for Cyber Incident Management, where there is a minimal amount of information for mitigation when software can no longer be patched.

Being able to update software or firmware in an IoT device is an essential security practice to fix security bugs [38]. Research conducted by Chen et al. [22], investigated firmware security across 42 device vendors. By analysing 23,035 firmware images against 74 known exploits they were able to determine at least 89 products were vulnerable to multiple exploits, interestingly 14 previously unknown vulnerabilities were discovered, emphasising the need for software and firmware updates on IoT devices. However, there are two major issues of firmware updates. The first is the issue of reliability, there has been a plethora of research papers which focus on applying trusted and authenticated solutions: Witkovski et al. [116], Huth et al. [47], Schmidt et al. [88], Choi et al. [26], but there has been little work on maintaining the longevity of IoT devices, it is not that there is a lack of security implementations but "instead the wrong firmware might be uploaded, the transmission of the new image failed, or the new firmware simply does not work as intended [87]." For instance, Fiat Chrysler Automobiles issued an update which caused their Uconnect infotainment system to go into an "endless loop of reboots," [32] causing the system to become unusable. Audi in 2018 had to recall 1.2 million cars after a software update had failed to fix a fuel pump issue [98]. It would seem there is a lack of testing which is directly effecting the reliability of these updates, an IBM [48] study found 80% of organisations do not test their IoT devices for vulnerabilities and this fact is made even worse when McConnell [66] highlighted in his book "Code Complete," the industry average of errors is "about 15–50 per 1000 lines of delivered code," with industries such as, e-medicine and the automotive industry adopting IoT as a solution, it will be only a matter of time before more fatalities occur due to an improperly configured software update. In 2016 a Tesla driver was killed whilst using autopilot, the car's sensor system failed to distinguish a large 18-wheeled truck with the bright sky [59]. As we have seen from the multiple papers referenced above, many of the current solutions to providing secure firmware updates rely on "software" level solutions and do not "consider the different usage patterns that IoT devices have [117]." This concern has been previously raised in the literature review where many of the events suggest IT best practices are being used but are clearly not sufficient when applied to IoT environments. The second, more alarming issue, is the inability for IoT devices to receive updates [97].

Raza et al. [83] provides a comprehensive overview of Low-power-wide-area-networks (LPWAN). The researchers express that over-the-air updates are a crucial feature for LPWAN networks and that there is a lack of support for OTA updates. They express the prominent LPWA technologies as being, SigFox and LoRaWAN. As of 2017 SigFox covers 45 countries and the number of connected devices stands at 2.5 million [96]. LoRaWAN on the other hand is expected to be 14% of the 27 billion installed bases by 2024, whilst cellular will account for 8% [61]. Raza et al. [83] studied both technologies and discovered alarming properties. SigFox does not provide the ability to update over-the-air, whilst LoRaWAN does. However, LoRaWAN is limited by factors such as, payload length, data rate and the band it uses for transmissions. Raza et al. [83] also expresses that there are few studies that have been conducted to see whether LoRaWAN can carry out OTA updates. One issue of LoRaWAN is the use of the unlicensed ISM band, which means devices must obey duty cycle implementations. The European EU 863–870 MHz band is controlled by the TTN Fair Access Policy, which applies a duty cycle of 1%. Adelantado et al. [2] defines duty-cycle "as the maximum percentage of time during which an end-device can occupy a channel," these limitations, enacted by the TTN Fair Access Policy are designed to remove the congestion of many devices being used on the same ISM band. Jongboom and Stokking [53] are one of the few researchers to give a working solution for providing firmware updates over LoRaWAN. They expressed that a packet over LoRaWAN could take up to 9 h, using the fastest data rate. One limiting factor to updating firmware is the problem of downtime whilst the device is being updated, some users depend on devices such as pacemakers which require 100% uptime always, at the current data rate stated by the researchers LoRaWAN would not be able to provide a reliable update and thus jeopardising the safety of its consumers. Jongboom and Stokking [53] further state that LoRaWAN transmissions are uplink orientated, meaning, their strength is transmitting packets and lack resources in receiving. Furthermore, uplink orientated means the device must make an uplink transmission to receive a downlink transmission. And therefore, suggesting that this is not a reliable or efficient method for OTA updates. To counter this issue, they incorporate multicasting, where a temporary session is made in which the devices share the same session keys. Commands are then sent down to the device that indicate when they should start listening, in turn, this will allow all devices to receive an update at the same time. Although they show a working practical solution there are, however, issues. For instance, many devices will have different configurations as to when they transmit data and thus the time chosen for the session to occur may interfere with the devices intended uplink time, and for this reason, it would be nearly impossible to achieve this feat in the real world. Also, studies by Bor and Roedig [17] and Georgiou and Raza [41] have expressed that the current architectural structure of LoRa networks are not sufficient in being able to scale with the current projections of IoT. An Interesting point both studies discuss is the lack of academic attention on LPWANs and their effectiveness on applying a scalable approach. One further limiting factor that can be inferred from their work is that congestion will have a direct effect on the reliability of data transmissions therefore an interesting point can be supposed, do regulatory bodies need to also account for the freedom of industries which have many limiting factors but are left to carry on installing the infrastructure, what are the potential effects on safety if this is feature is not considered.

The literature review above has shown there is a real problem of enforcement in IoT. Many organisations are seeing to take the approach of availability before software sustainability and safety. The idea of best practices is a step in the right direction, but it clearly is not forcing a much-needed cultural change, instead it is merely offering a different path, one that incorporates more time and costs into the production of an IoT platform or device; which, combined with the already low costs of IoT devices, is not a sustainable model. Interestingly the criteria above can be condensed into the term software sustainability. Manchester University [64] proposed a framework similar to the concept of "dependability: a measure of a system's availability, integrity, maintainability, reliability, and safety [10]." Instead Manchester University [64] included: "Extensibility, interoperability, maintainability, portability, reusability, scalability, usability and efficiency." Although this model was designed for the development of scientific and engineering software, it begs the question as to whether this can be applied to the IoT environment, more specifically regulations and standards.

4 Criteria for Case Studies

Prerequisites for the case studies which were based on the attributes set out by Manchester University [64]:

1. The case had been reported on, which is defined as having a disclosure from the OEM or a reputable news source
2. The case must have happened within several years prior to this study

3. The case included, two or more of the following factors that were decided from the literature review:
    i. Known vulnerabilities, past or present
    ii. Failed update
    iii. Known leakage of user's personal data [17]
    iv. Includes a form of legal contractions
    v. It is dependent on a service
4. The cases were unique to a specific industry

The researcher required that the cases had been within several years prior to this study because it reflects current implications, rather than incorporating old standards and regulations. We needed cases that have had an impact on consumer safety, or software sustainability. For these types of cases, which are published quickly, it is best to gather all outcomes otherwise key aspects are left out and the investigation can have insufficiencies. The researcher required the case to be substantially reported on, because all outcomes must be known, and as many cases have shown, more information becomes apparent after its public disclosure.

5 Criteria for Statistical Analysis

Part of the discussion section will have statistical data, formed from analysing 29 IoT devices. The number of devices was affected by the criteria set out below:

i. Must have a clear start and end date
ii. Must have clear indication through the OEM or a reputable news source that the product has been discontinued
iii. Start dates, if difficult to find, can be acquired from the initial sell date on the Amazon store
iv. IoT devices must rely on a service
v. IoT device must have an application

Statistical analysis is the collection of data which can then be analysed where inferences can then be made. Statistical analysis is often used to support hypothesis or discussions on the topic [100]. The above criteria will be used to analyse potential IoT devices which can be used in one of the sections of this study. It is important to follow this criterion as it will support part of the discussion in the later sections of this work. Overall 29 devices were deemed to match the criteria and the usage of a graph will be used to determine the average time of support for an IoT device. To find the relevant IoT devices I used Google which provided websites that listed IoT devices.

6 Case Studies

6.1 Case Study 1—Nest

Nest is a smart thermostat, designed to control the standard heating, ventilation and HVAC system based on heuristics and learned behaviour from the activities of a person. The thermostat includes a Wi-Fi module, supporting 802.11 b/g/n, which can connect to the consumer's household or business and interface with the Nest Cloud. Interestingly the device is fitted with a ZigBee module that can be used to communicate with other Nest devices within the household. The Nest thermostat uses a Texas Instruments (TI) Sitara AM3703 microprocessor, 64MiB of SDRAM, 2Gibit of ECC NAND flash [7, 49]. The device will collect statistics, environmental data which it then uses to "learn," when the device connects to the cloud all data, such as settings, logs and location data is uploaded. In 2016 The learning Thermostat suffered a software bug that drained its battery and sent thousands of people's homes into the cold and caused users without the means to warm their houses up. The bug had been found in late December but did not become apparent till early January when users were woken up to a cold household [15]. Although all Nest owners are now patched, originally, for some users, the fix required a daunting nine step manual update. Some of the procedural steps needed the user to detach the pre-installed device from the wall, charge the device for 15 min, reattach to the wall, pressing a series of buttons and then charging the device for another hour.

6.2 Case Study 2—Lockstate Smart Locks

Hardware business Lockstate, had a disconcerting problem with their $469 LS6i smart lock. An original error caused the device to become inoperable. Lockstate were able to issue a FOTA update but the new update caused the lock to fail when connecting to their web service making the lock unable to accept FOTA updates. They offered two solutions to the problem. Option 1 resorted in the user dismantling the lock and sending a portion of it back to LockState, so it could do their own update; this was estimated to take 5–7 days. Option 2 LockState offered to replace the interior of the lock but the user had to replace it themselves; this had the potential to take 14–18 days. Interestingly organisations such as, Airbnb, were mostly affected as it caused some users the inability to enter their rooms. A statement by El Reg identified their mistake as being a problem with version control, where they had mistakenly sent the firmware for the 7i model to the 6i model.

6.3 Case Study 3—Samsung Smart TV

Owners of the Samsung smart TV 2017 MU Series faced a disastrous firmware update which caused their smart TV's so much damage that they have to be repaired. The update caused the televisions to get stuck on a single channel, volume cannot be adjusted, and they were left inoperable. Interestingly many customers were left without support for some time, to then only be told that they must arrange a date where an engineer of Samsung could come out and fix the issue. The question is therefore do you install firmware updates and potentially leave your smart TV vulnerable or allow updates and potentially end up with a inoperable device [75].

6.4 Case Study 4—Abbot Pacemakers

The FDA addressed a genuine issue with Abbot pacemakers in 2017. Abotts pacemakers included the ability to include cardiac resynchronisation therapy and providing pacing for slow or irregular heart rhythms. The devices are implanted under the skin close to the heart area and contains electrical wiring that is inserted into the heart, many times, these devices are used to treat heart failure. The FDA provided a list of affected devices: Accent, Anthem, Accent MRI, Accent ST, Assurity, and Allure and stated clearly the effected audience are patients, caregivers and those treating the patients. Updating medical devices is a grey area as the update can change add unintentional features which mean the function of the device is now different as to when it was originally certified. The FDA approved a firmware update for 460,000 pacemakers, the devices contained vulnerable exploits and the connectivity of the devices meant action must be taken. Although no update failed the FDA raised the following concerns. The update would take 3 min to complete, during that time the device will not function properly. Other concerns such as the ability to reload the firmware if the update could not be completed, loss of programmed settings, loss of diagnostic data or the complete loss of functionality, were highlighted and this case is interesting as it emphasises not only the appropriate steps that had to be taken by FDA but also the impacts of firmware updates on safety-critical devices [49].

6.5 Case Study 5—Environmental Systems Data Controller

An environmental Systems Corporation Data Controller had been found as having a serious vulnerability, specifically in its ESC 8832 Version 3.02 and earlier. The vulnerabilities cause an authentication bypass which allows unauthorised persons to change the configuration settings, and unauthorised users are capable of bypassing privilege management by brute forcing other users accounts; the ICS-CERT noted that these exploits can be conducted remotely and with relatively low skill. Interestingly the ESC organisation notified ICS-CERT by saying, the Data Controller had no available space to make added security patches; therefore, a firmware update is no longer possible. The ESC replied by releasing an advisory which lists proper compensating controls. While the mitigation controls can help, it is notified that they are merely mitigation controls and will not completely stop peoples from exploiting the vulnerabilities found [49].

7 Identifiable Problems

There is clear requirement for industries to find a solution for providing reliable updates. This is emphasised with IoT devices because of their very nature. They are potential gateways to a persons or business infrastructure, and thus they are given an enormous amount of responsibility, which, as shown by the case studies, is not being taken seriously. All five case studies provide a different perspective, but they all equally prove the most simplistic actions, such as, poor version control, can cause the device to become inoperable; sometimes this can be for weeks if not months. The next section will find new challenges, that can be conveyed from the case studies above and will discuss and give potential solutions to already existing problems.

7.1 Extensive Remediation Steps

The procedural steps which were required to manually fix the issue with the NEST devices is extensive and difficult for those that may not have the knowledge to carry out such a task. The problem is these steps are currently the only remediation solution there is. This raises a further problem, some users may not have the technical knowledge to carry out such a task, and in that case many devices may be left vulnerable [46]. NIST [77] have interestingly highlighted the fact the need for more remediation effort when devices are abandoned, unfortunately as seen, this is not a clear-cut task. Some devices will inevitably be in geographical locations which are difficult to get too, and with many devices, who will pay a technician to attend all the locations. The underlying issue, it would seem, is the reliance on the user and the lack of motivation to properly discontinue products, therefore instead it must be implemented, and forced upon the OEM to drop their devices properly. However, to "properly" abandon a product is to brick the device which is no longer going to be supported, the concern here is the fact many users will be dependent on its service, especially when it's duty effects the health and safety of the user. Therefore, a sudden abandonment of the device is not feasible, as seen by the case with Revolv.

Solution. The idea of open-source has been used as a remediation for when devices are discontinued [7], but once again the idea has to be modelled to the IoT environment. Open-source works for when people want to actively take on the project, but the very nature of IoT, where it is dependent on a connection to a service, ensures this solution may not work, also it still does not help with updating the device and keeping them secure. Once again, this solution relies on the user to apply technical knowledge to maintain the device, with the update service being decommissioned the next step is to apply manual software updates. It would seem for IoT, open source would not be a sustainable system. Other solutions brick the device, for example, once Aether Cone had been discontinued, they decided to essentially brick the device, their daunting 10 step process still required technical knowledge from the user but at least this stopped some devices from being targetable. If an OEM however, is in control of a safety-critical system then bricking the device may cause harm to the user. For example, the NHS, in the UK, relies on Windows XP on much of its safety-critical equipment [73], it is therefore impossible to brick the service. The WannaCry ransomware effected the NHS but luckily Windows had the capital and the resources to distribute an update, even though they no longer support Windows XP. An additional question, as to who takes over, must also be addressed. IoT is unsecure by nature, is a gateway to the network, it can be deployed in safety-critical areas, and it generates large amounts of detailed data, therefore the question as to who takes over the open-source project is vital. Many consumers do not have technical knowledge, if a malicious attacker were to take on an open source project from a discontinued IoT device many consumers would not be able to detect malicious software, let alone where to look for it. Furthermore, it is not unknown that open source projects have had disastrous impacts on society, Heartbleed, Shellshock and the Debian OpenSSL fiasco are just a few examples, with IoT devices having less regulations than the IT ecosystem then it could potentially be very easy for these situations to happen again but for IoT, especially given the areas in which IoT devices are deployed and how insecure they are. Therefore, we must define the underlying issue and target it directly, with that in mind we can take insight from Ubuntu's history.

In 2005 the Ubuntu Foundation was created, which aimed to ensure the continued support for Ubuntu distributions. The foundation committed a first sum of 10 million dollars and ever since, this has been used as means for a sustainable operating system (OS) [106]. The problem stated by the literature review, the IoT environment is very different to the IT ecosystem and therefore revisions must be made. The revision is the money that can be set aside. Ubuntu is an established open source project, they make their money through support services, contracting services, Canonical Store and donations [8]. IoT devices on the other hand originate from Kickstarter projects (refer to Fig. 1), the majority do not have enough backing to set aside a lump sum of money for the continued support of their device. Furthermore, they tend to support far less staff and thus their running costs and the amount that must be set aside will be substantially less. With that in mind one solution could be to provide a government scheme which aims to dedicate some costs to the continuation of the device. Some factors which must be accounted for, the critical device must have already been established for $n = x$ amount of years, must be part of a critical infrastructure that

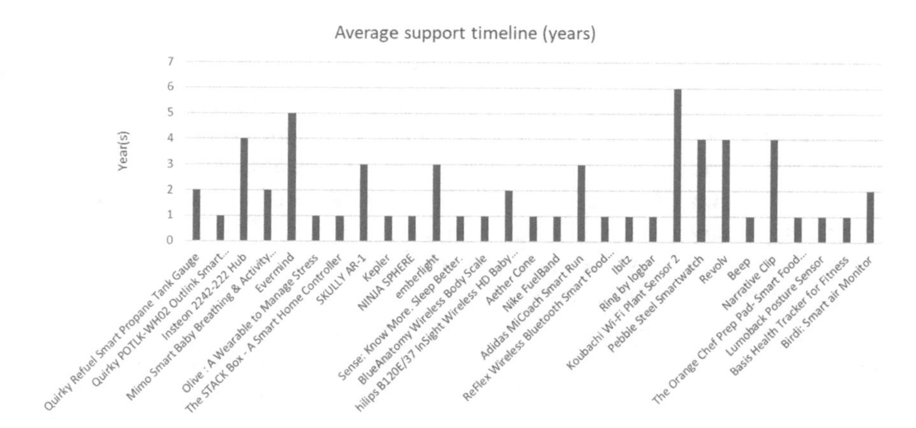

Fig. 1 Showing the average time organisations give technical support for their IoT products

has a direct effect on the health and safety of its users, and finally the organisation must prove they do not have the funds to set aside the full amount.

7.2 *Liability and Transparency*

We have previously spoken about Cisco, and its rich history of vulnerabilities and insecure devices, interestingly this raises another concern for the future of IoT. Currently large organisations such as Amazon and Windows are dominant in the IoT environment, as they have the freedom, resources and establishment to push their platforms to the masses. The problem is it seems many organisations are trying to push their agenda to the market too quickly without understanding the consequences. An example is the ACRN project, by the Linux Foundation, dedicated to applying a hypervisor in the automotive industry. A hypervisor is the method of separating a computer's OS and applications from the hardware it is on, acting as a middle-man by creating virtual machines [94]. ACRN have suggested that hypervisor can be used in scenarios which need to prioritise workloads which are related to safety [93]. If the entertainment system and the safety-critical system, such as, ECU or ABS, are using the same hypervisor then it would only take compromising the infotainment system to compromise the entire system, as can also be seen by Lv et al. [60]. Furthermore, the required safety between an infotainment system and a safety-critical system is magnitudes different, for infotainment, security is put last, and availability is preferred, while safety-critical systems, security and reliability is put first. The question arises as to whether we can currently rely on software to do its job, can we allow car manufactures to use only one OEM, such as, Intel, with their poor history record, Meltdown and Spectre are prime examples of what can happen, and this is only exaggerated when those same OEMs design the security for the safety-critical systems in the automotive industry or even medical equipment. Eventually devices and services stop receiving updates, Apple has already announced its latest IOS 11 operating system cannot be supported on the IPhone 5 model, which came out in late 2012. Apple have developed in car entertainment systems, such as, CarPlay, if they halt security updates within 5 years for their mobile phones, which are more popular than their car entertainment system, then we can safely assume they will not support updates for their car entertainment system either. If these systems become the weak security link in a car system, then they can be used as a target to gain full control of other more critical systems.

The shortage of regulations for organisations is also reflective in the descriptions of IoT legal contracts. Nest have a detailed legal items list. One section raises some concerns, the Dispute and Arbitration prohibits customers from suing the company or joining a class-action law suit, it would seem they are settled through arbitration instead. According to Bilton [15] who had contacted Sonia K. Gill, a lawyer for Civil Justice and Consumer Protection for public citizens noted that the terms of services for IoT devices "are inherently unfair to consumers." This is also emphasised by Diega and Walden [33] who deeply analysed the contractual terms of the Nest product. Interestingly, the researchers express in some circumstances, because the idea of "software" is not appropriately defined in the EU Liability Directive, some outlines set by Nest are inherently "unenforceable," and once third parties are included, the complexity of the legal contracts become ever greater, highlighting the potential future problems when multiple systems and services are incorporated in automotive vehicles.

Solution. As shown by NIST [77] Connected Vehicle solutions are focusing on applying cryptographic techniques and still need solutions for core areas, such as, hardware assurance, network security, software assurance and system security engineering. The problem is standards for cars have been designed based on physical access to motor vehicles, but as cars become more connected through wireless protocols, standards have not yet caught up. We end up with hackers capable of leveraging the out of date standards, where for instance, car attackers are able to break into a BMW without keys, by using a relay attack as the level of security and reliability is not yet sufficient [34]. The level of required reliability should be the same as that you would find in an aircraft, according to Jiang [52], a Boeing whitepaper, the average age of an airframe and its software is supported for at minimum 20 years, and at maximum 25 years. With that in mind we must learn from regulators such as the Civil Aviation Authority (CVA) and the Federal Aviation Administration, which have created one of the safest modes of transport seen today. One method is to adopt the "revisionary mode" as seen in planes. This is a mode in which all critical systems keep control through mechanical linkages. Organisations are creating software which will allow cars to communicate on a national scale, enhancing efficiencies in every factor that effects transport, cars are eventually going to be completely driver-less, we have seen the cloud unreliable, we have seen software unreliable, we must also make the assumption that software will fail and in that case we can use insight from other industries which have had years of experience in the development of reliable and redundant systems.

7.3 Support and Quality Assurance

Another common attribute between all case studies is the lack of support, or quality assurance from the organisations. NEST was able to apply a software update which drained their battery life, as seen many times in this study, IoT devices are continuously put into safety-critical areas, if an established organisation, such as, Google, can produce a disastrous software update which stops thermometers from working then it would suggest a serious lack of quality assurance, or motivation to apply reliable support for its user base. Furthermore, other case studies have shown the potential side effects of updating firmware, restating the fact that firmware updates can fail or can cause the device to malfunction. The incapability of receiving updates is a much more alarming factor, the literature review has already shown the prominent LPWAN protocols are not capable of applying updates to devices, and if this situation were to happen with a widely distributed IoT device, then not only does this add to the problem of abandonware, but it also allows the devices to be used un unintended ways. Additionally removing the ability of OTA updates from a poor software update could also spell just as much disaster as you are not able to update at all, at least the former had been designed without the intention, but as we have seen in other topics throughout this study it is difficult to determine whether Lockstate included remediation steps that allowed them to physically update the device. These are the types of topics which have yet to be discussed and a concrete solution is yet to be determined. We can empathise the abundance of standards and regulations, for instance, IEEE, Internet Engineering Task Force (IETF), 3GPP, LoRa Alliance, SigFox and many more but the question arises as to why they are not being followed.

The data above was collected from 29 chosen IoT devices, using specific criteria, that can be found in the methodology section. Using these statistics, we can develop some understanding as to why there is a lack of transparency, consistency, support and insecurity in IoT devices. The data showed the support life-time for an IoT device, on average, is 2 years. Furthermore, just under half of these devices originated from crowd-funding projects, such as, Kickstarter or Indiegogo. The common link between all crowd-funded projects is how they raise the money and the amount they raise. Larger corporate organisations tend to maintain a steady income through various means, however, crowd-funded projects are one off lump sums of money, and then the income stops after the device has been distributed, they are therefore left with two options; (1) produce a new version of the device and stop support for the old one, (2) discontinue support altogether. Both options add to the current problem of abandonware, but more importantly the data provides us with the understanding that many, potentially, run out of resources to support the devices, a problem arises concerning the consequences of ditching these devices, not just for the user but for the organisation. Interestingly just under half of the devices reviewed were still purchasable across major platforms, such as, Amazon. This leads on to the point of transparency. The biggest difficulty was determining the start and end date for the IoT devices, only 2 out of the 29 devices had statements issued by the original vendor stating their discontinuation of support for the device. Many of the websites were still functional and the applications that connected the device to the user were also still on the OS stores. It would seem the lack of motivation, support and security of IoT devices could be narrowed down to the lack of consequences, and therefore the lack of motivation to apply the industries best practices.

Solution. UL [108] issues a type quality mark, known as the Recognised Component Mark among organisations. It is placed on components which have been tried and tested using the criteria issued by UL. It is important however, to be aware of the difference in regulations among the countries, US and the EU hold different values and therefore providing a flexible dynamic framework for the certification of IoT devices will have to be different for each country. However, the concept of certification is to confirm a product is operationally safe and secure, features which are globally similar.

Although a full review of UL certification and implementation methods are out of scope for this work, it is important to understand which concepts can be taken from UL certification, so it can be used for future related work. The biggest issue is the cost of applying certification, some UL certifications cost upwards of $1500 [107] and therefore we must adapt this to IoT. As Fig. 1 has shown, cost is a direct factor to a sustainable device, by committing adjustments which align with the lower costs of IoT we will see improvements to quality assurance [35]. UL provide two factors to a standard, they are either "Required" or "Optional," with this in mind it would be beneficial if this idea was transferred for use in the IoT environment. Some IoT devices are not safety-critical and thus do not require as substantial certification, for instance, a device which controls the amount of water a plant pertains does not require the same security implementations or life-cycle information as a pacemaker would. But how would you distinguish between optional and required for IoT devices. Fortunately, risk management is an entire industry with years of experience, risk registers have been incorporated into organisations as a means for cost reduction and more notably, areas that are most vulnerable. Fortunately, NIST have recently developed an IT cybersecurity framework which finds the potential risk level of a product, software or service. By incorporating the reputation of UL certification and NISTS experience in cybersecurity a number could be given to a device depending on a strict set of factors, that number can then be used to define the severity or the impact it could have on the health and safety of its user base. However, enforcement is still an issue. It is noted by Leverett et al. [58] and Cohen and Rubin [27] that governments lack the ability to retain experts in specific fields, and standards should be left for those who are actively advancing the technology. With that in mind it is crucial that standards are not designed by the government but instead they should be enforcing laws when standards and best practices are not followed.

8 Conclusion

To summarise, the Internet of Things has the potential to do remarkable things, the digitalisation of society brings everything connected, business and people alike. It is easy to understand why IoT is becoming an important future technology as stated by Gartner [40]. The ability to connect anything and everything allows new areas of analysis and research, that, once incorporated into businesses and households, provides an enormous benefit to a wide range of industries. But as IoT grows and society becomes more digitalised and connected, we must start to incorporate safety with security. IoT has an immense amount of responsibility when it comes to the safety of its users, the incorporation of IoT in e-medicine, automotive, smart cities and smart homes will see a new development of security issues which have yet to be addressed. Much of the studies have been on the security aspect of IoT and have failed to discuss the importance of safety, reliability, and sustainability. Many IoT devices are insecure by nature and hold a plethora of vulnerabilities, for example Rios [84] discovered over 8000 vulnerabilities in third party libraries which had been used in four models of pacemakers. Other events have seen the banning of children's toys, as they were capable of spying on children [79]. IoT devices are typically seen as the weakest-link as they are embedded into infrastructure, and their poor security makes them an attractive target. This is a very real concern, as can be seen in 2018 when GitHub suffered a 1.3TB DDoS attack, the largest ever recorded, and whilst organisations release devices which have failed to be applied with industry standard best practices the DDoS attacks will only grow in size, it is therefore only a matter of time until critical infrastructures, such as energy grids, are targeted. Currently technology relies on a system which deploys monthly updates, such as "Patch Tuesdays" by Microsoft, at present this system is insufficient for the IoT environment. The limitations in IoT devices, such as the lack of battery life, memory size and bandwidth, cause security to have a complex meaning when it comes to developing security for IoT devices. Furthermore, the lack of regulations means organisations do not suffer consequences when they suddenly discontinue a product [39], or cause fatalities [59], and are therefore not motivated to apply secure-by-default into the design of their IoT device. The problem stems from crowd-funded projects, they receive a lump sum of money and can then no longer support their product until its end-of-life. Whereas large established organisations such as Ubuntu, have a wide range of enterprises which generate their income and apply it to the sustainability of their product [8]. However, that does not necessarily mean the established organisations are coherent in their ability to give security and safety. Cisco have a webpage dedicated to their vulnerabilities, Intel, Arm, Microsoft, Amazon, and Google have all been affected by Meltdown and Spectre, the biggest impact on security and safety today. But the inability to reliably fix these issues is a matter that has largely been unaddressed. Areas, such as, the automotive industry, are adopting IoT into safety-critical areas, and the inability to provide support to digital devices for long periods of time is only adding to the unprepared future of IoT. Organisations, such as, Apple have already decided to discontinue support for the iPhone 5 model, only 5 years after its introduction, these same organisations are helping develop systems inside vehicles. The average age of a vehicle in the US is 11.4 years, if organisations can only support a system for 5 years before they are incapable then the very notion of security and safety is forgotten. There is a plethora of bodies which produce detailed standards and best practices for IoT, but they are not enforced and lack in areas which effect the safety of the user. This study looked at highlight the underlying issues of IoT, instead of creating another solution which focuses on security. This study looks to develop and apply frameworks and strategies which allow enforcement, such as certification, to protect the user and create a sustainable future.

9 Future Work

Whilst this study provides solutions to existing and new challenges, the design of a proof of concept is lacking. This study has highlighted the underlying reason for the insecurity of IoT is the absence of enforcement, and currently the best method, highlighted by this study, is to create certification standards such as, UL, in which governments can then design laws which enforce the certification standards. The idea of combining NIST cybersecurity framework and UL certification is suggested but it will be a challenging task to complete. This study has shown the insufficiencies in the current standards in being used with IoT, therefore future work would take the core functions from NIST [76] cybersecurity framework; identify, protect, detect, respond, recover, and instead define them for IoT. For a first suggestion the core functions of IoT are: Reliability, Maintainability and Portability, moreover, each function has a set of sub-categories which must be defined: availability, fault tolerance, reusability, testability, adaptability, install ability and replaceability. By defining these terms, we can then use them as a basis for which we can apply "optional" or "required" standards, so governments can easily understand which standards and best practices to enforce and make laws on, but also so organisations have an idea on how much a project will cost to support.

References

1. Ackerman S, Thielman S (2016) US Intelligence chief: we might use the internet of things to spy on you. https://www.theguardian.com/technology/2016/feb/09/internet-of-things-smart-home-devices-government-surveillance-james-clapper. Accessed 20 Feb 2018
2. Adelantado F et al (2017) Understanding the limits of LoRaWAN. arXiv, s.l.
3. Anderson R (2018) Making security sustainable. Commun ACM 61(3):24–26
4. Antonakakis M et al (2017). Understanding the Mirai Botnet. USENIX, s.l.
5. Apple (2017) Repair file sharing after Security Update 2017-001 for macOS High Sierra 10.13.1. https://support.apple.com/en-us/HT208317. Accessed 26 Apr 2018
6. Apthorpe N et al (2017) Spying on the smart home: privacy attacks and defenses on encrypted IoT traffic. arXiv, s.l.

7. Arias O, Ly K, Jin Y (2016) Security and privacy in IoT era. UCF—University Central Florida, Florida
8. askUbuntu (2011) How does Ubuntu make money?. https://askubuntu.com/questions/21730/how-does-ubuntu-make-money/21744. Accessed 20Apr 2018
9. Atlas (2017) The average cost of IoT sensors is falling. https://www.theatlas.com/charts/BJsmCFAl. Accessed 2 Mar 2018
10. Avizienis A, Laprie J-C, Randell B, Landwehr C (2004) Basic concepts and taxonomy of dependable and secure computing. IEEE, s.l.
11. Batchelder D et al (2014) Microsoft security intelligence report. Microsoft, s.l.
12. Beall A (2016) Watch hackers control a Tesla Model S from 12 MILES away: firm issues fix after discovering the dangerous flaw. http://www.dailymail.co.uk/sciencetech/article-3799158/Tesla-fixes-security-bugs-claims-Model-S-hack.html. Accessed 20 Apr 2018
13. Bellavista P, Corradi A, Zanni A (2015) Integrating mobile internet of things and cloud computing towards scalability: lessons learned from existing fog computing architecutres. University of Bologna, s.l.
14. Bilge L (2012) An empirical study of zero-day attacks in the real world. ACM, s.l.
15. Bilton N (2016) Nest thermostat glitch leaves users in the cold. https://www.nytimes.com/2016/01/14/fashion/nest-thermostat-glitch-battery-dies-software-freeze.html. Accessed 26 Apr 2018
16. BITAG (2016) Internet of things (IoT) security and privacy recommendations: a broadband internet technical advisory group technical working group report. BITAG, s.l.
17. Bor MC, Roedig U (2016) Do LoRa low-power wide-area networks scale?. ResearchGate, s.l.
18. Bundesamt fur Sicherheit in der Informationstechnik (2014) Die Lage der IT-Sicherheit in Deutschland 2014. Bundesamt fur Sicherheit in der Informationstechnik, s.l.
19. Carey S (2017) Centrica shuts down UK data centres in move to the cloud with Microsoft. https://www.computerworlduk.com/cloud-computing/centrica-shuts-down-uk-data-centres-in-move-cloud-with-microsoft-3666318/. Accessed 3 Mar 2018
20. Carman A (2016) A security developer wrote such a scathing Amazon review that the product disappeared. https://www.theverge.com/circuitbreaker/2016/7/5/12096520/bad-amazon-review-product-pulled-auyou. Accessed 20 Apr 2018
21. Chau BK, Markel Z, Man D (2015) Liability for home IoT. MIT, s.l.
22. Chen DD, Egele M, Woo M, Brumley D (2016) Towards automated dynamic analysis for Linux-based embedded firmware. dcddcc, s.l.
23. Chen QA et al (2018) Exposing congestion attack on emerging connected vehicle based traffic signal control. Internet Society, s.l.
24. Chirgwin R (2018) Cisco casts an eye over IoT protocol landscape: everything the light touches is ours. https://www.theregister.co.uk/2018/04/18/cisco_intent_based_networking_for_internet_of_things/. Accessed 19 Apr 2018
25. Chirgwin R (2018) You publish 20,000 clean patches, but one goes wrong and you're a PC-crippler forever. https://www.theregister.co.uk/2018/01/29/malwarebytes_patches_patchy_patch/. Accessed 13 Feb 2018
26. Choi B-C, Lee S-H, Na J-C, Lee J-H (2016) Secure firmware validation and update for consumer devices in home networking. IEEE, s.l.
27. Cohen MA, Rubin PH (1985) Enforcing government policy: the evolution of efficient regulations. Working Papers, America
28. Coles C (2017) Overview of cloud market in 2017 and beyond, s.l.: Gartner
29. Cranor LF (2008) A framework for reasoning about the human in the loop. USENIX, s.l.
30. CVE Details (2017–2018) Cisco security vulnerabilities. https://www.cvedetails.com/vulnerability-list/vendor_id-16/product_id-19/Cisco-IOS.html. Accessed 3 May 2018
31. Daley J (2016) Insecure software is eating the world: promoting cybersecurity in an age of ubiquitous software-embedded systems. Stanford.edu
32. Denholm T (2018) Another failed update: connected cars, over-the-air updates, and what happens when it goes wrong. https://www.datalight.com/blog/2018/02/21/another-failed-update/. Accessed 26 Apr 2018

33. Diega GNL, Walden I (2016) Contracting for the 'internet of things': looking into the Nest. Eur J Law Technol 7(2)
34. Duffin C (2017) Shocking moment car hackers are able to steal a £60,000 BMW simply by holding a bag up to the front door of a house (so is the only solution to keep key fobs in a metal box or the fridge?). http://www.dailymail.co.uk/news/article-4456992/Shocking-moment-car-hackers-steal-60-000-BMW.html. Accessed 18 Apr 2018
35. Evans M, Maglaras LA, He Y, Helge J (2015) Human behaviour as an aspect of cyber security assurance. aRxiv, Leicester, UK
36. Fagan M, Maifi M, Khan H, Buck R (2015) A study of users experiences and beleifs about software update messages. Comput Hum Behav 51:504–519
37. Farahani B et al (2018) Towards fog-driven IoT eHealth: promises and challenges of IoT in medicine and healthcare. Future Gener Comput Syst 78(2):659–676
38. Fernandes E, Rahmati A, Eykholt K, Atul P (2017) Internet of things security research: a rehash of old ideas or new intellectual challenges?. arXiv, s.l.
39. Finley K (2016) Nest's hub shutdown proves you're crazy to buy into the internet of things. https://www.wired.com/2016/04/nests-hub-shutdown-proves-youre-crazy-buy-internet-things/. Accessed 26 Feb 2018
40. Gartner (2017) Leading the IoT. Gartner, Inc., s.l.
41. Georgiou O, Raza U (2017) Low power wide area network analysis: can LoRa scale?. IEEE, s.l.
42. Google Statistics (2018) Distribution dashboard. https://developer.android.com/about/dashboards/. Accessed 1 May 2018
43. Greenberg A (2017) MacOS update accidentally undoes Apple's "root" bug patch. https://www.wired.com/story/macos-update-undoes-apple-root-bug-patch/. Accessed 26 Apr 2018
44. Hewlett Packard (2014) Internet of things research study. HP, s.l.
45. Hill K (2013) 'Baby monitor hack' could happen to 40,000 other Foscam users. https://www.forbes.com/sites/kashmirhill/2013/08/27/baby-monitor-hack-could-happen-to-40000-other-foscam-users/#5ee0862658b5. Accessed 15 Mar 2018
46. Huq N, Hilt S (2017) US cities exposed—a Shodan-based security study on exposed assets in the US. Trend Micro, s.l.
47. Huth C, Duplys P, Guneysu T (2016) Secure software update and IP protection for untrusted devices in the internet of things via physically unclonable functions. IEEE, s.l.
48. IBM (2017) Ponemon Institute's 2017 state of mobile & internet of things (IoT) application security study. ARXAN, s.l.
49. ICS-CERT (2016) Advisory (ICSA-16-147-01B). https://ics-cert.us-cert.gov/advisories/ICSA-16-147-01. Accessed 20 Apr 2018
50. IHS Markit (2014) Average age of vehicles on the road remains steady at 11.4 years, according to IHS automotive. IHS Markit, s.l.
51. In L, Lee K (2015) The internet of things (IoT): applications, investments, and challenges for enterprises. Elsevier, s.l.
52. Jiang H (2013) Key findings on airplane economic life. Boeing, s.l.
53. Jongboom J, Stokking J (2017) Firmware updates over low-power wide area networks. https://www.thethingsnetwork.org/article/firmware-updates-over-low-power-wide-area-networks. Accessed 26 Apr 2018
54. Khan M, Bi Z, Copeland JA (2013) Software updates as a security metric, passive identification of update trends and effect on machine infection. IEEE, s.l.
55. Kovacs E (2014) Hackers attack shipping and logistics firms using malware-laden handheld scanners. https://www.securityweek.com/hackers-attack-shipping-and-logistics-firms-using-malware-laden-handheld-scanners. Accessed 26 Feb 2018
56. Lambert F (2017) Tesla is unlocking extra power in older Model S and X 75D vehicles—cutting a second off 0–60 mph acceleration. https://electrek.co/2017/10/21/tesla-unlocking-extra-power-older-model-s-x-75d-acceleration/. Accessed 20 Feb 2018

57. Leonhard W (2018). Microsoft Patch Alert: April patches infested with bugs, but most are finally contained. https://www.computerworld.com/article/3216425/microsoft-windows/microsoft-patch-alert-april-patches-infested-with-bugs-but-most-are-finally-contained.html. Accessed 27 Apr 2018
58. Leverett E, Clayton R, Anderson R (2017) Standardisation and certification of the 'internet of things'. University of Cambridge, s.l.
59. Levin S, Woolf N (2016) Tesla driver killed while using autopilot was watching Harry Potter, witness says. https://www.theguardian.com/technology/2016/jul/01/tesla-driver-killed-autopilot-self-driving-car-harry-potter. Accessed 3 Apr 2018
60. Lv S, Nie S, Liu L, Lu W (2016) Car hacking research: remote attack Tesla motors. Keen Security Lab of Tencent, s.l.
61. Machina Research (2015) Global M2M market to grow to 27 billion devices, generating USD1.6 trillion revenue in 2024. https://machinaresearch.com/news/global-m2m-market-to-grow-to-27-billion-devices-generating-usd16-trillion-revenue-in-2024/. Accessed 26 Apr 2018
62. Maggi F (2017) The crisis of connected cars: when vulnerabilities affect the CAN standard. https://blog.trendmicro.com/trendlabs-security-intelligence/connected-car-hack/. Accessed 2 Mar 2018
63. Malwarebytes (2018) Root cause analysis—web protection false positive. Malwarebytes, s.l.
64. Manchester University (2014) Defining and measuring software sustainability: towards an empirical framework for evaluation at the architectural level. Manchester University, Manchester
65. Markey E (2015) Tracking & hacking: security & privacy gaps put American drivers at risk. EDMarkey, s.l.
66. McConnell S (2004) Code complete, 2nd edn. Microsoft, s.l.
67. Microsoft (2016) Addressing ROI in internet of things solutions. Microsoft, s.l.
68. Microsoft (2017) CVE-2017-8529 | Microsoft browser information disclosure vulnerability. https://portal.msrc.microsoft.com/en-US/security-guidance/advisory/CVE-2017-8529. Accessed 20 Mar 2018
69. Microsoft (2017) June 2017 security updates. https://portal.msrc.microsoft.com/en-us/security-guidance/releasenotedetail/40969d56-1b2a-e711-80db-000d3a32fc99. Accessed 20 Mar 2018
70. NADA (2014) The impact of vehicle recalls on the automotive market. NADA, s.l.
71. National Audit Office (2017) Investigation: WannaCry cyber attack and the NHS. Department of Health, s.l.
72. Nazario J (2016) The anatomy of an IoT botnet attack. https://www.fastly.com/blog/anatomy-an-iot-botnet-attack. Accessed 26 Apr 2018
73. NHS (2018) Lessons learned review of the WannaCry ransomware cyber attack. Department of Health & Social Care, England
74. NHTSA (2015) NHTSA timeline. https://one.nhtsa.gov/nhtsa/timeline/index.html. Accessed 5 Mar 2018
75. Nichols S (2017) Nasty firmware update butchers Samsung smart TVs so bad, they have to be repaired. https://www.theregister.co.uk/2017/08/24/samsung_tvs_botched_firmware_patch/. Accessed 20 Apr 2018
76. NIST (2014) Framework for improving critical infrastrcuutre cybersecurity. NIST
77. NIST (2018) Interagency report on status of international cybersecurity standardization for the internet of things (IoT). NIST, s.l.
78. Novet J (2016) Pogoplug unlimited cloud storage service is shutting down on September 28. https://venturebeat.com/2016/09/21/pogoplug-unlimited-cloud-storage-service-is-shutting-down-on-september-28/. Accessed 2 Mar 2018
79. Oltermann P (2017) German parents told to destroy doll that can spy on children. https://www.theguardian.com/world/2017/feb/17/german-parents-told-to-destroy-my-friend-cayla-doll-spy-on-children. Accessed 20 Feb 2018

80. Palanca A, Evenchick E, Maggi F, Zanero S (2017) A stealth, selective, link-layer denial-of-service attack against automotive networks. In: Detection of intrusions and malware, and vulnerability assessment, vol 10327(1), pp 185–206
81. Perlow J (2015) IoT abandonware: when your cloud service leaves you stranded. https://www.zdnet.com/article/iot-abandonware-when-your-cloud-service-leaves-you-stranded/. Accessed 26 Feb 2018
82. Pinkstone J (2018) TomTom owners are furious after the promise of a 'lifetime' of sat-nav map updates for 66 older models is SCRAPPED. http://www.dailymail.co.uk/sciencetech/article-5329405/TomTom-scraps-lifetimesupport-map-updates.html. Accessed 20 Feb 2018
83. Raza U, Kulkarni P, Sooriyabandara M (2017) Low power wide area networks: an overview. arXiv, s.l.
84. Rios B (2017) Security evaluation of the implantable cardiac device ecosystem architecture and implementation interdependencies. WhiteScope, s.l.
85. Rolfe D (2017) Five foreseeable failures that will hit IoT. https://internetofthingsagenda.techtarget.com/blog/IoT-Agenda/Five-foreseeable-failures-that-will-hit-IoT. Accessed 22 Mar 2018
86. Rose A, Ramsey B (2016) Picking Bluetooth low energy locks from a quarter mile away. DEFCON, United States
87. Schmidt S, Tausig M, Hudler M (2016) Secure firmware update over the air in the internet of things focusing on flexibility and feasibility proposal for a design. ResearchGate, s.l.
88. Schmidt S, Tausig M, Hudler M, Simhandl G (2015) Secure firmware update over the air in the internet of things focusing on flexibility and feasability. ResearchGate, s.l.
89. Schneier B (2013) Our security models will never work—no matter what we do. Wired, s.l.
90. Schneier B (2014) The internet of things is wildly insecure—and often unpatchable. https://www.wired.com/2014/01/theres-no-good-way-to-patch-the-internet-of-things-and-thats-a-huge-problem/. Accessed 3 Mar 2018
91. Senrio (2017) Devil's Ivy: the technical details. http://blog.senr.io/devilsivy.html. Accessed 25 Mar 2018
92. Shah SH, Yaqoob I (2016) A survey: internet of things (IOT) technologies, applications and challenges. IEEE, s.l.
93. Sharwood S (2018) Linux Foundation backs new 'ACRN' hypervisor for embedded and IoT. https://www.theregister.co.uk/2018/03/19/linux_foundation_acrn_hypervisor_project/. Accessed 19 Apr 2018
94. Shaw K (2017) What is a hypervisor?. https://www.networkworld.com/article/3243262/virtualization/what-is-a-hypervisor.html. Accessed 20 Apr 2018
95. Sicari S, Rizzardi A, Grieco L, Coen-Porisini A (2015) Security, privacy and trust in internet of things: the road ahead. Tran Song Dat Phuc, s.l.
96. SigFox (2017) Sigfox presents 2017 results and 2018 roadmap. https://www.sigfox.com/en/news/sigfox-presents-2017-results-and-2018-roadmap. Accessed 26 Apr 2018
97. Simpson AK, Patel NS, Roesner F, Kohno T (2017) Securing vulnerable home IoT devices with an in-hub security manager. IEEE, s.l.
98. Smith LJ (2018) Audi recalls 1.2 million cars after software update fails to fix pump issue. https://www.express.co.uk/life-style/cars/951526/Audi-recall-2018-A4-A5-A6-Q5-coolant-pump-car. Accessed 26 Mar 2018
99. Stergiou C, Psannis KE, Kim B-G, Gupta B (2016) Secure integration of IoT and cloud computing. Elsevier, s.l.
100. TaylorHarris M (2007). Why you need to use statistics in your research. Kerrypress Ltd
101. Thomson I (2016) Wi-Fi baby heart monitor may have the worst IoT security of 2016. http://www.theregister.co.uk/2016/10/13/possibly_worst_iot_security_failure_yet/?mt=1476453928163. Accessed 26 Feb 2018
102. Toll N (2018) Using Shodan as a tool to find vulnerable devices | GRI Blog. https://globalresilience.northeastern.edu/2018/02/using-shodan-as-a-tool-to-find-vulnerable-devices/. Accessed 20 Mar 2018

103. TRAC (1999) Component acceptability for CE product safety. TracGlobal, Worcestershire, UK
104. Tsidulko J (2017) The 10 biggest cloud outages of 2017 (so far). https://www.crn.com/slide-shows/cloud/300089786/the-10-biggest-cloud-outages-of-2017-so-far.htm/pgno/0/10. Accessed 2 Mar 2018
105. UAW v Chao (2004) District Judge. Randy S. Rabinowitz
106. Ubuntu (2005) New Ubuntu foundation announced. https://insights.ubuntu.com/2005/07/01/new-ubuntu-foundation-announced. Accessed 20 Apr 2018
107. UL (2010) Standard for inverters, converters, controllers and interconnection system equipment for use with distributed energy resources. https://www.shopulstandards.com/ProductDetail.aspx?UniqueKey=20941. Accessed 2 May 2018
108. UL (2018) Specific guidelines and rules. https://www.ul.com/marks/ul-listing-and-classification-marks/promotion-and-advertising-guidelines/specific-guidelines-and-rules/. Accessed 2 May 2018
109. Vaniea K (2016) Tales of software updates: the process of updating software. ACM, s.l.
110. Wash R, Rader E, Vaniea K, Rizor M (2014) Out of the loop: how automated software updates cause unintended security consequences. USENIX, s.l.
111. Weber RH (2017) Liability in the internet of things. Kluwer Law Online, s.l.
112. Welch C (2017) Major Apple security flaw grants admin access on macOS High Sierra without password. https://www.theverge.com/2017/11/28/16711782/apple-macos-high-sierra-critical-password-security-flaw. Accessed 26 Apr 2018
113. Williams M, Nurse JRC, Cresse S (2017) "Privacy is the boring bit": user perceptions and behaviour in the internet-of-things. 15th International Conference, s.l.
114. Williams PA, Woodward AJ (2015) Cybersecurity vulnerabilities in medical devices: a complex environment and multifaced problem. PMC, s.l.
115. Williams ZD (2017) IoT platforms company list 2017 update. https://iot-analytics.com/iot-platforms-company-list-2017-update/. Accessed 26 Feb 2018
116. Witkovski A, Santin A, Abreu V, Marynowski J (2016) An IdM and key-based authentication method for providing single sign-on in IoT. ResearchGate, s.l.
117. Wurm J et al (2016) Security analysis on consumer and industrial IoT devices. cs.ucf.edu, s.l.
118. Yuchen Y et al (2016) A survey on security and privacy issues in internet-of-things. IEEE, s.l.
119. Yu T et al (2017) Handling a trillion (unfixable) flaws on a billion devices: rethinking network security for the internet-of-things. Carnegie Mellon University CECA, s.l.
120. Zhou W, Zhang Y, Liu P (2018) The effect of IoT new features on security and privacy: new threats, existing solutions, and challenges yet to be solved. IEEE, s.l.
121. Zurko ME (2005) User-centered security: stepping up to the grand challenge. IEEE, s.l.

Printed in the United States
by Baker & Taylor Publisher Services